AF394736

THE BLUE CLIFF

Climbing tales from the margin between land and sea

Edited by Grant Farquhar

This book is dedicated to Damian Cook (1969–2004) and Jonathan 'Woody' Woods (1976–2011).

First published in 2023 by Atlantis Publishing.

ATLANTIS PUBLISHING LTD, 19 Morar Place, Dundee, DD5 3HL. www.atlantis.bm Copyright © Grant Farquhar and individual authors 2023.

The authors have asserted their rights under the Copyright, Designs and Patents Act, 1988 to be identified as authors of this work.

ISBN: 978-1-9999600-6-3 (hardback) ISBN: 978-1-9999600-7-0 (ebook).

British Library Cataloguing-in-Publication Data. A catalogue record for this book is available from the British Library.

Designed by Atlantis Publishing. Printed and bound by Cambrian Printers, Llanbadarn Road, Aberystwyth, Wales, SY23 3TN. www.cambrian-printers.co.uk.

Distributed by Cordee Ltd, 11 Jacknell Road, Dodwells Bridge Ind Est, Hinckley, Leicestershire, England, LE10 3BS. www.cordee.co.uk. Cordee Code AP0002.

Other books by Grant Farquhar:

The White Cliff
Climb de Rock
Crazy Sorrow
A' Chreag Dhearg

Front cover: Rob Lamey on Davey Jones' Locker, F7b+ S2, at Ogmore, Wales. Photo Simon Rawlinson.

Back cover: Peter Biven at work on the Overhanging Bastion crux pitch of the Magical Mystery Tour on Berry Head, Devon, in September 1969. Photo John Cleare.

Facing page: Jonathan 'Woody' Woods on Barbecuing Traditions, F7b S2, at The Castle, Pembroke. Photo Dave Pickford.

We all have our favourite places, made special by moods and memories, atmospheres and conditions. A more open mind could point out that we are already deeply conditioned, but if I am colour-blind, I am most fortunate that it is my favourite colour that I can see clearest.

In a land-locked city, my eyes are drawn, my attention and imagination captured by a natural uncut boulder nestled unassumingly beside a more widely appreciated tourist spectacle. As if that boulder had burst through the surface of my thoughts, my memory is cast back to that place where the hills meet the sea, where my naked body is submerged in the most vivid blues and greens, greys and silver.

When the orchestra stops, the soloist begins. It's the time to put aside all life, all thoughts but one. The safety net is gone. The celebration of certainty and independence is an enormous gamble. One wrong note, one misplaced finger, and the curtain will fall. But the unthinkable happens, the hand fumbles and the price of failure must be paid. Yet then, facing the abyss, the dark veil is lifted. This isn't the abyss. There are the most vivid colours – here are our favourite greens and blues and silver.

Beside the ocean we are never alone. From here all life originates. Here we are protected from the worst excesses of gravity, from urban development, and from taking ourselves too seriously. At our favourite places are our friends, and it is our friends that make life colourful. Even in exile, my life is flavoured thus, and as I walk through this land-locked city, I can taste the salt spray on my lips and hear the waves in my friends' laughter.

A Letter from Budapest by Damian Cook

Facing page: Damian Cook diving the Rio Point
at Connor Cove in 1994. Photo Joff Cook.

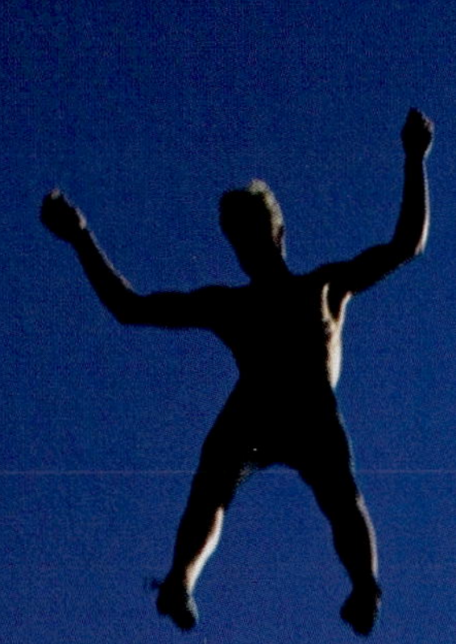

Now would I give a thousand furlongs of sea for an acre of barren ground, long heath, brown furze, anything.

William Shakespeare, *The Tempest*, Act 1

Foreword: The Liminal Game by Dave Pickford

We're going to teach you to fly the F-14 right to the edge of the envelope; faster than you've ever flown before. Faster and more dangerous.

Viper, *Top Gun* (film)

It's a blue summer's day somewhere at the beginning of things.

The beach carves out into the sun; silence and light merge where the shingle meets the English Channel. An offshore breeze ruffles the wide water of Worbarrow Bay. I run into the sea and swim straight out with the easy confidence of a nine-year-old boy. The water isn't cold, and refracted sunlight warms my body as I move through the sea. Gulls are flying overhead, diving and turning. The cliffs to the south, over Worbarrow Tout, are gold and topped with tuffs of yellow grass bleached by the summer heat. To the west, the chalk of Flower's Barrow shines in the sun, a great white shield held up against the blue arc of the bay. I can see the land behind, feel the sea all around, and sense the wind moving across the water. Out at sea, away from the noise and haste, I can dream of the world ahead of me. It's beautiful here.

After a while, I look over my right shoulder and notice that the beach doesn't look the same. I've moved to a different point in the bay to the place I set off from. My idea – the idea of many adventurous swimmers – was to go as far out as I could before heading back to shore. The cross-offshore wind has created a steady current on the surface of the bay, which is moving me westwards towards the chalk of Flower's Barrow. Quickly working out that the wind current is running due west down the beach, I know that as long as I swim in a diagonal line towards the beach I will reach the land eventually. The bay is a huge sickle-shape, forming a massive crescent over a mile wide. I've sailed across it before and feel I know this place. Children are less likely than adults to experience the counterproductive reaction of panic which, of course, can have catastrophic results for open-water swimmers.

I can see that I'm not in any danger, providing I stick to my plan of just swimming across the direction of the wind and not trying to return to where I started. Sure enough, that distant bank of rust-coloured shingle slowly becomes larger as I close in. My plan is working, even though the wind is stronger here away from the shelter of the cove. I can see a figure on the beach now, his arms waving in a fan-like motion, like one of those people who guide aircraft as they taxi on the runway. As I come into the shore, I roll through the gently breaking surf and feel the slow pressure of the shingle under my toes.

My dad gives me a towel and a good telling-off: don't swim out like that again in an offshore wind. But I don't get why he's scolding me. The situation was under control, and I never felt in any danger. Children sometimes know more about the facts on the ground than their parents, as any kid who gets up to mischief already knows. I understood what was happening in the sea and felt somehow that the sea understood me, too. That day was the beginning, in a way, of a lifetime of explorations on the edge of land.

That swim, more than 30 years ago, in Worbarrow Bay prefigured my life as an explorer of sea cliffs and of the sea itself. It also set out why it's a very good idea to respect it. The sea, like the mountains, doesn't care about anyone or anything. It's a primal force at the heart of nature. It's been there for a long time and will be there long after we're gone.

Facing page: Dave Pickford off Mark of the Beast, F7c S2. Photo Joff Cook.

A potentially life-preserving attribute for any sea cliff climber is a certain level of strength and skill as a swimmer. If you're going to climb above the sea, you're likely at certain times to enter it both by necessity and by accident. It is only because of our ability to swim that we might participate in that most pure and minimalist of climbing disciplines: the dark art of deep water soloing. The practice of climbing without ropes above the sea is one of the most powerful experiences we can have in a natural environment; it's also one of the least understood. Why is the sea itself so fascinating? In his brilliant, indefinable book *The Sea Inside,* the writer Philip Hoare explains why coastal areas exert a powerful existential draw:

'The sea defines us, connects us, separates us. Most of us experience only its edges, our available wilderness on a crowded island... And although it seems constant, it is never the same. One day the shore will be swept clean, the next covered by weed; the shingle itself rises and falls. Perpetually renewing and destroying, the sea proposes a beginning and an ending, an alternative to our landlocked state, an existence to which we are tethered when we might rather be set free.'

Hoare's proposition configures the sea as a space in which we might attain some kind of spiritual awakening. At the same time, the coast becomes its ritual ground, a place simultaneously sacred and profane where life and death might be blended, reordered, and magnified. Why, for example, is it so often the case that old people park their cars in certain quiet places facing the sea in order simply to observe it? They will drink tea from hot flasks – and possibly eat some sandwiches – whilst merely staring out into the distance, scanning the horizon for signs, like the lookout in the crow's nest of a sailing ship. Those elderly people are looking, I think, at what happens at the edge of land; the place where the light changes. They're also looking out beyond the land itself, observing the way the light glances off the waves and how the horizon begins to fade with the fall of evening.

And at the same time, they're thinking about the way life itself rises, shifts, changes – and ultimately disappears – with age. If there is one reason why elderly people seem to enjoy staring at the sea, it's because it is a mirror through which we might observe ourselves and somehow better come to terms with our mortality. A stretch of wild coastline, in this sense, is the ultimate liminal zone; a boundary between land and water, and between life and death.

One of the best ways of entering this zone is through sea cliff climbing, and through one style of sea cliff climbing in particular. Deep water soloing is a very special dimension of the sport of rock climbing. It's the proverbial cherry on top of the cake; the safety net of the water below allows us to climb alone and unroped in the most unencumbered style possible. Of course, you can do it only if the sea conditions and rock conditions allow it, which doesn't always happen. To be successful at deep water soloing, just as with mountaineering, you have to be in the right place at the right time. Achieving this can sometimes be extremely complicated, contributing to the beauty of the pursuit.

As the groundbreaking British climber Ben Moon once memorably remarked, 'If it was easy it wouldn't be hard, would it?'

The great thing about deep water soloing is that if you screw something up high above the sea, you'll likely get a chance to fight another day – if you are doing everything right. At least most of the time. Some of the most powerful memories in climbing are not the things you did first try; easily within your limit. Rather, they're the climbs that you either just managed to do by the skin of your teeth or the routes that spat you off in spectacular style, and you had to return to realise another day. This is never more true than in deep water soloing, which is to climbing what single-handed sailing is to seafaring: it's one of the most demanding and also one of the very best things you can do in the vertical world.

Anyone can have too much of a good thing, and there's a caveat in all this, too. Even in good conditions, deep water soloing – particularly if you are alone – is a very serious business. You're in trouble if you have a bad fall and lose consciousness in the sea or end up struggling to exit the water. In that situation, without a quick rescue, even the strongest swimmer can drown.

In early 2004, just after I spent a few weeks establishing the first routes on the tiny island of Laoliang in Thailand's Andaman Sea, I got an email from a friend saying that Damian Cook, one of the pioneers of deep water soloing in the UK, had drowned in Mallorca. We'd shared many climbs together, and I was truly shocked. He'd fallen off a route at Cova del Diablo and had been caught in a rip current. It was the end of winter, and the sea was cold; he couldn't exit at the usual place and tried to swim around the point to another landing place, but the current swept him away before he reached it. Damian was a really strong swimmer, and his death was a sobering reminder of the awesome power of the sea and the fact that deep water soloing is only as safe as the water that lies underneath the route you're climbing. If the tide is falling and it's too shallow, or if the swell is too strong to safely exit the water, then deep water soloing can easily be as dangerous as regular soloing. In such circumstances, it could even be more dangerous than soloing above the ground, as you may be tempted to try a line that's harder than something you ever consider soloing with a hard landing below it.

It's a matter of judgement as to what sea conditions you personally judge to be safe, but one thing's for sure based on my own experience: it is usually much easier to tell if the water is too shallow than to determine if the swell is too big to safely swim out. The sea, by its very nature, is unpredictable. And at the base of a sea cliff, where the swell meets the rocks, it's even more unpredictable than usual. All it takes is a big set of waves, and what looks like a safe exit point from the clifftop can turn into a surging maelstrom when you're in the water. The well-known American professional climber Michael Reardon – who was an ardent soloist – was swept off the base of Fogher Cliff in County Kerry in the west of Ireland by a rogue wave in the summer of 2007.

Dave Pickford on Wreck of the Zephyr, F6c+ S0, in Barrel Zawn, Pembroke. Photo Mike Robertson.

His body was never recovered. This tragic incident shows that even for the most experienced sea cliff climber, the sea itself may be much more dangerous than the wall you're going to climb.

My friend Grant Farquhar and I once had a lucky escape of this nature whilst attempting to make the first continuous crossing of the gigantic, 17km long Exmoor Coast Traverse, Britain's longest climb. We were 11 hours deep into our attempt, about three-quarters of the way along, and both mildly hypothermic. It was an hour after high tide, and I suggested swimming a narrow, flooded chasm to the west of Red Cleave to save time. Grant looked at me quizzically. 'Maybe best not to, mate,' he said. A minute afterwards, a huge set of waves smashed into the chasm, completely filling it with surging whitewater. If we'd been in the chasm at that time, we could easily have drowned.

I've used up more than a cat's nine lives in my climbing career and don't like to take so many chances these days. The potential for a serious mistake whilst soloing is known as a high-impact, low-probability event, like a Category 5 Hurricane hitting a particular island in the Caribbean. In deep water soloing, the impact level is likely lower, as you have a softer landing, but the impact sure is still there. Every time you go soloing, even above the sea, the risk of that 'high impact event' taking place rises in the same way that if you sail back and forth across the Atlantic Ocean continuously – the risk of sailing into a hurricane rises.

After my friend Damian drowned in 2004 – an accident which was a definitive example of a low-probability, high-impact deep water soloing event – I realised that a vital rule of thumb for safe deep water soloing is that you should never, ever set out on a climb if you think the sea below it looks too wild to go swimming in. It's a good idea to look at the sea under the route you're climbing from the perspective of being a weaker swimmer than you actually are. If you take a big splashdown from 50ft after a desperate battle with a crux sequence, you'll be hyperventilating for sure. So your swimming ability will be considerably worse than if you went for a swim off the rocks in the normal way.

Despite all its potential seriousness, deep water soloing is also one of the most magical experiences in climbing. It's just the rock and the sea, above which your body forms an improvised dance as it moves through the liminal terrain between earth and water. Like an astronaut breaching the atmosphere, crossing the shadow line between planetary orbit and space, the deep water soloist enters a new configuration of experience when pushing hard on a sea cliff.

When you're climbing well, the feeling of flow can be powerful. This is amplified several times over when deep water soloing. It's as if a switch is flicked, and you're suddenly on autopilot. The edge of the envelope is pushed a little further out. You eyeball that distant hold, tense, commit, and make the next move. Then you carry on, climbing higher above that deep blue water, the golden evening light flashing off the waves. It's always beautiful up there, high above the waves.

And if it goes wrong, the sea invites you; at once mesmerising and strange, familiar and utterly mysterious, it is the ultimate landing zone. The sea, as the writer Philip Hoare has pointed out, is where we came from. By falling into the sea, then, we might be returning home.

Deep water soloing evolved hard and fast during the noughties. Glamorous images of honed guys – and girls – on the famous DWS cliffs of the UK, Mallorca and Thailand adorned the glossy pages of climbing magazines. People began exploring well off the beaten track; exotic places like Halong Bay in Vietnam and Palawan in the Philippines now harbour some of the most exciting deep water soloing in the world, not least because of the paradisal setting, perfect rock and amenable sea temperatures. Other, even more far-flung, destinations will be discovered in time.

One of the most appealing things about deep water soloing is that it can take place on cliffs with absolutely no access other than by sea. In fact, some of the best deep water solo venues must be accessed this way. In the most extreme cases, such as some of the routes in Halong Bay, the only way to descend the climbs is simply to jump off at a particular point, for example a large stalactite protruding from the wall at 15m. There's a unique beauty about this; it's almost a perfect geometric reversal of mountaineering, where a pointy summit marks the top of the climb.

Back home, as the years went by, I returned to deep water soloing at some point every summer, lured back by that magnetic thrill of the dance in the liminal zone between land and water. I pushed the boat out – way too far out sometimes – but always managed to get away with it. Once, halfway along the mega 800ft traverse of Zodiac at St Govan's Head in Pembrokeshire, I looked down at the 30ft waves booming into the boulder-filled cave 100ft below. Clearly, this was not deep water soloing just because the sea was below me: it was conventional free soloing, where the consequences of a fall are inevitably fatal. (Zodiac fell down in a huge rockfall a few years later, but the point remains).

Other deep water solos on the sea cliffs of Pembrokeshire, one of the best places in the world for this style of climbing, were equally memorable. One September day in 2001, I raced over to the west face of Mowing Word, alone with just an abseil rope and harness, plus rock shoes and my chalk bag. I soloed 10 routes on that shining, sun-bleached wall of golden limestone that afternoon, culminating in the mega E5 that takes on the challenge of the steepest, smoothest part of the cliff head-on: the appropriately-named All at Sea. The tide was falling fast, but I was climbing strongly and confidently. In any case, I had no intentions of falling off into the sea from a route with a crux at over 25m anyway. Once I reached the headwall, I glanced down. Big waves were slamming into the base of the cliff, sending spray flying. But what was that yellow colour in the spray? Sand! The tide had fallen so far that the waves were churning up the sandy beach that's only exposed at low spring tide at the base. This cliff is barely a deep water solo venue at high tide; right now, I was

Emma Madeira on Octopus Weed, F6c S0, at Cave Hole in the mid-90s. Photo Joff Cook.

most definitely free soloing. The edge of the envelope, once again, was being pushed out. I refocused, concentrating hard on the short blank wall that forms the crux. A few sharp pulls on a series of crimps led to a final stretch for a good incut, and the apex of the headwall was in sight. Up here, well over a hundred feet above the surf, I was completely free as I moved alone across this beautiful wall in a silent, solitary dance under the late summer sun. Deep water soloing becomes conventional free soloing pretty easily on a cliff as big – and as tidal – as the west face of Mowing Word. As long as you can work out where the shadow line falls between the two, you'll probably be alright.

The art of climbing exploration is, at least partly, the art of getting off the beaten track. Motivated by an intense personal enthusiasm for getting away from the crowd, I've discovered quite a lot of new routes – and sometimes whole new crags – for deep water soloing. Over the years, I've found great lines all over the place: that perfect 50ft arête in Oman's Musandam Peninsula, rising up from the crystal-clear water like the protruding bow of an Arabian dhow; that crazy, sea-scalloped wall of ultra-compact overhanging limestone west of Loutro on the south coast of Crete, the strange, conch-shell holds forming a natural rising rightwards traverse; the beautiful, gently overhanging pink wall at Clarence Cove in Bermuda that climbed like a waking dream and became, inevitably, Fifty Shades of Pink; or the utterly bizarre and brilliant route I found near the North Light on Lundy Island, where I traversed about 30m horizontally along the retaining wall inside a massive cavern-arch to emerge into the sunlight at a single, protruding jug on the arête on the far side. The climbing had been quite sustained and superb, and this single hold was literally the end of the road, a lone feature on a canvas of blank, almost jet-black granite. The only options were to reverse the line or jump in; I chose the latter. The Ungraspable Phantom was born as I took a refreshing plunge into the cool, blue water of the Celtic Sea. It remains one of the most thrilling and unlikely deep water solos I've climbed: a line without an obvious beginning, traversing the interior of a giant cavern, and ending in a no-mans-land of blank granite like a spacecraft adrift in a sea of dark matter.

Deep water soloing enables cliffs that would otherwise never be climbed upon to be explored. During my long-distance voyage around southwest England by standup paddleboard that was completed over several summers between 2017 and 2022, I found dozens of new, virgin crags along Cornwall's Atlantic coast that would be ideal for deep water soloing, particularly with the benefit of an approach boat. Many are invisible from the clifftop, and only one or two of these crags currently have any documented routes. It simply isn't true that the UK has been 'climbed out' in terms of new route potential; it just requires imagination and some serious hunting around in the liminal zone between land and sea.

Being creative reaps rewards for the exploratory climber. Back in 2006, a team of us drove over to North Pembrokeshire on a rumour early one Friday morning, just before I was due to fly out to India to lead a mountaineering trip. A certain Mr Robertson, one of the key pioneers of DWS on the south coast, had found a perfectly-formed barrel of high-quality sandstone above a small inlet near St Davids. It was christened Barrel Zawn, and we climbed around a dozen new routes that day, from some short and sweet bouldery climbs to a long traverse line that I climbed into the heart of the cave that featured a bold and sketchy exit, made more memorable by the half-dead pigeon (probably the victim of a recent peregrine falcon attack) flapping in its death-throws on the sloping ledge at the end of the crux. I pulled through the last moves, and the pigeon squawked one final time before plunging into the sea. Terrapigeon thus came into the world, the route's name immortalising the poor bird's fate.

Deep water soloing, by the very nature of the craft, creates potential for climbing where otherwise there would be none. This particular cliff is a textbook case. It would be useless for trad climbing, as the base is almost always a sea-filled trench, and there's not much natural gear. There's certainly no bouldering there, and people don't put bolts in British sea cliffs as a matter of course. Only with the mindset of the deep water soloist can a cliff such as this come into its own.

In 2017, I found a small zawn whilst out paddling on the Pembrokeshire coast, close to Raming Hole, that had no recorded routes at all. At high tide, I noticed it had at

Dave Pickford on Overexposed, F6c+ S2, at Stennis Ford, Pembroke. Photo Pickford Collection.

least three metres of water underneath it; just enough. I did four new deep water solos here of varying difficulty, all of them poised on the seaward edge of the zawn. The lefthand one was particularly good, but I didn't record any of them; I felt they were more special somehow left that way. The cliff is so tricky to find from the clifftop if you don't know exactly where it is that few people would venture there, in any case. Such is the way of these kinds of cliffs: they are elusive prizes that reward the dedicated and the crafty.

There's a fascinating tradition in Norwegian trad climbing of not recording (or only minimally recording) new routes on certain crags well off the beaten track as, so the tradition goes, the very record of them would dilute the adventure they might offer to another team, another time. There's an alluring beauty, I think, about this approach to climbing. At the same time as leaving no trace of your presence on the rock itself, you also seek to leave no trace of the route you climbed as a piece of recorded information. As the American poet Wallace Stevens remarked of the beauty of the blank page: 'The sombre pages bore no print / except for the trace of burning stars / in the frosty heaven.'

Perhaps it's easier to undertake this approach to climbing through the present-tense, visceral aesthetic of deep water soloing? The very notion of climbing without ropes above the sea stands in complete contrast to the classic mountaineering aesthetic: a quest for a summit to be conquered. In opposition to this acquisitive mindset, the deep water soloist might discover a perfect wall of limestone overhanging a secret sea cave whilst swimming beneath it, climb a new route up the centre of it completely onsight, and leave the place unmarked and just as mysterious as when they arrived. This is a more appealing aesthetic state, of course, than that of a climber who brandishes a flag on the top of a mountain and takes a selfie to upload to their social media profile.

I had just such an experience myself in Menorca in 2017 when I found a gorgeous wall of ochre-coloured rock on the northwest tip of the island near Cala Morell. I traversed into it and climbed it onsight; it was the only route I climbed that day. It wasn't particularly hard, but the wall had this special feeling of solitude, forming the edge of a small zawn on the outermost edge of a tiny island in the middle of the Mediterranean, completely impervious to the James Bond vibe of high-end yachts and glamorous people in the sheltered cove just around the corner.

Sometimes, deep water soloing takes you right out beyond the 'edge of the envelope' that fighter ace Viper refers to in the iconic movie *Top Gun*. The medium of climbing without ropes above the sea is, by definition, an uncompromising one. It doesn't take prisoners. And it certainly doesn't suffer fools. You can learn more about your own relationship to fear, and about how much risk you're prepared to take, from an afternoon of deep water soloing than you can learn in a year of bouldering or sport climbing, for example. That's not to say that you can't experience fear and risk in those disciplines. It's just that when you're high over deep

Above: Dave Pickford on Splendid Isolation, F6c S2, at White Hole in the 90s. Photo Joff Cook. Below: Jonathan 'Woody' Woods making the second ascent of The Bandersnatch, F7a+ S2, at Hollow Caves Bay, Pembroke. Photo Dave Pickford.

water, everything becomes a whole lot more involved. The consequences of action are always pretty high if you're going to fall 15m into 15°C water.

Gavin Symonds, one of my oldest friends and a regular climbing and watersports companion, was another key pioneer of hard deep water soloing in the UK. In 2008, I hung on an abseil rope in Hollow Caves Bay in South Pembrokeshire, photographing his attempts to climb San Simeon (an E8 trad route) as a deep water solo. The thing about this route is that although it's entirely above deep water, the crux comes at a cool 19m above the high-tide mark. That's over 60ft. It's a very long way to fall into the sea from a dynamic, powerful move. Gavin is a master at the art of the body-angle correction technique needed to enter the water safely in a fall from this height.

I watched him, once again, set up for the big crux lunge up and right to a shallow pocket. I could feel the humidity in the August air; as usual for this cliff, conditions were sub-optimal. His right hand flicked out, hovered in a moment of hesitation as his left foot slipped a fraction of a centimetre, and he was off. Pirouetting perfectly in the air, he flapped his arms rapidly to attain a vertical 'pole' posture and locked both hands down by his sides as he entered the water like a human bullet. That's it, I thought; that's exactly how to fall safely into the sea from a hard move a really long way up. If, that is, you've got the stomach for stuff like that.

The height limit I'm personally comfortable with whilst pushing hard on a deep water solo is definitely lower than the crux of San Simeon, and I'm quite happy with that. It's a good idea to work out your own personal parameters for a 'safe' deep water solo at your earlier stages of experimentation with it and carefully monitor what you're comfortable with. Some people don't like to swim more than 50m to exit the water; others, such as myself, prefer to solo no higher than about 15m above the water.

In the same way, a proficient deep water soloist knows how to fall into the sea safely from height, a skilled racing driver will know how to balance a car in a high-speed corner. The two activities are closely linked through a complex interplay between speed, skill, and judgement. There are a number of interesting links between climbing and motorsport; both disciplines involve using high-level technique, intense concentration, intelligence, and perfect timing in a fast-moving, high-risk environment. The groundbreaking British climber Johnny Dawes has said that a fast car moving through a high-speed corner is 'an unresolved work of art in a physical form'. The same could be said, I think, of a deep water soloist setting up for a crux move high above the sea. Some of the reasons that fast cars are my personal guilty pleasure are probably linked to the reasons I have a natural attraction to deep water soloing.

As the American philosopher Matthew Crawford has explained in his book *Why We Drive*: 'To drive is to exercise one's skill at being free.' Exactly the same thing is also true of deep water soloing. People with a high-risk, high-reward psyche and a desire to master difficult tasks in challenging environments are likely to be drawn to high-octane adventure sports for the very reason that Crawford so clearly sets out.

There's a lot that deep water soloists can learn from sailors, too, about operating with a reasonable margin of safety. One of the greatest mistakes in the history of seafaring was the choice of the route taken on the maiden voyage of the *Titanic* in April 1912. At the time, the vessel was the world's largest ship and (unwisely) considered unsinkable. The *Titanic's* captain, EJ Smith, reportedly made the following statement before the voyage: 'I have seen but one vessel in distress in all my years at sea. I never saw a wreck and never have been wrecked, nor was I ever in any predicament that threatened to end in disaster of any sort.'

Captain Smith's sense of hubris ultimately led to – or at least contributed to – the *Titanic's* famous demise. Sailing much too far north in the North Atlantic for any reasonable margin of safety, she hit a massive iceberg at her cruising speed of 21 knots and sank in less than three hours. One thousand five hundred people lost their lives. Captain Smith's belief that shipwrecks were not a problem at sea was possibly a major factor in the most spectacular disaster in maritime history.

What does the story tell us about deep water soloing? Well, if you think you might be smarter than the sea – like Captain Smith of *RS Titanic* – or if you think your knowledge or skill is going to somehow protect you from it, the sea is probably going to take a shark-sized bite out of your arse, and possibly help itself to your head as well.

From decades of sea cliff climbing, and also more recently from my long-distance standup paddling expeditions around the British Isles, I have learnt that the ocean is one of the wildest environments we can enter. Have you ever been caught in a rip current with triple-overhead surf breaking on the outer reef? Well, I have and managed to get out of the situation before it became critical. But I can tell you that the feeling of an eight-knot underwater current pulling you towards breaking waves the size of a double-decker bus tells you a lot about not messing around with the sea. (The key to surviving this, by the way, is to swim sideways across the current). The sea is an almighty beast, and it takes no prisoners whatsoever.

The American sociologist Diane Vaughan formulated a very interesting concept about risk management in her book *The Challenger Launch Decision*, a forensic study of the causes of the 1986 Challenger Shuttle disaster. Her idea is important to think about, in my view, if you go deep water soloing. Vaughan identifies a process she calls 'the normalisation of deviance' as crucial to the evolution of this most spectacular of exploration-related disasters. During the developmental phase of the Space Shuttle Program, Vaughan shows how the

acceptance of deviance within the team of engineers and rocket scientists resulted in a critical, dangerous design flaw in the design of the joints on the rocket boosters of the spacecraft. The design team conducted an analysis to find the limits and capabilities of joint performance, but evidence initially interpreted as a 'deviation from expected performance' was then later reinterpreted as 'within the bounds of acceptable risk'.

To interpret Vaughan's concept more widely and generally, the normalisation of deviance can be any process by which we do something that does not follow the accepted rules of good safety procedure – for example, climbing without a rope above a double-overhead swell or sailing a very large ocean liner through iceberg-infested waters – which we then get away with. Then, believing it's safe to make the same safety shortcut a second time, we do the same thing again. And again. Repeat this process indefinitely, and something will eventually go wrong; the ship sails into the iceberg and sinks, or the climber falls off the route – and drowns.

The notion of climbing competence is rooted in the concept that you and your partner will be able to extricate yourself from any given situation. If you can't do the route, you have to abseil off it, and so on. In deep water soloing, this notion is complicated by the fact that in order to safely extricate yourself from the sea after a fall from height, you might have to get out of the water in a difficult environment whilst fatigued. It's always worth thinking about before you set off into the unknown. Deep water soloing is not necessarily the 'safe' style of climbing as it's sometimes assumed to be. If the water underneath the route isn't deep enough – a common problem in tidal waters such as the British coast – then you're in trouble. And if the water is deep enough, but the sea is too rough to exit safely, you're also in trouble. Other complicating factors have arisen, too; for example, the sudden presence of a swimmer, kayak, or other vessel in the water directly underneath the deep water soloist. This is a situation that can be avoided with a good ground-crew, but it's a key issue to think about in busier areas.

Mountains and deserts can be very wild, often inhospitable, and sometimes dangerous places. But they don't have the same dynamic quality that the sea has, which makes it both so alluring, so magnificent, and also potentially deadly. Tracy Edwards, skipper of the first all-women team to compete in the Whitbread Round-the-World yacht race, memorably remarked that 'the ocean's always trying to kill you... and it doesn't let up'. From her enormous experience of ocean sailing, Tracy Edwards instinctively knew why the sea was to be deeply respected, even feared if necessary, for a requisite level of competence on the water. I think that if your starting point is this perspective, rather than Captain EJ Smith's, then you can become proficient at the great game of exploring on the edge of land.

Climbing is much more than just a sport. It's a kind of ritual, a spirit quest, a journey into the heart of your own life. It could even be an art form. Deep water soloing is perhaps the style of climbing that comes closest to all these things. Like an astronaut experiencing weightlessness for the first time, with the gift of water beneath their feet, the soloist enters a new realm in which the conventional laws of physics no longer apply.

Like Peter Pan in Neverland, climbers moving above the deep may transcend place and time, remaining ageless and powerful as they strive to master the shadow world between the land and the sea. The great polarities of life begin to swirl around us in this most sublime of earthly regions: love and loss, success and failure, innocence and experience, joy and sorrow. They are written down, these stories – all of them – in vanishing print on those imagined lines in the rock we climb, in the faces of our friends on and off the wall, and in the constant churn and surge of the sea beneath it all.

Here, amid the briny edge-land between earth and water, we are possessed by the action of wave over stone, of wind against tide, of refracted light flooding the darkness of the chasm we enter. Here, we can find out what it means to be truly alive. And here, too, we can grasp with crystal-clear certainty that one day, some day, we will die.

This is the liminal game.

Introduction

It may be too early to say that sea mountaineering has sufficient potential for development on separate lines, having its own traditions and literature. Some aspects, such as the long traverses, are different enough to justify speculation that it may one day come to be looked upon as a separate sport, in much the same way that British rock climbing became finally independent from Alpine mountaineering.

Peter Biven in *Tor*, The Journal of the Exeter Climbing Club, 1973

The style of this book will not be appreciated by everyone. It does not try to be a general history of sea cliff climbing in the UK. Instead, this book is focused on the more aquatic sea cliff exploits, i.e. those where you stand a serious chance of getting wet, and the sea is by necessity engaged with, not just a pleasant backdrop to the action. So, it's mostly about sea-level traversing, sea stacks and, above all, deep water soloing.

This book takes you on a clockwise circumnavigation of the British coastline, starting from the chalk of the White Cliffs of Dover. Along the way, the book deals with some particular questions: Did Aleister Crowley invent sea-cliff climbing, and was he the best climber in the world in the 19th century? Did the St Kildan natives really evolve prehensile feet? What is the truth about the famous story of commandos watching Menlove Edwards deep water soloing and then drowning while trying to emulate him?

What has been included and what has not may seem whimsical to some, and I make no apology for that. During research for this book I have encountered many rabbit holes that make for interesting stories, such as the under-appreciated climbing career of Mr Crowley.

For the early history of sea cliff climbing in the British Isles we can turn to Peter Biven:

'Sea cliff climbing as a sport, and distinct from activities such as mining or collecting birds-eggs, is of Victorian origin. Most of the earliest references are to the chalk of South-East England. The pioneers – famous alpinists such as Mummery, Tyndall, Whymper and the notorious black magician Aleister Crowley – were practising for Alpine mountaineering and treated the chalk as though it were ice or snow. Crowley climbed Etheldreda's Pinnacle on Beachy Head – a remarkable achievement and, in concept, years ahead of its time. Others were not unaware of the peculiar hazards of chalk climbing. As early as 1869, Tyndall wrote of Swanage: "These cliffs provided me with quite sensational risks – there are bits on these places as dangerous as anything in the Alps."

'But after the turn of the century, the mountaineering world unaccountably seemed to lose interest, and for many years hardly anyone climbed on our chalk cliffs. There are, however, signs that the wheel may be turning full circle. Beachy Head has been climbed, mainly with pegs and with the protection of a very long top-rope. Robin Collomb climbed one of the Needles on the Isle of Wight from a clinker dinghy in 1951. There have been sporadic attempts on the other Needles and Old Harry on the other side of the Solent. Devon climbers have recently completed some high-grade chalk traverses at Beer in East Devon. But, so far, the technique required for dealing with flints which pull out precipitately and the crumbling nature of the chalk has proved elusive.'

As you will discover from the chapters that follow, sea-level traversing has a long and rich history. Prior to his Himalayan exploits, Dr Tom Longstaff (1875–1964) climbed on, and traversed, Baggy Point in 1898. Another famous Himalayan climber, Albert Frederick Mummery (1856–1895) had earlier, in 1894, described 'traverses of great interest and no slight difficulty' on the chalk cliffs.

Traverses are the peculiar property of sea cliffs, which are often awkward or impossible to approach from their base for more direct ascent. Whole days, in fact, if it were desirable, a whole holiday lifetime of delightfully varied traversing at different levels may be had along the cliffs, high or low, of the west coasts of Scotland and of its islands, of Ireland, of England, and of parts of Wales.

Geoffrey Winthrop Young in *Mountain Craft*, 1920

Arthur Westlake Andrews (1868–1959) declared his ambition to 'make a complete traverse of the shore of the British Isles between low and high water mark'. He did not manage this, rather ambitious, project (and it has probably not been done to date), but he did climb extensively on the sea cliffs in Cornwall.

Clement Archer (1902–1967) pioneered the remote and serious Exmoor Coast Traverse in the 1950s. Investigating this massive sea-level traverse, which lies below the highest cliffs in England and has few escapes, took him at least five years and included such techniques as swimming while fully clothed and roped up, and the use of ice axes. In 1965, he fell and broke his leg at the base of the cliffs, resulting in an epic rescue.

The era of 'modern' sea-level traversing was ushered in by the efforts of John Cleare, Rusty Baillie, Peter Biven and Frank Cannings, amongst others, in the 60s.

This married contemporary technical difficulty to the significant objective dangers of sea-level traversing, resulting in some very long, and difficult, traverses. More recently, Coasteering – a term coined by Edward Pyatt in the 60s – in the inter-tidal zone has become a popular commercially guided recreational activity.

One of the earliest descriptions of what sounds like deep water soloing (DWS) in climbing literature referred to one of the many exploits of psychiatrist John Menlove Edwards (1910–1958). In the early 1940s, he developed a technique of using strong waves off the Cornish coast to lift him onto a cliff, whereupon he would pick out a line to the top. Three Commandos who were watching were said to have tried to emulate this feat but were tragically drowned.

DWS was developed independently by Miguel Riera in Mallorca, in the late 70s, and British climbers in the south of the UK around the same time. Pat Littlejohn soloed the early Torbay traverses in the late-60s and all four sections of The Magical Mystery Tour, in one push, in 1982. Nick Buckley soloed the perfect DWS line of The Conger (E2) in 1983, followed by a wave of DWS exploration from climbers, notably Crispin Waddy and Mike Robertson, who then exported the techniques to other areas.

Sea cliff climbing was revolutionised when DWS took off in the UK – followed by the rest of the world – in the 90s. The boom resulted from a change in climbers' mindsets about how best to approach steep cliffs with deep water underneath.

DWS is a branch of rock climbing that involves climbing without any equipment, except shoes and chalk, above water that provides a safe landing zone. It is a joyous type of climbing that enables solo climbing without the encumbrance of ropes, gear and belayer while, mostly, removing the fear of injury and death that is the constant companion of the solo climber otherwise. The style of ascent is, usually, necessarily pure – there is no shouting 'take', resting on the rope or dogging the moves. It's up or off, and the climber has to figure out the moves as they climb.

Most DWS routes are started along a sea-level traverse. A chalk bag is carried, and the climber stays dry until they fall in the water, at which point it's usually game over. Soloing above water was not 'new', but the systematic application of DWS techniques was. Suddenly, all those 'inaccessible' overhanging crags above water and lacking convenient belay ledges were open for business. To the uninitiated, DWS sounds and looks very dangerous. So there are, inevitably, a number of questions to be answered about the safety of DWS.

The first question is, what height above the sea is it safe to fall or jump from?

Traditionally in DWS, so far, up to 40ft has been regarded as almost completely safe, up to 60ft as mostly safe, 60–80ft as pushing the safety margin into the danger zone, 80–100ft as definitely involving serious injury, and possibly death, and above 100ft as simply soloing: no longer 'deep water' soloing, and definitely death if you fall off.

Many injuries and some deaths have, indeed, occurred during DWS. Injuries that seem to be relatively common are back injuries, ruptured lungs, perforated eardrums and involuntary enemas.

Deaths are, thankfully, rare but have happened – four DWS-related deaths have been recorded in Mallorca. Damian Cook (35) was one of the original Dorset DWS pioneers and tragically drowned at Cova del Diablo in 2004. He reportedly got into difficulties after falling into the sea on a day of bad conditions. Philip John Hawkes (25) died in 2012 after falling into the sea at Cala Serena while 'loaded with rope and carabiners'. In another incident in 2021, two American climbers, Brandon Scott Burns (25) and Mark Anthony Bacolod Stiles (35), both died at Cala Sa Nau. This double fatality was attributed to rockfall.

There are many variables that relate to the propensity for injuries, but the strongest variable is the physics of hitting the water. According to the online Splat Calculator, a 70kg person falling from 30ft will take 1.4 seconds to hit the water and will do so at a speed of 30 miles per hour. From 60ft this increases to 1.9 seconds and 41mph, and so on to 2.47 seconds and 54mph from 100ft.

Hitting the water is obviously a softer landing than hitting the ground, but there are still forces at play that can cause injury or death. Impact forces on the surface of the body can cause blunt trauma and fractures. Rapid deceleration of the body in water can cause rupture of internal organs and death.

The theoretical height limit for DWS probably lies above what is generally regarded as 'safe' currently, but in order to operate at such theoretical heights in practice, the climber would have to be highly skilled at altering their body position in mid-air in order to enter the water with the optimal streamlined position to minimise surface area, maximise water penetration and avoid injury. Some data regarding this theoretical limit can be gleaned from various sources, some scientific studies (often post-mortem) and some less rarefied.

The Golden Gate Bridge in California is the most notorious suicide spot in the world, with an estimated 1,500 takers for the 245ft drop. There are only 34 recorded survivors, and the death rate from jumping is therefore estimated to be around 98%. Suicide nets are supposedly in the process of being constructed at the time of writing. What is amazing is that 2% of people actually survive this fall which shows that it is humanly possible, albeit very unlikely, to survive a jump into water from 245ft.

In the 1980s, ABC's *Wide World of Sports* TV show featured world-record-attempt high diving. Their rules required contestants to execute at least one somersault and exit the water without the assistance of others. Dives from higher than 170ft were regularly shown, as were many serious injuries such as broken femurs and backs; by the way, these 'high dives' all involved entering the water feet-first. The highest head-first dives occur from similar heights to those of the famous divers at Acapulco, Mexico, from around 120ft above the water.

Facing page: Neil Gresham falling off The Flying Dutchman, F7b+ S2, on Lundy. Photo Dave Pickford.

The *Guinness Book of World Records* informs me that the world record for the highest dive is 192ft, 10 inches achieved by Swiss/Brazilian high diver Laso Schaller at Cascata del Salto, in Switzerland, in August 2015. At the time of writing, he is the current record holder for the highest dive from a diving board, simultaneously holding the highest cliff jump record for the same jump. Wearing a wetsuit, Schaller jumped feet-first from a specially built diving board platform next to the waterfall into the pool below, which was aerated with scuba tanks. In the video, his water-entry position looks good, perhaps leaning back slightly, but he had to be assisted from the water and appeared to be in pain, although his injuries were later described as 'slight'.

Døds or Norwegian Death Diving is a form of extreme freestyle diving from height, initially with extended arms and belly, landing hands and feet first in a foetal or cannonball position. This discipline may more closely resemble what could reasonably be expected to be achieved in the water entry position from a high DWS. The current world record for this, at the time of writing, is 112ft and is held by Frenchman Côme Girardot.

So – theoretically – a climber who also happened to have some of the skill set of someone like Schaller (who has a background in gymnastics, trampolining, canyoning and professional diving) or Girardot could contemplate taking on deep water solos in the 100ft-plus range, although bear in mind that in a fall from, say, 120ft the falling climber will only have 2.71 seconds of free-fall time within which to orientate themselves to the optimal body position before hitting the water at a speed of 60mph. This would fit anyone's definition of extreme.

The next question to be answered in DWS is what water depths are safe to land in?

Traditionally, in DWS, anything more than 20ft water depth has been regarded as deep and completely safe from any height, and anything less than 10ft as shallow and not safe. There is, of course, the 10–20ft grey area in between, which may be markedly affected by the tide. For example, Olympiad is a viable DWS at high spring tides, whereas you can walk around at the bottom of it and climb it as a trad route on a low spring tide.

The Olympic diving rules stipulate that the pool for 10m diving competitions must be at least 5m (16ft) deep, and a diver holding a streamlined position will, indeed, stop at around this depth. However, ongoing rotation during the diver's water entry will result in stopping at a shallower depth – typically around half of that. The 5m water depth rule for 10m height, one would suspect, is a conservative estimate. So, how shallow can you go?

Once again, some data can be gleaned from elsewhere. Shallow diving is a recognised extreme sport whereby enthusiasts attempt to dive from the greatest height into the shallowest depth of water without sustaining injury, and it has a long history. Roy Fransen was a British stuntman best known for his displays of acrobatic diving into shallow pools. And if that is too boring for

Above: Disclaimer posted at the 2003 DWS festival in Dorset. Photo Grant Farquhar. Facing page: A climber touching down on the bottom after a DWS fall into shallow water in Bermuda. Photo Andrew Burr.

you, these high dives were usually performed with both the diver and the water surface being set ablaze with burning petrol. His 1948 record high-dive, from a height of 110ft into a depth of eight feet, remained unbroken worldwide for 49 years. Fransen continued to perform his 'On Fire Into Fire' dives until he died during a performance in 1985 – at the age of 69 years.

At the time of writing, the Guinness World Record for the shallowest dive is held by Darren Taylor, aka Professor Splash, after he dived from 37ft,11 inches into 12 inches of water in 2014. Taylor's technique is best described as an extreme belly flop. Taylor is said to have sustained eight concussions during his career but never broken any bones.

What conclusions can be drawn from these statistics for application to DWS? Perhaps not many, except that it reveals the limits of what is possible, and there is not much else to go on. So the old 10ft and 20ft chestnuts are probably a good rule of thumb. If in doubt, it's best to snorkel the landing area first and inspect the water depth and sea bottom for obstacles. Sand is much more forgiving to hit than rocks or reef, for example. If the water is shallow, then the safest thing to do is wait for a high spring tide or look for another project.

If you are determined to climb the route in question anyway, then you can attempt to minimise water penetration by using such techniques as an armchair landing during which, on feet first striking the water, the falling climber will sit back – as if in an armchair – to maximise surface area and water resistance. This will, of course, increase the impact forces with the water surface and above 60ft may cause injury by itself. This devil's calculus of height and water depth is part of what makes DWS so interesting and challenging as a pursuit.

Above: Kelvin Briscall about to fracture his spine jumping off the top of Diablo, Mallorca. 'You are fine,' I delivered my diagnosis to a nonplussed Kelvin adding, 'you wimp' for good measure. Except he wasn't fine and when the pain failed to subside he sought a second opinion at the hospital where an X-ray revealed compression fractures to his vertebrae. In my defence, I'm a psychiatrist not an orthopaedic surgeon. Below: Tim Emmett making the 70ft jump off the top of Feist Queen, Mallorca. Photos Grant Farquhar.

A sea surface state with pronounced waves puts the climber at increased risk of striking the bottom if they fall into a trough, as I discovered one day in Bermuda while attempting the first ascent of a 45ft route above six feet of water. Over the years of DWS in Bermuda, I'd become more and more comfortable with falling into increasingly shallow water. You don't know where the line is until you step over it; I didn't realise it, but I was about to do precisely that.

Full-moon spring tide that day happened to be one of the highest of the year at 1.3m, but what I failed to take into account in my calculations was category 4 Hurricane Teddy, which was passing Bermuda and creating massive swells. In my enthusiasm to climb, I turned a blind eye to this inconvenient fact.

On my first attempt, I fell off and touched down on the reef, but not too hard. So I had another go and fell off again. This time touchdown came with a very hard and painful impact directly onto my heel – I had fallen into the same spot, but this time my fall coincided with a trough in the waves which meant that I'd landed into waist-deep water from 45ft.

Swimming and then exiting the water without further injuries was tricky due to the very rough swell surging in and out and up and down the sharp exit rocks waiting to impale the unwary. On dry land, I was not surprised to find that I couldn't put any weight on my right foot. Thirty minutes of crawling later, I had collected all my gear and got back to my bike. A sketchy one-footed ride and then an X-Ray at the hospital confirmed a fractured calcaneum and a premature end to the DWS season for that year.

***

Other questions to be answered besides sea bottom, body orientation, and water entry described above include sea-surface state and water temperature. Air in the water, whether caused by accident through waves or by design through aerated pools, reduces impact forces on hitting the water. You might factor this in when contemplating a DWS above a quarry pool versus the sea. Water temperature is most likely a factor in some deaths recorded during sea-level traversing and deep water soloing. If you are injured or struggling with gear in the water, then colder water will reduce your chances of survival.

What other objective hazards should the sea cliff aficionado take into account? For example, what are the chances of being swept away by a rogue wave?

Rogue waves (also known as freak waves, monster waves, killer waves, and sneaker waves) are unusually large, unpredictable and suddenly appearing surface waves. There is not a universally accepted definition of rogue waves, but one that has currency defines them as waves whose height is more than twice the significant wave height (which is itself defined as the mean of the largest third of waves in a wave record). Rogue waves probably don't have a single distinct cause but occur where physical factors such as winds, tides, currents and interference patterns converge to create a single exceptionally large wave.

Rogue waves were once considered mythical. Early eye-witness accounts such as damage to the RMS Lusitania in 1910 were taken with a pinch of salt – literally fishermen's tales – but through the more modern accumulation of hard evidence from satellite imagery, surface radars, pressure transducers, research vessels, oil platforms and wave buoys, rogue waves have now been proven to exist. The Great Wave off Kanagawa by Hokusai is probably the most famous example of a rogue wave.

Rogue waves are distinct from tsunamis which are caused by a massive displacement of water from earthquakes or landslides after which they propagate at high speed over a wide area. They are nearly unnoticeable in deep water and only become dangerous as they approach the shoreline and the ocean floor becomes shallower. Rogue waves (and tsunamis) are clearly potentially catastrophic events for the sea cliff climber but are so rare that they are probably not worth factoring into your risk calculations. You are more likely to encounter problems from more mundane and predictable tidal phenomena and boat wakes, such as the well-known wave generated by the regular Holyhead fast ferry at Gogarth. However, Joe Brown has a tale from an attempt on Spider's Web in Wen Zawn that may be an example of a rogue wave:

'It was a nice day, very calm with no wind. We had traversed across from Genuflex, and Fred had taken a stance at the base. I had climbed most of the first pitch of Spider's Web when I realised that we couldn't communicate; the noise was too great. I started to traverse back, and when I reached the arête, the first of a whole lot of big waves was just coming in – God knows what caused them – there was no wind. Anyway, this one completely buried Fred the Ted. I almost fell off laughing, saying, "Never mind Fred, I'll probably get wet next," not for one minute thinking that I would.

'The traverse back to the ledges by Genuflex was being covered by the waves, so we went 50ft up Archway, which we had climbed a few days earlier. Our plan was to abseil down, to above the waves, and then rush across the traverse when there was a lull. I started to do this, but when I was quite a long way above the traverse, about 30ft, I saw a big wave coming in. By the time I had realised just how big it was, it was too late to do anything. It came sailing into the zawn, right over the top of me; I was completely soaked.'

And, more chillingly, it may have been a rogue wave that killed American professional climber Michael Reardon on the west coast of Ireland in 2007, as Damon Corso witnessed:

'Without any warning, I saw Michael's eyes go wide and watched him scramble to the edge of the wall. A giant 15ft wave was coming right at him. He couldn't secure a good grip, and I watched in horror as he was sucked into the torrid ocean and rocks.'

Above: Under the wave off Kanagawa by Katsushika Hokusai (1760–1849). Sourced from the Metropolitan Museum of Art metmuseum.org. Open Access Public Domain licensed under Creative Commons zero licence. Below: Grant Farquhar exiting a DWS in Vietnam. Photo Neil Gresham.

An important question is, how do you access DWS routes?

DWS routes are very rarely started directly from the water. The goal is to stay dry as long as possible. Ideally, they have a relatively easy sea-level traverse to the base of the chosen route, which gives a good warm-up while remaining dry for the route. Sometimes, it's necessary to access a ledge or hanging stance by either abseiling or swimming in with a dry bag containing a towel, shoes, chalk, etc. Another, rather high-class, option is to gain the rock directly from a boat.

Other safety considerations include the water exit. Does it require a long swim? If so, are you a strong swimmer? Should you preplace a knotted rope to facilitate egress? Climbing with a partner is likely to be safer than climbing alone in most situations. The sea state and water temperature will be major factors at play once the water has been entered and the climber is swimming to an exit and are probably the most crucial factors when it comes to the fatalities described above.

What about hypothermia?

The seawater temperature in the UK varies widely from 6–10°C in the winter to 15–20°C in the summer. But, even in the summer, hypothermia is a significant risk. There is no set time for when hypothermia will set in, but generally the colder the water, the faster it happens (duh).

An illuminating study conducted by the Department of Sport and Exercise Science at the University of Portsmouth in 2019[1] found that the majority of their subjects could not complete a two-hour swim in 16°C water without becoming hypothermic.

That studies conclusions were: 1. Humans vary widely in their ability to maintain their body temperature in cold water, but the main factor is having less body fat (which describes most climbers). 2. Heat loss in water is increased by exercise (for example, a long swim in choppy sea to an exit below a cliff). 3. Swimming performance deteriorates before core temperature reaches the stage of hypothermia (so you may well have less time than you think). 4. Swimmers cannot be relied upon to assess their own state of hypothermia while swimming (you need to be wary of your judgement after a period in the water). Take home message: Even at the height of summer in the UK, you are at significant risk of becoming hypothermic if you spend two hours in the water without a wetsuit, and that risk is exacerbated by having a climber's physique.

What grading systems are used for DWS?

Climbers love grading systems, and DWS is no exception. 'S grades' were invented by the Dorset DWS pioneers in the 90s and are still in use. S0 is safe at most tides and not particularly high, with a low crux (up to 30ft). Care may be required on an S1 due to water depth or height

Facing page: Martin Atkinson on Mark of the Beast, F7c S2, DWS festival, 2002. Above: DWS festival 2003 at Connor Cove. Photos Grant Farquhar.

(crux up to 40ft). S2 is likely to have shallow-ish water, a high crux (up to 50ft) or objective hazards such as ledges or rocks in the landing zone. S3 will have a high crux and/or shallow water, and you most likely can't afford to make an uncontrolled fall off one of these without a high possibility of injury. S3s are connoisseur's routes for experienced soloists only who will need to do their homework. SX denotes a very dangerous solo that you can't afford to fall off and probably no longer 'DWS'.

French sport climbing grades are used to give a measure of the difficulty of the climbing on a DWS. English grades are also used but usually apply to the difficulty of climbing the particular route on ropes with trad gear, which is often a harder proposition than simply soloing the route.

For beginners, it is not recommended to go deep water soloing alone, to climb close to high tide, to work out the exit point from the water before embarking on a solo, and to climb only S0 and S1 graded routes.

What equipment do you need for DWS?

Multiple pairs of rock shoes and chalk bags, towels, and dry bags for any swimming approaches. Some routes finish on a feature part way up a bigger cliff, which means that the descent involves jumping. A small dry bag can be useful in this situation to put your chalk bag in prior to jumping.

1   Saycell J, Lomax M, Massey H, et al. *Br J Sports Med* 2019;53:1078–1084.

I had climbed and rambled for some 10 years in the Alps and the British Isles but till then had never realised the possibilities of what seemed at first glance a rather dull and low coast. I had had the ambition to make a complete traverse of the shore of the British Isles between high and low tide marks, as far as practicable, but have not yet finished it. We began our pilgrimage with a traverse from St Ives to Pendeen in which, though much was easy, we met unexpected obstacles.

AW Andrews quoted in *A Climber in the West Country* by EC Pyatt

Andrews's dream of the complete intertidal zone traverse of Britain has never been realised, and, in case you are contemplating it, there are significant obstacles to this challenge. Not least, how long is it and how much time would it take?

The entire coastline has, of course, been walked, and the times taken are an indication of the scale of this challenge. Quintin Lake, a self-employed photographer, completed it in sections. He started his 'Perimeter' photography project in 2015 and followed a clockwise route from London, returning home to edit, print and sell the images before setting out on the next leg. It took him 454 days of walking over five years. Christian Lewis, a former paratrooper, set out on a charity trek around the coastline in August 2017 and eventually returned to his starting point in July 2023, six years later.

'How Long Is the Coast of Britain? Statistical Self-Similarity and Fractional Dimension' is a paper by mathematician Benoît Mandelbrot. Published in *Science* in 1967[2], this is one of the first publications on the topic of fractals, although Mandelbrot does not use the term in this paper, as he did not coin it until 1975.

The paper examines the coastline paradox: the property that the measured length of a stretch of coastline depends on the scale of measurement. Mandelbrot argues that the smaller the increment of measurement, the longer the measured length becomes. If one were to measure a stretch of coastline with a yardstick (a three-foot ruler), one would get a shorter result than if the same stretch were measured with a (one-foot) ruler. This is because one would be laying the ruler along a more curvilinear route than that followed by the yardstick. This suggests a rule that, if extrapolated, shows that the measured length increases without limit as the measurement scale decreases towards zero.

Facing page: Mike Robertson DWS Perfect Pitch, F6a+ S1, in Pembroke with a coasteering party at the base. Photo Mike Hutton. Above: DWS fest 2003 at Connor Cove. Photo Grant Farquhar.

Coasteering

The virgin summit, even the little visited summit, has long since disappeared among home mountains. To do genuine exploration, it is necessary to travel very far afield indeed, while the making of a new climbers' route on a mountain crag face in these islands is only possible for the very expert. Coasteering opens up a new field here, providing in the shape of islets, pinnacles and so on, the unclimbed summits, which were the original objectives of mountaineering.

EC Pyatt. *A Climber in the West Country*

The first company in the UK to offer coasteering as a commercial activity was TYF in Pembrokeshire. Andy and Sarah Middleton bought the old temperance Twr Y Felin Hotel on the outskirts of St Davids, which became Twr Y Felin Outdoor Centre and started trips along the local coast in 1986. Since then, the activity has exploded in popularity. Over 150 companies now offer similar activities along the UK coastline under an advisory body called the National Coasteering Charter.

Therefore, it is difficult to accurately measure the UK coastline. Estimates vary from 7,723 miles in the *CIA Factbook* to 12,251 miles according to the World Resources Institute. Taking 8,000 miles as a nice rounded figure along with EA Newell Arber's rule of thumb of progress in the intertidal zone as one mile per hour gives us 8,000 hours which would take around three years to circumnavigate the UK intertidal zone assuming traversing for eight hours per day. The logistics of this would be a nightmare, and an unsupported solo effort would take a lot longer.

2   *Science.* 156 (3775): 636–638, 1967.

The Devil's Rejects by Grant Farquhar

I set my standards pretty low, so I'm never disappointed.

Otis B. Driftwood in *The Devil's Rejects* (film)

The global birthplace of deep water soloing is Mallorca, specifically Porto Pi. This venue had the primordial soup – it's highly accessible, lying on the outskirts of Palma, and a popular place for locals to jump and swim in the warm water. Bouldering already existed there in several sectors above platforms. The vital spark came in the shape of Mallorcan native Miguel Riera. He extended the bouldering from the platforms to above the sea on the neighbouring DWS area, which at around 12m is an ideal height for this. The route Cuquets, climbed by a teenage Miguel in 1978 at F6b, is possibly the first recorded DWS in the world. Miguel described this in an interview in *Climbing* magazine in 2016 with Kevin Corrigan:

'Psicobloc has been practised since the first time there was a climber, a sea cliff, and it was hot. I've seen old American videos of the Stonemasters playing and fighting over a lake and even older pictures of French climbers in Marseille doing long traverses over the sea. But it wasn't until we began to establish routes, name them, and grade them in Mallorca in 1978 that anyone considered it a sport. In Mallorca, Psicobloc arrived before sport climbing. We were still aid climbing up mountains at the time. Then George Meyers's *Yosemite Climber* book fell into our hands.

'Imagine: shirtless hippies bouldering. We wanted to do that, too. In Mallorca, the best place to do it was on the sea cliffs. To us, we were simply bouldering like the Americans in Yosemite. In England and Germany, the sea is cold and rough. It's not a good place for kids. For us, it's an exciting game to jump from 15m into the sea, nothing more. For climbers who come from outside the island, that's very psychological. When we saw that, we stopped calling it "bloc" (bouldering in Catalan) and began to call it Psicobloc (psychological – or psychotic – bouldering).'

DWS in the UK took off in the 90s with Tim Emmett and Neil Gresham being keen proponents. They met some Austrian climbers, including Klem Loskot, at the Ice Climbing World Cup in 2000 in Pitztal. The UK climbers got the Austrians excited about DWS and they decided to book a trip to the UK for a long weekend. After showing them around Lulworth and Connor Cove, Klem mentioned a guy who was doing some DWS in Mallorca. Tim had also heard of this guy who worked in a climbing shop. Tim contacted the shop, and it turned out to be Miguel Riera. Miguel sent Tim a picture of himself, taken in 1992, in fluorescent-yellow shorts high above the water on what looked like the most amazing DWS cliff Tim had ever seen. This was Diablo. Miguel had climbed there in the 1980s, but his Mallorcan amigos had apparently stopped DWS, and he was the only one still interested in it. So, in October 2001, a UK team and an Austrian team converged on Porto Cristo to meet up with Miguel and be shown around.

***

We waited to meet Miguel Riera at a cafe in Porto Cristo. There was a tank of lobsters in the cafe, one of which was a truly gigantic specimen. We looked at it. It looked back at us. We were thinking about climbing. It was probably thinking: don't eat me. After coffee, we dragged ourselves away from the comfort of the cafe, and the lobster, and drove the short distance out of town so that Miguel could show us the promised crag: Diablo.

Left top: Tim Emmett and Gav Symonds. Left middle: Mike Robertson. Left bottom: Kirilee Wood and Lauren Matthews at Diablo. Photos Grant Farquhar. Facing page: Grant Farquhar on Afroman, F7b+ S1. Photo Neil Gresham.

As Miguel pointed out the three routes that he had already climbed 20 years previously, our DWS expert, Mikey Roberston, wasn't waiting for the briefing. Like James Bond ignoring Q's instructions on a lethal new toy, Mikey was immediately off and, before long, could be seen clearly enjoying himself dangling off jugs high on the right-hand arête of the crag.

The immaculate central sweep of the crag appeared, amazingly, to be unbothered by routes. There were some in situ sun-bleached-white threads visible, which Miguel had apparently placed but never climbed the route. Neil abbed down to have a look and then quickly soloed it, to our surprise, as it looked a lot harder than the F7a grade he gave it. This was the route Lobster, named after our friend in the cafe. Over the next few days, we carried out the devil's work and threw ourselves at the routes which sometimes threw us off hence the name of the route Ejector Seat. The routes were, without exception, fantastic. Calamares was, literally, named after lunch. Kirilee Wood was Superwoman. Afroman was a rapper in the charts at the time and also described Tim's (and Ken's) hairstyles.

Speaking of Afroman, I watched Tim make the first ascent of this, which was an impressive display of stamina and his ability to recharge his guns on steep ground. This impression was reinforced when it took me two days of attempts to eventually get up it. Afroman, however, would be in my personal list of the top 10 routes of the world. It is a perfect DWS route in terms of rock quality, height, water depth and, for me, in 'the zone' when it comes to grade.

Loskot and Two Smoking Barrels was a superb name, made up by Neil, although it took Klem, the Austrian dynoing machine, a while to understand the pun based on the name of the popular film at the time for his first ascent. Klem was an outstandingly strong climber and an interesting, somewhat extreme, character.

One morning, between coffee and climbing, I went with him to the shop. He wanted to buy a stamp to send a postcard. The little old lady behind the counter smiled benignly at him and appeared keen to make small talk following his attempted request for the stamp in Catalan.

'Where do you come from?' she asked in broken English, obviously trying to start a dialogue.

'Does it matter?' he barked.

Small talk was probably not his strong point, I thought, but climbing was. Maybe at that point in his life, nothing else mattered.

Left top: Miguel Riera, Kirilee Wood, Gav Symonds and Neil Gresham at Diablo. Left middle: Neil Gresham, Lauren Matthews, Ken Palmer, Gav Symonds, Mike Robertson, Grant Farquhar, Kirilee Wood and Tim Emmett at 'the lobster' cafe. Left bottom: Tim contemplating the jump off the top of Feist Queen. Facing page: Mike Weeks enjoying Lobster, F6c+ S2, at Diablo. Photos Grant Farquhar.

Neil and I watched him DWS the first ascent of In the Night Every Cat is Black. He moved lithely through the crux section while Neil (who had abbed it) commentated:

'That hold is shit, so is that one, and that one. Blimey!'

At night, we slept out on the parapet of the Tower of Falcons, which was located above one of the crags. The ancient spiral staircase inside this local tourist attraction had crumbled away to the point where you had to basically use chimney-climbing techniques to get up it. This would have been very intimidating for non-climbers but pretty straightforward for climbers, even when drunk. This arrangement lasted for a while until early one morning, the Guardia Civil, no doubt alerted by late-night music and partying, showed up and ordered us to come down. They were not, it seemed, amused by our antics.

There was a brief standoff. They strutted about and huffed and puffed in that Spanish way, like frustrated matadors, but it was clear that there was no way they were going to climb up the tower. It looked like they weren't going to go away, though, and they had guns. We also wanted to go to the cafe for breakfast, so we gave up and descended. Hands in the air, we promised not to use and abuse the tower anymore, and with that, they appeared mollified and let us go on our way.

Meanwhile, the ever-affable Miguel carried on climbing with us and showing us around with his local knowledge and generally going out of his way to assist. For example, one day the wind was strong out of the east, so he took us to an obscure Mallorcan west-coast DWS venue that was in the lee and in good condition to climb.

One day, I watched him abseil down the middle of Diablo, bouncing in and out to arrive neatly on the ledge below Lobster and thereby avoid the long traverse in. That looks easy, I thought, and tried it only to find myself somewhat red-faced and stranded in space before, with no other option, abseiling into the sea: harness, GriGri, chalk bag and all.

On the last night, Miguel took us out for an extended visit to the shanty town of unofficial bars and clubs in the hinterland of Porto Cristo before driving us bleary-eyed in the early morning to catch our flight and for him to start work. Miguel worked at the airport at the time driving the baggage-handlers truck, about which he liked making the joke that it was a harder job, 'for the real experts', than driving the aeroplane.

Josh and Brett Lowell produced the 2003 climbing film *Dosage 2* using footage they filmed at Diablo of Miguel, Tim and Klem, along with later footage of Chris Sharma, which brought Mallorcan Psicobloc to the attention of the masses. Miguel featured in a number of other films and TV programs, and he wrote the *Psicobloc* climbing guide to Mallorca. Tragically Miguel died in 2019, aged only 56 years, from lung cancer.

Above and below: Strong Oz climber Chris Jones attempting and then being ejected from Ejector Seat, F7c S1. Facing page: Neil Gresham on Morning Glory, F7a+ S1. The Tower of Falcons is visible above. Photos Grant Farquhar.

There is a cliff, whose high and bending head

Looks fearfully in the confined deep.

Bring me but to the very brim of it,

And I'll repair the misery thou dost bear with something rich about me.

From that place I shall no leading need.

Shakespeare. King Lear

Chalk

Climbing on Beachy Head has a place of its own among the fine arts. But let no devotee seek to penetrate the shrine of this spotless deity when she forbids – I mean on wet and windy days. On the former, the soddened chalk is slippery and dangerous, and too unpleasant, in fact, to be indulged in by the most enthusiastic. Dry, windy days, on the other hand, when chalk particles, varying from fine dust to large nuggets, are being driven about, are fatal to the eyes, which may be bloodshot and sore for days afterwards. In external appearance, also, the worshipper must face the infidel fashion of Eastbourne and be scorned for a miller or a baker; but then, these are small drawbacks to the glorious sensations that await him who is careless of mockers. But the mighty goddess reserves a terrible doom for the profane who would impiously violate her sanctuaries; he will be a foolish man who treats her otherwise than with reverence and respect. Go then, with true worship, undaunted, and your reward shall be joy unspeakable in the glorious divinity of sun-glistening altitude and towering whiteness.

Aleister Crowley

The Victorian age (1837–1901) saw the creation of most organised modern sports: The Football Association was formed in 1863, and the league began in 1888. The first cricket championships were held in 1873, and the first Wimbledon tennis championship was in 1877. Golf, rugby, boxing and amateur athletics were all spawned in the same era. Others, presumably craving more adventure, turned to Alpine climbing. The Golden Age of Alpinism (1854–1865) saw the formation of the Alpine Club in 1857 and was bracketed by Sir Alfred Wills's ascent of the Wetterhorn and the tragedy on the Matterhorn. Some Alpine peaks, including Mont Blanc, had been climbed prior to 1854, but the majority were first ascended over the subsequent decade.

The prominent white cliffs on England's southern shore have long been associated with its island identity. The North and South Downs outcrop where they reach the sea near Folkestone, Dover and Eastbourne. At these points, and for some distance in both directions along the coast, steep and often vertical faces form the most familiar cliffed shoreline in the land. They are often taken as a symbol to represent the country as a whole and are from which the name 'Albion' is derived. Chalk is a fine-textured – and very soft – type of limestone that is notoriously unreliable to climb on.

Nevertheless, the white cliffs are a spectacular and very obvious climbing challenge that was not overlooked by some of those living in the Victorian age. Despite being porous, soluble, crumbly and extremely slippery in the wet, the chalk cliffs, along with other UK crags, were regarded as good training for alpinism as Frank Smythe noted in his book *Mountaineering Holiday*: 'Anyone who can gaze unmoved upon the sea from the edge of Beachy Head need have no qualms as to his reactions on the most sensational of Alpine precipices.'

Facing page: Beachy Head.
Photo John Cleare.

Chalk, and the methods of dealing with it, form a study by themselves. Chalk climbing provides the missing link between rock and ice technique. Those who frequent its cliffs use big claws and ice methods and pronounce it to be an unrivalled training for ice work. Its occasional hard surface and its abundant projecting flints, whose security is in inverse ratio to their graspability, have to be treated with the measures of precaution proper to unsound rock.

Geoffrey Winthrop Young in *Mountain Craft*

So, in the early days of Alpine mountaineering, chalk was used for practice, being soft enough to enable steps to be cut in places and steep enough to require them. The friable rock and lack of good belays were not of such importance to the early alpinist as they were to the later rock climbers.

The white cliffs of Old England have proved themselves too much for many of her sons. Chalk is a very troublesome material to climb; it is loose, breaks away, and, if wet, either with rain or the draining of the land above, forms a sticky paste, which, lodging between the boot nails, renders them of little use. Doubtless, many of those who get into difficulties are only shod for a walk on the promenade or pier, but chalk cliffs must not be treated carelessly even by the well equipped.

Charles Pilkington in *Mountaineering*

Albert Frederick Mummery (1855–1895) grew up in Dover, where his family resided at Maison Dieu House, an old Jacobean mansion adjoining Dover Town Hall. Mummery became famous for his ascents in the Alps. He died during an expedition to Nanga Parbat and was later described by its first ascensionist, Hermann Buhl, as 'one of the greatest mountaineers of all time'.

Mummery started climbing at the age of 16 years and was one of the earliest to sample the merits of the nearby white cliffs. Mummery found that the poor quality of the chalk forced him to explore horizontally, rather than vertically, to make sea-level traverses between Dover and St Margaret's Bay. These take a series of ledges and steps from beach to beach and thus produced the sort of traverse which may have influenced the young AW Andrews and other coasteering pioneers. Mummery contributed the notes on chalk to Walter Parry Haskett Smith's 1894 magnum opus *Climbing in the British Isles* which described the explorations by Alpine climbers on the 'training' crags to be found in the UK, including the sea cliffs:

'Though this can hardly be regarded as a good rock for climbing, much excellent practice can be gained on it. As a general rule, it is only sufficiently solid for real climbing for the first 20ft above high-water mark, though here and there, 40ft of fairly trustworthy rock may be found. These sections of hard chalk are invariably those which, at their base, are washed by the sea at high tide; all others are soft and crumbly. Whilst any considerable ascent, other than up the extremely steep slopes of grass which sometimes clothes the gullies and faces, is out of the question, traverses of great interest and no slight difficulty are frequently possible for considerable distances. A good objective may be found in the endeavour to work out a route to the various small beaches that are cut off from the outer world by the high tide and cliffs.

'The best instances of this sort of work are to be found along the coast to the eastward of Dover (between that town and St Margaret's). Between the ledges by which these traverses are in the main effected, and the beach below, scrambles of every variety of difficulty may be found, some being amongst the hardest mauvais pas with which I am acquainted. Owing to the proximity of the ground, they afford the climber an excellent opportunity of ascertaining the upper limit of his powers. Such knowledge is a possession of extreme value, yet in most other places it is undesirable to ascertain it too closely.

'Chalk, it must be remembered, is extremely rotten and treacherous, very considerable masses coming away occasionally with a comparatively slight pull. In any place where a slip is not desirable, it is unwise to depend exclusively on a single hold, as even the hardest and firmest knobs, that have stood the test of years, give way suddenly without any apparent reason. The flints embedded in the chalk are similarly untrustworthy; in fact, if they project more than an inch or so, they are, as a rule, insecure. The surface of the chalk is smooth and slimy if wet, dusty if dry, and does not afford the excellent hold obtained on granite. As a whole, it may be regarded as a treacherous and difficult medium, and one which is likely to lead those practising on it to be very careful climbers.'

Edward Whymper (1840–1911) was an English mountaineer, explorer, illustrator, and author best known for the first ascent of the Matterhorn in 1865, during which four members of his climbing party were killed. One of the earliest references to climbing the 500ft Beachy Head comes from his 1855 diary: 'July 21st to August 9th: I went to Eastbourne, which is a rather large country town on the south coast near Brighton. I did not like the place itself at all. I went from thence to Beachy Head, where I nearly broke my neck trying to climb the cliff.'

Whymper wrote several books on mountaineering and exploration including his classic 1871 book *Scrambles Amongst the Alps* which mostly concerns his attempts on the Matterhorn. However, Whymper also remembered his chalk escapade in the book: 'On the 23rd July 1860, I started for my first tour in the Alps. As we steamed out into the Channel, Beachy Head came into view and recalled a scramble of many years ago. With the impudence of ignorance, my brother and I, schoolboys both, had tried to scale the great chalk cliff. Not the head itself – where sea birds circle and where the flints are arranged in orderly parallel lines – but at a place more to the east, where a pinnacle called the Devil's Chimney had fallen down. Since then, we have been often in dangers of different kinds, but never have we nearly broken our necks than upon that occasion.'

Whymper included an engraving of Beachy Head in his book. Over the next few years, various publications including *The Leisure Hour* magazine, *The Graphic* and *The Standard* newspapers carried reports of several rescues from the head by the local coastguard. Various uncorroborated claims of successful ascents were also sensationally reported, but the first widely accepted ascents on Beachy Head were made in 1894. 'Insensate folly takes various forms. One of the most dangerous is that of attempting to climb chalk cliffs on the southern coast' is how the *Eastbourne Gazette* described the ascent by an 18-year-old Edward Alexander Crowley who, at this period, was known by 'the dreadful name', Alick, but he would soon change it to Aleister.

Much has been written about Aleister Crowley (1875–1947), not least by himself. The English occultist became infamous as 'the wickedest man in the world' and remains an influential and polarising figure today. Crowley was clearly aware of climbing from an early age, and this features prominently in his 'autohagiographical' *The Confessions of Aleister Crowley*. He recalls posing a *Touching the Void*-esque riddle to his uncle at around 12 years of age: 'Suppose a climber roped to another who has fallen. He cannot save him and must fall also unless he cut the rope. What should he do?'

Crowley lists mountain climbing amongst various other childhood adventures feeling that it was 'natural' for him to find ways up mountains which looked interesting and difficult. In April 1894, Crowley had been sent from boarding school at Tonbridge to live in Eastbourne with a tutor from the Plymouth Brethren Christian sect that his family belonged to. During the day, he visited Eastbourne College. The college website, today, is at pains to point out that 'despite various myths', Crowley did not attend as a pupil nor teach at the college. At the time, he 'assisted' the Head of Science, Professor Robert Edward Hughes, in the chemistry laboratory. In his spare time Crowley was variously 'hunting flappers on the front, playing chess or climbing Beachy Head'.

Chess was something that Crowley had a natural talent for, and he edited a chess column in the *Eastbourne Gazette,* but his 'grand passion' was Beachy Head, as he described in his *Confessions*:

'Most chalk cliffs are either unbroken precipices, unclimbable in our present stage of the game, or broken-down rubble; but Beachy Head offers rock problems as varied, interesting and picturesque as any cliffs in the world. I began to explore the face... The fantastic beauty of the cliffs can never be understood by anyone who has not grappled them. Mountain scenery of any kind, but especially rock scenery, depends largely on foreground.

Right top: Aleister Crowley c1925. Commons.wikimedia. org. Creative Commons Public Domain Mark 1.0.
Right middle: Aleister Crowley as Baphomet X° OTO c1919. Photo Arnold Genthe. Commons.wikimedia.org. Creative Commons Public Domain Mark 1.0.
Right bottom: Crowley on the lower Baltoro Glacier during his 1902 expedition to climb K2 in the Himalayas. Photo Jules Jacot Guillarmod. Letemps.ch. Creative Commons Public Domain Mark 1.0.

'This is especially the case when one has acquired an intimate knowledge of the meaning, from the climber's point of view, of what the eyes tell one. The ordinary man looking at a mountain is like an illiterate person confronted with a Greek manuscript. The only chalk in England which is worth reading, so to speak, is that on Beachy Head.'

Crowley made a number of ascents of conspicuous features on the head with his cousin, Gregor Grant, including Etheldreda's Pinnacle, Devil's Chimney, and attempted the line that he christened 'Cuillin Crack' and was later first ascended by Mick Fowler. He had to be rescued from the latter, but these routes were futuristic and possibly the hardest rock climbs in the world at the time. This seems like an outrageous claim: Aleister Crowley, the best rock climber in the world? Ironically, despite his penchant for hyperbole and publicity, this is a claim that Crowley never made for himself, but by examining his climbing history and the difficulty of his routes as discovered by repeat ascensionists, we can evaluate this.

The teenage Alick started off with ascents of various Munros in 1890 and made an ascent of the Pinnacle Ridge of Sgurr nan Gillean in September 1892 in 'wet and windy' conditions. Crowley had met Sir Joseph Lister at the Sligachan Inn, who persuaded 'some real climbers' to include him in their party. Crowley found himself 'up against it' and realised 'that there was something more to be done than scrambling'. Crowley visited Snowdonia for the first time in April 1893. It is, perhaps, unsurprising that he should have been drawn to the feature called the Devil's Kitchen. The embryonic sport of rock climbing was not as advanced in Snowdonia at the time as it was in the Lake District, but two unclimbed features were recognised as the outstanding challenges of the time: Slanting Gully on Lliwedd and Twll Du aka the Devil's Kitchen. The latter is described by William Russell Reade[1] in the *Climbers' Club Journal,* 1898:

'A huge cleft in the rocks on the southwest flank of Llyn Idwal. A stream flows through it to the lake below. Roughly speaking, the chasm is about 450ft long, 300ft deep, and 18 or 20ft wide, the sides being either vertical or overhanging. The floor rises, not in a gradual slope, but in a series of pitches.'

In his 1894 application to join the SMC, Crowley documented the 'first ascent' of the Devil's Kitchen in May of 1893. This coincides with the Great Drought, which prevailed over the spring and early summer of that year and would, presumably, have resulted in favourable conditions. In his *Confessions* he wrote:

'The Devil's Kitchen or Twll Du, which I had climbed by taking off my boots. I had no idea that the place was famous, but it was. It was reputed unclimbable.

Almighty Jones himself had failed. I found myself, to my astonishment, the storm centre. Jones, behind my back, accused me flatly of lying. Quite unconsciously, however, I put myself in the right. I have always failed to see that it is necessary to make a fuss about one's climbs. There is a good reason for describing a first climb. To do so is to guide others to enjoyment.'

Crowley is referring to Owen Glynn (OG) Jones (1867–1899) who was the leading rock climber in the country at the time having taken up the sport, as legend has it, after seeing one of the Abraham brother's photographs of Napes Needle in a London shop window. Jones's later partnership with the two photographers, along with his skilful self-promotion, was to have a huge influence in popularising the sport. He was known for his gymnastic feats and climbing style which led to the first ascents of some of the hardest routes in the country in the Lake District such as Kern Knott's Crack, VS 4c, in 1897, possibly the first 4c.[2]

Jones was criticised by some, notably Crowley, for his use of a top rope and his publicising of the sport through photography. Maybe he was the Neil Gresham of his day? It's interesting to see that top roping was seen as not sporting from the very beginning of the sport – the games climbers play. In his 1911 book *Rock Climbing in the English Lake District*, OG Jones introduced a grading system for rock climbs that formed the basis of the system we still use today. He divided the climbs into a classification system of four 'sets': Easy, Moderate, Difficult and Exceptionally Severe.

OG Jones was presumably not the only climber to dismiss Crowley's claims. Therefore, his ascent is at odds with the official chronology of ascents which, according to the guidebooks, is as follows: The Devil's Kitchen was first ascended in March 1895 by James Merriman Archer Thomson, the leading climber in Snowdonia at the time, in exceptional winter conditions during the Great Frost of 1895.

Believing the kitchen to be unclimbed, OG Jones attempted it on 31st May of that year but failed. George and Ashley Abraham climbing with Oscar Eckenstein also failed in 1897. Geoffrey Winthrop Young retreated from Jones's high point in May 1898. The day after Young's attempt, the route was ascended as a summer rock climb by WR Reade and WP McCulloch who are duly credited with the first ascent in the guidebooks. OG Jones is credited with making the second ascent in 1899. Crowley's claimed first ascent in 1893 appears to have either simply been overlooked or, more likely, not believed by the cognoscenti and therefore ignored.

Later, in 1893, Crowley first visited what was then, in the closing years of the 19th century, the cradle of British rock climbing, Wasdale Head in the Lake District. There he met the pioneers of the day, including the man who was to be the greatest influence upon him as a climber, Oscar Eckenstein (1859-1921): 'Eckenstein was a man 20 years older than myself. His business in life was mathematics and science, and his one pleasure mountaineering.

---

1   William Russell Reade (1874–1945) became known to all his mountaineering friends by his initials 'WR' to distinguish him from his contemporary Herbert Vincent Reade (1870–1929), who was often a member of the same party.

2   This now gets HVS 4c on UKClimbing.com.

'He was probably the best all-round man in England, but his achievements were little known because of his almost fanatical objection to publicity… The combination was ideal. Eckenstein had all the civilized qualities, and I all the savage ones. He was a finished athlete; his right arm, in particular, was so strong that he had only to get a couple of fingers onto a sloping ledge of an overhanging rock above his head, and he could draw himself slowly up by that alone until his right shoulder was well above those fingers… In the actual technique of climbing, Eckenstein and I were still more complementary. It is impossible to imagine two methods more opposed. His climbing was invariably clean, orderly and intelligible; mine can hardly be described as human. I think my early untutored efforts, emphasized by my experience on chalk, did much to form my style. His movements were a series; mine were continuous. He used definite muscles; I used my whole body. Owing doubtless to my early ill-health, I never developed physical strength, but I was very light and possessed elasticity and balance to an extraordinary degree.'

This opinion about Crowley's climbing style, perhaps influenced by the chalk, was echoed by Tom Longstaff, who described Crowley as 'a fine climber, if an unconventional one'.

Napes Needle had been first ascended in 1886 by WP Haskett Smith, an event widely regarded as the birth of the sport of rock climbing, with climbers such as Haskett Smith, Mummery, Geoffrey Hastings, Collie and William Cecil Slingsby at the forefront of this new sport. Repeats of the needle followed in 1889, including the second ascent by Geoffrey Hastings. The needle was first climbed by a female, Miss Koecher, in 1890, thus, presumably, completing Mummery's infamous three stages of a climb: An inaccessible peak, the hardest climb in the Alps, and, finally, an easy day for a lady.[3] This route nowadays gets a grade of Hard Severe but was originally graded Very Difficult. This may be an example of 'grade creep', but it has also got harder due to the polish from thousands of ascents.

In June 1893, Crowley made a solo ascent of Napes Needle by the SW route and returned in September 1893 to make the first ascent of his 'crack and traverse' route on the needle. The 1926 guidebook credits him with the first ascent of this 'Variation II', and the FRCC *Gable & Pillar* 2007 guide describes this as 'The Crowley Route', and it gets a grade of Hard Severe.

In 1894, Crowley made further ascents in the Lakes including Professors Chimney, Deep Ghyll, Steep Ghyll, Pillar Rock, and the third ascent of Kern Knott's Chimney making the first ascent of 'the sensational step' at the top without the aid of a shoulder. Crowley was the first to climb this unaided:

'Jones had been blowing his trumpet about the first ascent of Kern Knotts Chimney; the top pitch, however, he had failed to do unaided. He had been hoisted on the shoulders of the second man. I went to have a look at it and found that by wedging a stone into a convenient crack, and thus starting a foot higher up, I could get to the top and did so.[4] The following day a man named HV Reade, possibly in a sceptical mood, followed in my footsteps. He found my wedged stone, contemptuously threw it away, climbed the pitch without it, and recorded the feat. That was a double blow to Mr Jones. It was no longer a convincing argument that if he couldn't do a thing, it couldn't be done.'[5]

Crowley also climbed Mickle Door Chimney South Face exit 'which was always thought impossible' and the first ascents of the Ridge West of Arrowhead Ridge and the Direct Climb from Mickle Door, 'one of the original impossibilities of the district':

'This time, the fat was in the fire. My good faith was openly challenged in the smoking room. I shrugged my shoulders but offered to repeat the climb the following day before witnesses – which I accordingly did. I suppose I am a very innocent ass, but I could not understand why anyone calling himself human should start a series of malicious intrigues on such a cause of quarrel. I must admit that my methods were sometimes calculated to annoy, but I had no patience with the idiotic vanity of mediocrities.'

Crowley roped up with OG Jones, of whom he formed a low opinion: 'I was only once on a rope with Jones. It was on Great Gable; the rocks were plastered with ice, and a bitter wind was blowing. In such conditions one cannot rely on one's fingers. Our party proposed to descend the Oblique Chimney on the Ennerdale face. Robinson led the way down. The second man was a Pole named Lewkowitch, who was generally known as "Oils, fats and waxes" because of his expert knowledge of them and the personal illustration of their properties which he afforded. He had no experience of climbing and weighed about 16 stone. It was up to me, as third man on the rope, to let him slowly down. I had, of course, to descend little by little, the rope being too short to allow me to lower him from the top.

'I soon found myself in the most difficult part of the chimney, very ill-placed to manipulate a dangling ox. I looked up to Jones, the last man, to hold my rope so that I could give full attention to Lewkowitch and saw to my horror that he was maintaining his equilibrium by a sort of savage war dance! He was hampered by a photographic apparatus which was strapped to his back. Robinson had urged him to lower it separately. As neither Einstein nor the Blessed Virgin Mary was there to suspend the law of gravitation, I have no idea how we got to the bottom undamaged, but when we did, I promptly took off the rope and walked home, utterly disgusted with the vanity which had endangered the party.

---

3 Mummery's quote from *My Climbs in the Alps and Caucasus* is often referred to. However, Leslie Stephen had previously described this progression of climbs in *Playground of Europe* (1871): 'Since the first summer I spent in the Alps, more than one excellent mountain of my acquaintance has passed through the successive stages denoted by the terms inaccessible, the most difficult point in the Alps, a good hard climb, but nothing out of the way, a perfectly straightforward bit of work, and, finally, an easy day for a lady.'

4 Crowley recorded this in the book kept at the Wastwater Hotel which was started for the use of climbers in January 1890.

5 According to Mike Cocker, Reade dispensed with the stone six months later (not the following day, as recollected by Crowley).

'Of course, there could only be one end to that sort of thing, and Jones ended by killing himself and three guides on the Zinal side of the Dent Blanche a few years later… Jones obtained the reputation of being the most brilliant rock climber of his time by persistent self-advertisement. He was never a first-rate climber because he was never a safe climber. If a handhold was out of his reach, he would jump at it, and he had met with several serious accidents before the final smash. But his reputation is founded principally on climbs which he did not make at all, in the proper sense of the word. He used to go out with a couple of photographers [The Abraham brothers] and have himself lowered up and down a climb repeatedly until he had learnt its peculiarities, and then make the "first ascent" before a crowd of admirers… This conduct seemed to me absolutely unsportsmanlike. To prostitute the mountains to personal vanity is, in fact, something rather worse.'

These feelings appear to have been reciprocated by Jones, who (according to Crowley) accused him of lying about his ascent of the Devil's Kitchen. OG Jones referred to his practice of climbing with a camera sack in his book, *Rock Climbing in the English Lake District*, as 'a frequent source of annoyance in a tight place' and described the same incident: 'I remember making a mistake when descending the great overhanging pitch at the bottom, in assuming that it was an easy matter to climb down with a camera sack on my back.'

When Crowley moved to Eastbourne in April 1894, he applied his new-found climbing skills to the obvious challenge of the nearby chalk cliffs. In Crowley's opinion, chalk was probably the most dangerous and difficult of all kinds of rock:

'Its condition varies at every step. Often, one has to clear away an immense amount of debris in order to get any hold at all. Yet indiscretion in this operation might pull down a few hundred tons on one's head. One can hardly ever be sure that any given hold is secure. It is, therefore, a matter of the most exquisite judgment to put on it no more weight than is necessary. A jerk or a spring would almost infallibly lead to disaster. One does not climb the cliffs. One hardly even crawls. Trickles or oozes would perhaps be the ideal verbs.'

Crowley regarded his chalky predecessor, Mummery, as 'a genius' and 'the greatest rock climber in England', but thought that Mummery was incorrect in his stated view that chalk was only good for traverses. He wrote to Mummery pointing this out. Mummery was initially sceptical but warmed up to Crowley after he was sent pictures of the routes. As well as writing to the local newspaper, Crowley contributed an illustrated article on the subject to the 1895 *Scottish Mountaineering Club Journal*:

'In wet weather the chalk forms a paste which clogs the boots and makes foothold impossible. In dry weather the dust takes possession of the eyes and throat. But for all that, many of my happiest days have been spent on the face… My associations with Beachy Head possess a charm which I have never known in any other district of England.

Above: Crowley's application for membership of the SMC in 1894. Courtesy Scottish Mountaineering Club.

Tom Patey traversing out towards Etheldreda's Pinnacle on 8th June 1969. Photo John Cleare.

'My climbs there fulfilled all my ideals of romance, and in addition, I had the particularly delightful feeling of complete originality. In other districts I could be no more than primus inter pares. On Beachy Head I was the only one – I had invented an entirely new branch of the sport.'

This claim of Crowley's is not without foundation. The chalk cliffs were, indeed, the first sea cliffs in the UK to have rock climbing routes on them. On 4th April 1894, Crowley made the first ascent of Etheldreda's Pinnacle by the Castor chimney on the east side of the pinnacle with his cousin Gregor Grant: 'Directly I saw this magnificent pyramid I determined to climb it at once. Two chimneys, side by side, and since named Castor and Pollux, presented the most obvious route to the ridge, joining the pinnacle with the mass of the cliff. The north one (Castor) looked easy, and so I determined to try it. The chimney was almost entirely filled with chalk dust of the consistency of fine flour, caked on the top and having blocks of various sizes in the middle. All this, at the touch, came down, and the whole weight jammed on my legs, which were well into the chimney. A convulsive series of amoeboid movements enabled me to get out over the debris when it immediately thundered down, leaving me in a very comfortable gap. I was soon over the jammed stone and onto the ridge.

'My friend refused to follow, but as he was roped, I put on sudden pressure, and he, well, changed his mind. When he reached the ridge, I went on for the north face. This has several natural steps, but the first two yards required a few gentle touches with the magic wand, the chalk being very hard. Thence the route lay westward to the NW corner and then back again to the shoulder, only one step being at all awkward. The whole climb, however, requires a method of climbing which may be described as a chemical combination of the writhe, the squirm, and the slither. This indeed applies to all difficult chalk.'

Crowley described Etheldreda's Pinnacle as his 'first great triumph' and claimed to name it 'after my dog, or a schoolgirl with whom I had stolen interviews, I forget which'. In June and July 1894, Crowley made a number of other ascents on a feature called the Devil's Chimney west of Etheldreda's Pinnacle, which he named Dent de Voleur and Aiguille du Bois and made a repeat ascent of Etheldreda's Pinnacle from the west:

'On July 13th, Grant having arrived, we made another attempt upon Etheldreda's Pinnacle and climbed it by a difficult chimney on the opposite side of the cleft from Castor and Pollux. This chimney is 50ft high, is shaped like a bow, with the concavity facing outwards, and is closed towards the middle by several large jammed stones. The angle above these is not so severe, and another block forms a fine natural arch. Grant took the lead, and after some arduous climbing, which involved a good deal of going outside at the jammed stones, emerged victorious on the top. From this col, my previous route was followed to the top of the pinnacle, and in descending we made a new variation by way of the Pollux chimney.'

Crowley then attempted his 'Cuillin Crack' but had to be rescued by a rope from above: 'This chimney is nearly 200ft high and affords the finest and most difficult piece of climbing that I have yet found in the whole neighbourhood. It is broken at two places, one near the bottom and the other about 100ft higher up. The first break presents terrible difficulty, but after incredible exertion it yielded, and then I got a leg and an arm jammed, and managed to wriggle up about 60ft higher. At this point, the rope and my strength were alike exhausted, some four hours without any sort of rest having already passed.

'Foothold and handhold there were none that could be relied upon to support Grant's additional weight if any pressure were to be put on the rope. There was nothing for it but to let down a rope from above or to descend with ignominy and much toil. So Grant sped away to invoke the assistance of the coastguard, and meanwhile, I sat wedged in a most uncomfortable position, at the bottom of the second break, up to which I had struggled while awaiting Grant's return. Presently a rope was let down from above so that by tying my own length of 60ft to it, I could scramble to the top, climbing for the most part, but receiving now and then aid that was more moral than any description of it as such. Whoever first climbs the Cuillin Crack will have no reason to be displeased with his trophy and will be able to reflect with satisfaction that the superior character of the chalk renders this climb more justifiable than a great many others.'

It took nearly 90 years for Crowley's routes to see repeats. In 1969, John Cleare and Tom Patey tried to repeat Etheldreda's Pinnacle but failed to reach the access chimneys. A photo of Patey is reproduced in Edward Pyatt's *Climbing and Walking in South East England*, where he can be seen traversing out towards the pinnacle. Arnis Strapcans recorded in his diary that he made the second ascent of Etheldreda's Pinnacle on 21/1/79. He abseiled into the notch between the pinnacle and the main cliff and found the ascent 'terrifying'. He gave it a 'chalk-ice grade' of III.

On 13th April 1980 and suitably armed with ice axes and using ice screw belays and runners, Mick Fowler with Mike Morrison and Brian 'Blob' Wyvill climbed the pinnacle via the second ascent of Grant's Chimney on the west side of the pinnacle. Fowler thought the grade of the route to be VS 4b and then turned his attention to Crowley's 'Cuillin Crack'. Fowler conceded that it looked 'horribly intimidating', but the route succumbed by incorporating modern ice climbing techniques. Fowler renamed the route Crowley's Crack and graded it XS 5a:

'Crowley, of course, would not have had such protection, and I could not help wondering if I would have succeeded had I been climbing with his equipment... Crowley was outrageous! He was obviously good; something of a star rock climber, in fact... He was light years ahead of his time in his attitude to tackling vertical chalk cliffs. The ground he covered was without a doubt amongst the most technically difficult in Britain, but his achievements were never really appreciated in his lifetime.'

Above: Etheldreda's Pinnacle and Crowley's Crack. Photo Brian Wyvill.

Below: Arnis Strapcans's diary entry describing his ascent of Etheldreda's Pinnacle. Photo courtesy Gordon Jenkin.

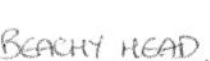

Chalk and Cheese by Brian Wyvill

A sunny Sunday, and Beachy Head was crowded with tourists moving aimlessly, sometimes watching brightly-coloured hang gliders and often lurching dangerously near the edge of the famous chalky precipice. The dog pulled me along, straining on her lead to find Messrs Morrison and Fowler.

Kim, the dog that is, was good at remembering people; she always remembered Morrison by the taste. A commotion at the top attracted my attention, and I caught sight of a blue uniform. That's where they must be, I thought and set off to see what sort of trouble this unorthodox pair of climbers had found.

'Good morning officer, anything wrong?' I used my best authoritative voice, and Kim growled menacingly.

'Some would-be climbers, sir, you can't be going down there. It's dangerous. I've sent for reinforcements.'

Pulling the dog away from the constable's trousers, I said, 'Actually, officer, I know them both. They are…' (I resisted the temptation of saying a couple of lunatics) 'very experienced alpinists.'

The policeman narrowed his eyes and studied my scruffy climbing apparel, 'I suppose you are an experienced alpinist too, eh?'

I couldn't tell a lie: 'Well, actually, yes.'

'Well, suppose you tell your experienced alpinistic friends to get themselves off the cliff before I call out the coastguard.' At that moment, reinforcements arrived with a megaphone.

'Good afternoon, sergeant, I have the situation well in hand here.' I said, getting in my bit before the constable could say anything to the contrary.

'That looks dangerous.' The sergeant's stance looked even more dangerous, balancing on the white crumbly edge peering at the two Mikes who were eyeing up Etheldreda's Pinnacle, first climbed in 1894 by Aleister Crowley.

'Do you know to whom you are speaking?'

'I don't care who you are. You ain't going climbing up these cliffs on my beat.'

'But, sergeant, we are Alpine Club members.'

'Oh! Alpine Club, you say?'

'That is Michael Fowler, a leading light of the British Alpine Club. Part of the new generation, whose predecessors climbed Everest for the first time in 1953.'

'Everest? Oi've heard of that.'

'For our training, we are attempting to climb Etheldreda's Pinnacle, which has not been attempted since Aleister Crowley climbed it in 1894. This is a historic occasion. Do you really want to stop us? Was this the spirit in which Everest was climbed? Would Sir John Hunt have succeeded if he was forbidden from training on British rock?'

'Well, since you put it that way, sir. Let me clear the crowd back from the edge to give you more space. Come along now, you lot. This is an official British expedition. Give them space there.'

'Thank you, sergeant, much obliged.'

By now, we had attracted a lot of spectators. At that moment, Kim ran between the sergeant's legs; he tripped over the lead and narrowly avoided plunging to chalky death.

'I wouldn't stand near the edge if I were you, sergeant,' I said. The law continued arguing for a few minutes before repairing to the safety of the Eastbourne nick, satisfied with my assurances that I would keep an eye on our heroes.

Meanwhile, I made my way down to the beach to observe the climb more closely. Kim and I started up to the base of the hundred-foot pinnacle, which the two Mikes ascended separately, not daring to belay on the rickety summit. Our ascent was barred by a steep crack which looked at least canine E3. Anyway, Kim's jamming technique was too paw (!) to make it, so we returned to the top by the easy path.

Down below, Mick Fowler emerged at the top of the pinnacle, looking like an advert for a white tornado. After a brief search for any satanic symbols left by Aleister Crowley, he descended amidst tons of rubble. Fowler managed to reach the safety of an ice-screw belay, and the performance was then repeated by Morrison, much to the amusement of the growing crowd of onlookers at the top.

'A fine new route', declared Fowler as he eyed a horrific-looking chimney line in the main cliff. This apparently was the same line as that attempted by Crowley when he had been rescued. Fowler set off, undaunted by the failure of so great a climber as Crowley, attacking the line with Terrordactyls, ice pegs and lunacy. The rumbling of discarded blocks grew louder as Mick became more at home with the 'chalky severity'. With frequent cries of, 'it's unjustifiable!' he pirouetted from block to block, often pausing to kick off the hold that he had previously been hanging on to. Anita, Mick's girlfriend, looked on, apparently unconcerned.

'If you marry him, you could be a young rich widow,' remarked one of the onlookers darkly.

'This is solid compared to Berry Head,' retorted Anita, whose comment only served to increase the doubt over Mick's continued good health.

At last, Mick squeezed through the final narrow chimney at the top. I looked down to see his totally white face framed against the grass, looking delighted with his efforts. The pitch was a full 150ft, so he was glad to use a second rope to belay. Morrison came up, and they both looked at me:

'Come on Blob, your turn now,' said Fowler enthusiastically.

Having originally agreed to go with them, I couldn't really back out, so I belayed Kim to Anita's leg and the rope to a grass stake. Down I went to the start of the pitch, where the full horror of the situation faced me.

Despite the fact that I had a top-rope, I was far from safe. This must be what it was like in the trenches, I thought as I suffered a full frontal bombardment of the Beachy Head chalk battalion.

With great courage and perseverance, plus the fact that there was no choice (and a big crowd of onlookers), I started to make doubtful progress up the chimney. The secret was to treat it as 'rice'; something halfway between rock and ice.

Every move caused a ton or so of what schoolmasters love best to rumble down the crack, and I soon became covered in white dust. I wondered if there was a chalk equivalent to silicosis; after all, I was breathing an atmosphere of only 50% air.

Near the top, I felt a growing panic. It was not the danger of the situation but the sudden realisation that I may be accused of breaking one of my cardinal ethical rules: there I was halfway up Beachy Head, but was anyone going to believe that it was a chalk-free ascent?

Although the *practice* of rock climbing was an important component of Victorian mountaineering in the Alps, the *sport* of rock climbing as a pursuit in itself gradually evolved simultaneously in the Lake District, Saxony and the Dolomites in the late 19th Century from an alpine necessity to an athletic sport in its own right. Haskett Smith's solo of Napes Needle in 1886, as publicised by the Abraham brothers' photographs, was the catalyst in England. Others were also pursuing rock climbing as an end in itself in the UK around the same time: Jim Puttrell in the Peak District, AW Andrews in Cornwall and, of course, Crowley on the chalk.

How do Crowley's routes compare in difficulty to the other routes of his day? We can't go and climb Crowley's routes to see for ourselves because a rockfall in early 2001 removed Etheldreda's Pinnacle and Crowley's Crack. However, after the repeats described above, we can arrive at a grade of VS 4b for Etheldreda's Pinnacle (in 1894). The Devil's Chimney routes were never repeated (except by Crowley) before they fell down. Members of the Alpine Club had climbed from the base of the Devil's Chimney to the gap between the tooth and the top of the cliff and thence to the summit of the cliff but had refused to attempt the tooth because they (correctly) believed that the pinnacle was 'likely to collapse into the sea without warning' (Crowley and Grant had climbed it from the gap). Crowley had to be rescued from 'Cuillin Crack', which was later climbed by Fowler at XS 5a, so that was an outrageous attempt for the time but doesn't count as he didn't get up it on the lead.

Why were Crowley's routes dismissed? Was Crowley a good enough climber to have made his ascent of the Devil's Kitchen? The answer to that, on the basis of his described and witnessed climbing skill, is definitely yes. Crowley was very well informed about the topography of the routes he was doing in the Lakes and Snowdonia.

In his application form to join the Scottish Mountaineering Club in 1894, he stated that he climbed the Devil's Kitchen 'out by left face (i.e. the southeast face)', which is the correct description for where the route goes. It's also clear that he was a very bold and talented climber. He was also keen to get full credit for his pioneering climbs. So why didn't he? There are many possible explanations for this.

Stephen Reid: 'In terms of the first VS in the Lakes, I think that is generally thought to be Eagle's Nest Ridge Direct (1892) on the Napes which was led on sight by Godfrey Solly – a very bold undertaking. Solly also led the Hand Traverse finish to North Climb in the same year, having climbed it on a rope once. Although this is graded Hard Severe, that is taking into account that these days you can pretty much lace it with protection. In 1892, it would have been very serious – the handholds are good, but there are no footholds, so it's extremely strenuous, and you are climbing some 15ft above a nasty ledge.

'I think even as a young man Crowley made himself so unpopular that he found it hard to get people to go climbing with him, so much of what he did, or claimed to have done, was solo with no witnesses. My wife and I repeated a short route of his on Pisgah a couple of years ago that had never been in the guidebook, and that turned out to be Mild Severe, and very exposed, so I guess he could have soloed VS. One thing was, and this is not often realised, that proper belaying techniques did not come into use until around 1917 when George Bower wrote an article on using a shoulder belay in the *FRCC Journal*. Prior to that, the rope appears to have been just held in the hands, though the belayer was usually tied to a spike or a chockstone if they could find one. So, effectively, and considering the low breaking strain of hemp rope, all the leaders in those days were soloing.'

Crowley visited the Alps annually between 1894 and 1898, climbing without guides in contrast to the members of the Alpine Club before making audacious, but unsuccessful, expeditions to the then-unclimbed K2 and Kanchenjunga:

'Mountaineering, I saw, was primarily a scientific problem... I discovered independently the facts of the case. I found that I could go pretty well anywhere without the least danger or difficulty, whereas all the people I met were constantly on the brink of disaster. I began to think that solitary climbing was the safest form of the game... My experience with chalk helped to give me confidence. I was accustomed to estimating the breaking strain of rotten material.'

Whilst he was alive, Crowley was out of favour with his contemporaries and his achievements were dismissed. He did get elected to the Scottish Mountaineering Club in 1894 and was a member until 1902, but he was not able to join the Alpine Club in January 1896 despite being nominated by Collie and seconded by Conway. It appears that he was not formally blackballed, but his reputation had preceded him, and his application was withdrawn before a vote took place.

In death, Crowley has become a cult figure whose life story is intertwined with his own mythology which consists of 'magickal' feats, exaggerations, embellishments, half-truths and lies. Fantastic tales about his life have been widely disseminated in the popular press, and he remains an influential cultural figure.

So, was Aleister Crowley the best climber in the world in 1894? He certainly held his own with his contemporaries, who were considered to be the best climbers of the day. He made many first ascents, including the Devil's Kitchen which was recognised as a major challenge and rebuffed attempts from the likes of Jones and Eckenstein. He certainly was one of the first sea-cliff climbers and climbed up-routes on the chalk sea cliffs rather than the traverses that his predecessors had pioneered. However, at the time rock climbing was a new sport with vague ethics and few participants. Perhaps a good analogy is that of the Wimbledon tennis tournament.

The world's first official lawn tennis tournament was held in 1877 at the All England Croquet and Lawn Tennis Club in Wimbledon. The winner, Spencer Gore, was born and raised within a mile of the club. Was Spencer Gore the best tennis player in the world, or even England, in 1877? Probably not, but he was better than the 20 other entrants and won the championship for which his name is remembered for posterity. Crowley's achievements have been dismissed by the climbing establishment, but his routes on chalk at the grade of at least VS were, indeed, most likely the hardest rock climbs in the world in 1894. And how else do we measure ability in climbing other than by the grades of routes ascended? By this metric, Aleister Crowley was, arguably, the best climber in the world in 1894.

It appears that Crowley did not climb much once in his 30s as he was busily occupied with the occult, sex magick, drug addiction and notoriety. Of course, from a scientific perspective, magic is, to use a technical term, a load of bollocks, but in that respect no different from the organised magic that we call religion. In fact, Crowley founded his own religion – Thelema – self-identifying as the prophet entrusted with guiding humanity into the 'Æon of Horus'.

But this is not the last you have heard from him in this book. Similar to the horror movie monster or Bond villain that you think has been vanquished in the movie – like a Schizotypal Terminator: He'll be back.

Facing page: An attempt to make a sea-level traverse to the Needles, on the Isle of Wight, with support from a small inflatable on 23rd October 1971. Photo John Cleare.

The Isle of Wight by John Cleare

Attracted by the colour of the cliffs
Which here stands vertical and tell a tale
Of dire commotion when the level beds
Of this fair isle were wildly tossed on end.

Ballad of the Isle of Wight by John Tyndall, 1856

The complex geology of the island gives rise to many rock formations and correspondingly varied coastal scenery. The vertical folding of chalk is plainly evident on the western tip of the island. On the north side of this promontory, working out of Alum Bay from coloured sandstone structures towards the Needles, and on the south side in Scratchell's Bay, which can only be entered from the sea, ribs of rock rise in a perpendicular plane and alternate with horizontal and angular strata. The cliff in Scratchell's Bay is more imposing than anything on Beachy Head, and parts of the north-side cliff, tinged yellow and orange from sandstones and oxides, resemble Dolomitic walls. The rock is pitted with flints which offer unreliable holds.

Any climbing interest centres on the Needles – three distinctive chalk stacks about 100ft high running in a line out to sea, with a lighthouse poised on the base of the outer and tallest one. Strong currents and a tidal race surge through gaps between the stacks, making landing a tricky proposition. The central pinnacle was climbed by Robin Collomb in 1951, using an old clinker rowing boat, on the south face from the inner end, then along the razor edge. In a calm sea, it is possible to sail a dinghy under canvas through the gaps, though conditions good enough to achieve this are quite rare. In normal conditions, the only practical and safe procedure is to use a power boat and transfer to a rubber dinghy towed for landing purposes.

In 1971, a party of four attempted to make a sea-level traverse from Alum Bay all the way to the Needles. At low water, and after several duckings in the sea, they were stopped about 150 yards short of the target by two caves in the last section of cliff before the promontory. At this point, the sea heaves against the cliff all the time. The first assault party of two therefore took to a seven-feet rubber dinghy and rowed up to the stacks where they were promptly overturned by a huge wave crashing through the first gap. This episode proved the futility of attempting the Needles by any means except from a motor boat anchored at sea.

The Isle of Wight chalk is prolonged as a submarine reef in the Channel opposite Bournemouth. It reappears inland on Ballard Down. The promontory extending from this hill into the sea is the Foreland, which divides the popular bays of Studland and Swanage, just South of Poole Harbour. The tip of this point is composed of razor-edge foreshore stacks, undercut by small caves and tidal channels and known collectively as Old Harry Rocks. From the furthest stack, out to sea stands Old Harry itself, a slender shining-white pinnacle, one of the most impressive chalk stacks in Britain, and some 70ft high. It can only be reached by swimming 40 yards, or from a dinghy, and looks quite unclimbable by any methods known today with the possible exception of bolting. A successful landing itself would be quite an achievement as a rough sea continuously washes its base.

In the narrow channel between Old Harry and the outer razor-edge stands Old Harry's Wife, a dumpy little rock of 25ft, accessible at normal low water by wading up to the knees in a strong current. Ian Howell climbed it in 1971 and found carved initials on top. The foreshore stacks were reputed to have been climbed before 1971. However, no trace of an ascent of the outer stack was found by the 1971 party, which climbed it and built a cairn facing Old Harry. An attempt on the inner stack failed by 15ft due to approaching dusk and an impatient speedboat driver waiting to take the party off the rocks. The peg for the top-rope descent from this stack was removed by dint of two members at the bottom pulling on it with the rope. As if to have the last say in the matter, the peg hit one of them before landing on the beach.

Stacks To Go At by Mick Fowler

Chalk climbing is a dog with a bad name, but it does not deserve hanging – only see that it is muzzled.

Herbert Somerset Bullock

The average mountaineer will tell you that all summits in Britain have been climbed for centuries and that the country offers little in the way of exploratory excitement. It will come as a surprise, then, when I say that some of England's most spectacular tops only received their first ascents in the last five years, and many in Scotland are still virgin. In describing the ascents themselves, terms such as 'difficulty of approach', 'technically exacting climbing' and 'serious situations' come to mind. They are normally associated with mountaineering in the greater ranges but are equally applicable to something much closer to home.

I talk, of course, of sea stacks. Dotted around the British coast are many of the challenges of traditional mountaineering but in a very different setting. Few climbers appreciate the wealth of exploratory excitement still available. Before the challenge of the stack can be accepted in full, it is necessary to look at the sport of mountaineering in a new light; an affinity with the sea is desirable. It must be admitted that taking to the sea did not come naturally to the Fowler body. After a childhood of mediocre swimming ability, I never dreamt that I would ever be in a position to enthuse over the joys of sea stacks and associated seaborne epics.

Back in the early 1970s, when sea cliff climbing first became universally popular, I had a distinct preference for mountain scenery. It somehow seemed wrong – obliquely against the mountaineering ethic – to descend to the foot of a cliff in order to climb back up again. Nevertheless, alter initial sessions on the increasingly popular cliffs of Bosigran and Gogarth, real enthusiasm was gradually generated for the sea cliff pleasures of difficult approaches, serious retreats and variable rock. Many battles followed, firstly on Littlejohn's North Cornish gems and then on new ground, exploring the intimidating shale cliffs of North Devon and Cornwall. There are some sea stacks in the vicinity of these cliffs, but they did not really compare with the real challenges that I will come to later. Mind you, there were a few among them to whet the appetite.

On the North Cornwall coast, Long Island and Lye Rock (near Tintagel) provoked numerous character-building swims after their first routes in 1982, and Gunver Head near Padstow provided memorable moments on the first ascent in 1985 when a Tyrolean anchor pulled, and Paul Bingham took a spectacular fall into the sea. Nevertheless, exciting as they were, these exploits lacked the challenge of what I now refer to as the 'true sea stack'. Firstly, they were not always surrounded by sea, and secondly, they often had disappointingly easy routes to their summits. The combination of virginity, isolation and difficulty are now deemed essential for maximum enjoyment.

It was visits to the chalk headlands around the Needles on the Isle of Wight and to Handfast Point in Dorset that first drew my attention to the wealth of spectacular and obviously difficult unclimbed pinnacles or 'true sea stacks' around the coast of Britain. Action was precipitated by Chris the Newcombe's sudden urge to attempt the sheer and unclimbed 70ft chalk stack now known as Press Gang Pinnacle at Handfast Point near Swanage. A boat was judged necessary, and Chris aptly emphasised his enthusiasm for the project by tracking down a small, obscure boatyard in Basildon which was prepared to take the risk of hiring a 12ft inflatable and outboard engine to us totally inexperienced and potentially incompetent mariners.

Left: Three chalk stacks stand off Handfast Point – from back to front – Old Harry, Press Gang Pinnacle, and Tusk AKA The Wine Bottle. Photo John Cleare. Facing page: Rupert Hoare on top of the middle Needle, Isle of Wight. Photo Mick Fowler.

Ian Howell on the summit of Old Harry's Wife in October 1971 with Old Harry beyond. Photo John Cleare.

Thus, on 2nd September 1985, Chris, Lynn Newcombe, Lorrainne Smyth, and I found ourselves on a very public slipway on Swanage seafront pondering the problem of how to create a seaworthy item from 150 pounds of floppy rubber, five wooden boards and three keel sections. Much amusement was afforded the crowd of holidaymakers, but the end result was thumbs up all round as the hopeful quartet zig-zagged in a vaguely controlled fashion towards the stack itself. On the way we passed Tusk Pinnacle – the most impressive stack on the south coast and one whose presence we noted for the future.

The climbing on Press Gang Pinnacle proved not to be particularly memorable, it being a rather loose and unspectacular HVS. Nevertheless, the final few feet to reach the knife-edge summit gave considerable interest as the top proved to be so loose and narrow that it was necessary to substantially change the skyline profile in order to stand on the highest point. The name was derived from the peculiar, continuous mantel shelving technique required to climb this type of rock. However, the quality of climbing was secondary; the important thing was that an obsession had been born. Even in the immediate vicinity of Press Gang Pinnacle, true sea stacks such as Tusk Pinnacle and the well-known landmark of Old Harry remained unclimbed. Across the Solent, the Needles on the Isle of Wight bad no recorded ascents. It was clear that the Basildon boatman would have to be persuaded to part with his craft.

Six months later, after an outlay of £400, *Deflowerer I* (as we christened her) was ours, and the assault proper could begin. As most of the impressive stacks in Southern Britain are made of chalk, the experience I'd gained in climbing on chalk cliffs was to prove invaluable in the search for objectives. Having acquired the boat, our attention immediately became directed towards the extremely spectacular 120ft-high knife edge of Tusk Pinnacle at Handfast Point. May 1987 had the boat back in action and three of us – Andy Meyers. Sonya Vittoris and myself – belayed beneath the soaring crackline splitting the extremely narrow West Face.

The face is so narrow that at the top it seemed obvious that any attempts to fist jam the crack would substantially alter the summit structure. Taking note of this probability, Andy made a start up the spectacular ground on the North West arête. In sharp contrast to nearby Press Gang Pinnacle, the rock was excellent, and two 5b pitches left the three of us perched astride the 20ft-long knife-edge summit, taking turns at standing gingerly on the highest point.

Epics were non-existent, and we felt well-satisfied with our efforts. In sharp contrast to this successful start, our second visit to the pinnacle resulted in the only occasion so far (!) in which both climbers have ended up in the sea together. Remarkably this admirably incompetent performance was in no way a reflection of our climbing ability but could be directly attributed to our non-existent nautical talents. Initially, the problem was one of no ledges at the foot of a prominent groove on the seaward face, which we had spied as a possible line.

Duncan Tunstall and I effected a precarious landing whilst Nicki Dugan, Lynn Newcombe, and Jackie Stead fended the boat gently away from the pinnacle. Unfortunately, much dithering over the best belay spot ensued, and 15 minutes later, Duncan was perched on a tiny ledge in the groove with me in a distressingly intimate position bridged out above him. At this point it became clear that the painter was tangled around my body. This realisation was prompted by being suddenly and violently pulled backwards (the result, I later discovered, of some rather overenthusiastic fending off by the female contingent). In my frantic efforts to grab anything vaguely secure, Duncan's jacket came to hand. Alas, Duncan was not belayed, and the ensuing swim added a certain piquancy to the day's experience. Hurling my waterlogged chalk bag in the direction of the boat and watching it fall wide and sink forlornly provided a fitting finale.

The climb itself could have been an anticlimax, but such words are rarely applicable on chalk stacks. The excitement culminated with strenuous and frightening 5b climbing up the groove and a final 5a knife-edge providing Duncan, Nicki and myself with the second ascent. The only remaining stack at Handfast Point was now Old Harry. Her virginity had remained intact despite the attentions of suitors for many years. Their attempts to win her over ranged from bolting to attempting to cast a line across from a nearby, easily accessible, stack. The stack itself resembles an ice cream cone sticking out of the sea. It is about 30ft long, 15ft wide and 75ft high and is, perhaps, unique in that it gently overhangs all the way around.

It's a formidably challenging obstacle which is complicated further by a 50ft-wide channel which can only be waded (in neck-deep water) at low tide. The discomfort factor and the dangerous Solent tides make an approach by boat desirable for all but the most adventurous. In fact, I had swum/waded to the base of the pinnacle in 1980 and remember classing it as 'impossible' for the free climber. On that occasion, a retreat was made to Swanage where the local ironmonger was dismayed to hear why we were purchasing his stock of six-inch nails. Perhaps, fortunately, the next day the weather was bad, our aided attempt never materialised, and the nails are still rusting in my garden shed.

So in June 1987, the rather leaky *Deflowerer I* was scraping against the seaward face. It was time to prove wrong my judgement of 1980. Chris Newcombe and I attempted to gain a small ledge on the high tide mark whilst Mark Lynden and Andy Meyers alternated between failing to manoeuvre the boat and successfully bailing out the rising water (it must be admitted that the rising water level in the boat was eventually attributed to nobody having put the drainage plugs in). Although our nautical skill still left much to be desired, good luck seems the most vital ingredient on such ventures, and a liberal dose on this occasion allowed Chris and I to lodge a precarious footing on the stack. An obvious belay ledge at 35ft gave us something to aim for, and after an initial dither on the first 10ft we were soon safely established beneath the final 35ft overhanging wall.

Technically difficult moves protected by good ice screws then led to the crux: a horrifically loose overhanging six-foot section leading to the top. Pulling over with a loose flint in one hand and an incredibly poorly-rooted grass tuft in the other was a pure delight. As I lay panting in the grass, it struck me that something was distinctly amiss. Fifteen feet away, a rusty reinforcement rod protruded from the foliage – it was the trademark of Simon Ballantine, the man famed for introducing the reinforcement rod to Dover chalk.

A phone call the following day confirmed my suspicions. In May 1986, he and John Henderson had waded out, climbed the stack and swum back. The south coast climbing scene is sufficiently mellow for him not even to have bothered to report such a major ascent. A sharp contrast indeed to what we have become used to in more traditional areas. In fact, I had to give it to them that their efforts showed considerably more willpower and imagination than ours. Noting a conveniently low spring tide, they had waded out at 5am wearing wetsuits and rock boots and carrying ice axes, ice screws and a four-pound lump hammer.

Having reached the top via the landward side with a combination of ice, rock and aid techniques, they then slept there for a few hours, intending to simultaneously abseil when the tide went out. However, these strong-willed young men soon realised that they were in with a chance of making the pub by lunchtime. A premature abseil resulted, and having pulled down and coiled the ropes whilst treading water, the intrepid duo did not let the weight of two axes, 10 pounds of hardware and the four-pound hammer prevent them from swimming 100 yards to the nearest beach and prusiking up to fence posts on the cliff top. A commendable performance indeed; successful in all respects.

By 1988, the only south coast sea stacks which I knew to be unclimbed were two of the three Needles on the western tip of the Isle of Wight. The inner one had been climbed in 1985, but the situation with regard to the other two was less certain. Rumours of previous ascents were rife, but nothing could be confirmed, and on seeing ascents recorded in his sea cliff climbing book, we contacted John Cleare. No, he said, he hadn't climbed them, but his friend Robin Collomb definitely had. We phoned Robin Collomb – no, *he* hadn't succeeded, but John Cleare definitely had. And so the situation stood: minimum confirmation, maximum confusion.

Having pressed the boat into action, the weekend rapidly developed the atmosphere of a children's holiday outing. Isle of Wight sun hats bearing the insignia 'Up the Needles 1988' were donned all round, and a variety of water wings and plastic lifebelts sprouting ducks' heads formed our safety equipment in the event of a capsize. Perfect weather induced some serious sunbathing as the trusty 3.5 HP engine throbbed against the tides. Our route along the coast gave us a fine view of Rupert Hoare's one-day sea-level traverse from Alum Bay to Freshwater Bay. Well over five miles long and with a lot of hard climbing and swimming. This traverse undoubtedly provides one of the most tiring outings in the south. Rupert, whilst ineffectually trailing a fishing line over the side, vividly recounted the various epics they had experienced to Andy Meyer, Jon Lincoln, Nicki Dugan and myself as we scorched ourselves en route to the action.

The first plan was for the middle Needle: The south face looked most appealing, firstly because it caught the sun and secondly because it obviously provided the easiest line to the summit. The lighthouseman later informed us that the only known previous 'ascent' of this stack was by the RAF, who apparently use the area as a training ground and periodically drop poor defenceless crew members onto the summit, where they are abandoned for many hours in the middle of cold, windy, winter nights. Presumably, this builds their characters immensely.

Having reached the summit by way of a pleasant Severe we, perhaps, imitated their actions by leaving Rupert stranded on top for him to photograph the four of us making an ascent of the outer stack. This one proved more problematic, not because of the difficulty of the climbing, but more due to the proximity of the lighthouse, which is actually built on the outer edge of the Needle. It appeared that it was not normal practice for the public to have access to the concrete platform on which the lighthouse stands and from which we wanted to start climbing. In fact, it was only because we had informed the coastguard of our intentions that we were able to land at all.

Acknowledging our obviously extensive nautical experience from the unusual array of safety equipment on display, the keepers soon became more relaxed, and our curious desire to climb the Needle became a source of real interest. Having enjoyed tea and biscuits, we duly climbed the seaward arête. It had seen more than its fair share of hold enlarging and was apparently first climbed in the 19th century by the workmen building the lighthouse. It now sports several reinforcement rods and, despite appearances, succumbs at about Difficult standard. It has probably been a trade route for generations of lighthousemen – so much for a first ascent.

And so to the current state of play. I know of no major stacks on the south coast which remain unclimbed, but most have had only one or two ascents and are ripe for new routes which can be guaranteed to provide excitement. Also, please note that my knowledge is far from exhaustive. Hidden gems are still coming to light – the explorations are far from over.

Skeleton Ridge by Mick Fowler

Climbers generally seem to have come to the conclusion that it was altogether too dangerous. It must be admitted that, at any rate, it is very unpleasant.

Aleister Crowley

'Dad, are there any good climbs on those cliffs?' I asked.

I was 13 years old, and we were crossing the English Channel on the way to the Alps. The grasslands of southern England that roll in small swells toward the south coast had suddenly ended in brilliant, sheer, vertical white. But although my father had introduced me to the short sandstone outcrops 20 miles south of our home in London, he wasn't a devoted rock climber, and he didn't know.

Three years later, I'd started visiting the sandstone crags on my own, and soon I acquired a group of core climbing friends. As avid readers of climbing literature, we were intrigued to learn that when Tom Patey had tried to repeat Etheldreda's Pinnacle, his team had found the steep grass and rock slopes so challenging that they'd failed even to reach the technical climbing. And yet, Patey was very much a leading adventure climber of the day. Clearly, there was something exceptional about this Crowley character.

I researched more and found his account of another chalk climb, Devil's Chimney, a wild pinnacle that has sadly subsided. 'I scooped a hole out of the east face', read the account, 'inserted my chin, and hauled. I had not shaved for a day or two, so was practically enjoying the advantages of Mummery spikes'. Mummery spikes? Turned out they were early crampons. This all seemed very interesting – a little too interesting to start out on, perhaps, but worthy of note for later.

By the time I reached my mid-20s, I'd discovered that I most enjoyed exploratory climbing away from the mainstream crags. And I knew that the only other truly vertical cliffs in England were the chalk ones jutting proudly over the channel, only 70 miles south of London. It seemed inconceivable that London climbers hadn't focused on them before. I quickly recruited a group of my sandstone friends: Mike Morrison, Andy Meyers and Chris Watts.

At Beachy Head, we did Crowley's routes first, finding the climbing serious but almost conventional: we wore rock shoes and protected the climbs with traditional nuts and slings along with the occasional drive-in ice screw. The razor-sharp flints made good hand- and footholds, although they tended to catch the rope. Later, we moved on to Dover, where the chalk was softer, and it was possible to hang two climbers from an ice axe placed in the sponge-like rock with just one blow. Gradually it became clear: there was a tremendous wealth of urge-giving, unclimbed rock not far from home. And as much as it was adventurous – sometimes loose, and sometimes requiring eccentric techniques – if tackled the right way, it need not be dangerous.

In 1985, Andy Meyers, Lorrainne Smythe and I strolled along the rolling grass-topped cliffs of the Isle of Wight in plastic boots, with crampons affixed to our ice screw-loaded packs and ice axes in hand. We'd based our foot- and hard-wear on Andy's reconnaissance; he'd declared the rock was most similar to the ice-climbing chalk at Dover. Since he was a fastidious man with a keen awareness of his surroundings combined with the willpower to build kit cars, we had no reason to doubt his judgment. The other weekend walkers studied us dubiously but offered no comment. At the bottom of a 300ft abseil, on the otherwise inaccessible beach of Scratchell's Bay, I experimentally swung my axe: 'Shit!'

It bounced back a bit as though it had hit a concrete wall. I scowled, Lorrainne collapsed laughing, and the normally loquacious Andy went quiet. It transpired that he had not actually taken his ice axe on the reconnaissance visit. However, between the extra time and cost of the ferry ride and our committed position on the beach, we were not in the mood to give up. I strapped my axe to my pack and stepped gingerly onto the rock, pleased to find solid flints, sound underlying chalk and straightforward climbing. To the west, the three Needles pierced the blue sky, presenting many obvious future objectives; to the north, cargo ships ploughed lanes up and down the Solent, which separated us from the British mainland; to the south, racing yachts dotted the sea; while to the east, the 400ft, knife-edge ridge soared up over white pinnacles resembling a giant's weathered vertebrae.

A coastguard lookout sat atop the final steep section. At the first belay, a keen north wind whipped across the crest until I was almost glad to be wearing my Himalayan mountain boots. Lorrainne and I huddled while Andy led a short pitch to a 20ft, near-vertical step. As I surveyed the impasse, I felt confident that if I'd had some sounder protection than ice screws I could have carried on.

'Tide's coming in,' announced Lorrainne while Andy and I dithered. The truth of her words prompted a unanimous and undignified abseil into the sea. A good saltwater dosing of our crampons, boots and screws helped mark the occasion, but the 300ft jumar out of Scratchell's Bay in strong wind and drizzle made the day clearly and suitably memorable.

The skeletal ridge nagged me as I sat at my Tax Office desk, as no doubt the Beachy Head objectives had pecked at Crowley's mind 90 years before. When Lorrainne rang, I was not difficult to persuade. Andy couldn't make it, so a few weeks later, in beautiful weather, Lorrainne and I were back with rock shoes and nuts. It appeared that we had misread the tide tables.

'A little sea-level traverse will be fine,' I announced. Soon I was wet, but the day was warm and drying out in the sun was pleasantly relaxing. Back at our previous high point, calm, sunny conditions and rock gear made all the difference. After placing some welcome protection in a crack on the right-hand side of the arête, I made a few delicate moves up an exposed, steep slab. Then, I udged along the crest with a leg on either side to gain a fine ledge. Lorrainne followed, whooping.

Meanwhile, we were creating much interest from passing leisure craft. I was glad we'd contacted the coast guard beforehand and thus avoided unwelcome rescue efforts. We signalled that all was well, and indeed it was: we were able to pass a steep and intimidating step surprisingly easily on the left. From the top of the step, a simple horizontal section and a short wall led to the lookout, and we were done.

After his Beachy Head climbs, Crowley felt carried away by the poetry of the moment, having 'fulfilled all [his] ideals of romance and in addition... the particularly delightful feeling of complete originality'. But I'm a taxman, not an occultist, so we just sat in the warmth of the sun and enjoyed our success.

Facing page: Bill Birkett on Skeleton Ridge. Above: Lorrainne Smythe on the central section of Skeleton Ridge. Photos Mick Fowler. Below: Mick Fowler on the central section of Skelton Ridge. Photo Bill Birkett.

Above: Phil Thornhill on the first ascent of Great White Fright, VIII, Dover.

Below: Victor Saunders in the Himalayas. Photos Mick Fowler.

A Battle at Hastings by Phil Thornhill

No! I won't! I can't! Oh God, no! I will not; I cannot get up. Not now. Staggering around in the cold, in the gloom, in the filth, decay and disorder of our 'short-life' house (a kind of glorified squat). Half alive in the half-light, still half-drunk but half hungover, jarred into consciousness just at the stage when that deeper kind of sleep was stealing me away into soothing oblivion. And here was Victor, like an evil spirit of the night. He hadn't overslept, and he was on time. Incredible! Why wasn't he still asleep, like the rest of the world?

Victor is driving me through the darkened streets of London. I want to surrender myself, as far as the car seat allows, to whatever I can regain of my stolen sleep. Empty streets. Strange in their lifelessness, like a different place. Suppose no one awoke, and it was always like this. 'Do you recognise these? I think we've been here already.' Groan, mumble, groan. 'I think we'll have to go back.' Victor goads me into route-finding activity as we get lost again, but I slip back into semi-consciousness at the first opportunity. I treasure every moment of slumber, every remaining instant of blessed inactivity. But it's a slumber in which something looms. We're getting closer. Soon, inevitably, we'll be there. Hastings. Yes, Hastings – where the battle was.

He was saddle-sore and foot-sore; his body ached all over. The wound in his forearm stung. And now this drenching rain, the mud clinging to his feet, as he tugged at the reins of his tired, limping mount. God curse all this marching! He'd rather be killing Norsemen. Up there in the meadows near York lay the bloody corpses of twelve of those Norse vermin, bearing the cuts of HIS sword, by God's truth, his sword alone! And now this bastard Norman – son of a tanner's wench! The pox on him! His sword would build up heaps of Norman corpses now.

The first stop at Hastings would usually be The Mermaid Cafe. Full breakfast, two extra teas AND toast and marmalade – to drag it out a bit longer. The condemned men ate heartily. Condemned? Well, it was my idea, as Victor would ceaselessly remind me: 'Oh God, this is so frightening; my God, I'm so scared. What are we doing here? It's all your fault!' Now, most climbers, I would guess, are scared to some degree (except for one Michael Fowler, mutual acquaintance, who is not capable of so sophisticated an emotion), but the difference with Victor is that he likes to let you know, in precise and vivid detail, exactly how scared he is. At all times. And of the state of his health: 'I do really feel ill this morning.'

'I know what makes you feel ill.'

'No. Yes, it is fear, but it's not only that. I think it's food poisoning it must have been that curry. I do actually, really, feel definitely, physically ill.' Thus spoke Victor the morning we started the monster traverse. No breakfast this time. Unnaturally early. A haunting freshness and silence along the desolate foreshore. A cold, lonely sea in the early glimmerings of the sun. A fox bounds out of the shadows like an omen. Victor and I tramp over tidal rocks and seaweed with ungainly plastic mountain boots. We have come to climb on mud.

My bizarre idea it was, to climb on the Hastings sea cliffs. Cliffs of mud, well, mud and, to be more precise, bands of a very soft sandstone gave the mud sufficient solidity to allow a certain verticality; to form a cliff rather than a slope. The idea had germinated from the seedbed of the North London Climbing Club, a collection of deranged individuals responsible for a variety of manic enterprises. And the mountaineer had come to Hastings, straight off the London bus that first time, alone. Plodding stolidly down the seafront with monstrous bulging rucksack festooned with ice picks, crampons, sleeping roll, ropes, and mountain boots swinging by their laces. All the paraphernalia of the alpinist incongruous amongst the cafes, chippies and ice cream parlours, the coloured flashing lights and the amusement arcades. I had bivouacked in one of the caves in the cliff and made a couple of experimental climbs using a back-rope, returning on several more weekends. Local reaction was not always understanding, and on one occasion, I inadvertently left my rucksack by the coastguard headquarters: 'You're not going to climb on the cliff, are you?'

'Er...'

'Look, we know these cliffs; they're dangerously unstable. Only the other week... etc, etc.'

'Actually, I'm not going to climb. I'm just going to retrieve some equipment by abseil.'

Which was true; it was a foul, windy day anyway.

'Yes, and what do you use to abseil off?'

'Er... bushes.'

'Bushes! When we abseil, we use five-foot iron stakes, hammered into the ground 10 yards back, reinforced with... etc, etc.'

'I've climbed at Dover and Beachy Head, and the coastguard allowed it there...'

'Right, we'll ring up the Beachy Head coastguard.'

A little later, rather reluctantly: 'Well, you're allowed to climb, but you'll have to pay for any rescue.'

Since I had no intention of being rescued, this was excellent news. Just as I was about to go, some evil spirit prompted the thought: 'Oh, I am insured for rescue costs, anyway.'

'Well, I hope you're insured for fucking funeral costs!'

Came the somewhat intemperate response. Now, climbers that I know cringed at the thought of climbing on mud or other less substantial substances like chalk. They called me – or Fowler, doyen of less stable climbing media – crazy. But everyone else knows that all climbers are just that anyway.

I started off at Hastings, just as a rock climb in PAs, on the more solid bands of sandstone. Holds on the faces were generally too friable to use, so routes took obvious weaknesses, cracks and chimneys, often lined with half an inch of clay so that once you stopped struggling upwards, you began to glide remorselessly back down again. All kinds of squirming, squeezing, oozing, bridging, off-widthing, chimneying, tunnelling... getting stuck for hours, wedged in terror. Shattering the brittle rock with useless pegs; abseiling off stacked Hex 9s in the rain; clawing desperately at disintegrating baked clay chockstones that fell apart as fast as you pulled on them; hacking away with an ice hammer at a loose block of sandstone poised above, destroying it gradually before it could fall on me in one big lump.

Then came the next stage. Further round the cliffs were bigger but more broken, more muddy so rock climbing would be impossible. Here, I was able to import the use of ice climbing equipment and techniques. Thus it was that dawn found Victor and I geared up as if for Ben Nevis or The Eiger but actually about to start a monster traverse of mud.

A curse on the gloom of this everlasting forest. God knows what stirred in those dark recesses – goblins, spirits, fiends? Fool! He was growing weak-headed with the marching. The endless broken stone road, stretching ever straight ahead, a road built by the giants, or maybe ancient armies, marching to the battles of a former time. Who were these men who fought on horseback? Must we keep moving so fast? They were Harold's men; they had been the housecarls of Harold, Earl of Wessex, and now they were the housecarls of Harold the King. They had brought victory to their Lord. He remembered the feasting in the mead hall, the longing and admiration in women's eyes, the sound of songs that filled the hall – songs about their deeds – and the telling of stories about their valour and daring... 'We have to stop!' it was Svein. 'We have to wait for the fyrd-men.' They wouldn't be out of the forest until evening then – that Norman son of whore would be ready for them by the morning... God, let the battle be soon. He wanted to be back in his mead hall, singing songs of another victory.

The first pitch looked easy, and really it was, but at the same time, distinctly disturbing. The angle of the slope was invitingly shallow, but the material that formed it was horribly insubstantial – a kind of clay mudstone, soft and gooey on top, brittle and fissured underneath. It did not take placements but shattered beneath the pick and crumbled ominously under crampons. I kicked my way steadily across, clearing a substantial platform for each insecure footstep until I reached an easy part of the slope, consolidated by vegetation. I attempted a belay, and soon Victor appeared, shuffling sideways with disconcerting ease, 'I still really do feel quite ill, you know. Honest, I'm not kidding.'

It was the first pitch of our monster traverse, probably the silliest route on the silliest cliff ever climbed. The cliff consists of horizontal bands of sandstone interspersed with sloping shelves of mudstone. The top sandstone band is the most impressive, and beneath it runs what appeared from below to be a ledge – actually a sloping band of mudstone – more or less continuously for three-quarters of a mile. Once you stopped thinking in a purely vertical plane, it was the most obvious line on the cliff, and some obscure hand of fate had determined that it would be Victor and myself who climbed it.

The early start that first morning proved to be over-ambitious. Easy ground followed the first pitch, a grassy ramble to where the cliff juts out in the form of a prominent clay ridge. There was more scraping around on crumbly horror as Victor led out around the prow and out of sight. The rope snagged as Victor waited for me to follow, and as I waited for Victor to take the rope in, we both dozed off in the sunshine. Two sleeping bodies – tethered to a mud face.

The traverse had to be done in stages. Many were the weekends spent completing it, and many were the tribulations. There was the time I left a message at Victor's workplace to meet me that evening in a pub called The London Trader. He spent all evening searching London for it but couldn't find it. That was because it was in Hastings.

The amazing thing about it all was how the dirt and grit penetrated everywhere, even into your very underpants. Tired and dirty, as (or after) the sun went down, we would abseil off a collection of ice screws hammered into the mudstone (which wasn't as bad as jumaring back up on them) or, on several occasions, perform marathons of bushwhacking in that cross between semi-arid thorny scrub and tropical rain forest that flourishes on top of the cliff. This was a time when shattered nerves could finally blossom forth into untrammelled fury and find expression in dazzling displays of colourful language. Tired and dirty, and maybe scratched and scoured by brambles, the mountaineers would settle into some haven of a cafe or pub to relax after the day's deeds.

The difficulties of the climbing varied according to the angle and narrowness of the mudstone shelf, but a certain insecurity pervaded all. This was due to the difficulty of finding good protection, which could only come from scraping away as much of the loose and friable surface material as possible and hammering one or several ice screws into the loose and friable material underneath.

Seconding was not a bonus, for after leading out in exposed and teetering, crumbly insecurity, it was a relief, after sundry awkward manoeuvres, to have placed an ice screw that might, while you were close enough to it, hold your weight. For the second, the exposed, crumbly, teetering and the awkward, devious manoeuvres only resulted in your having removed the one point of questionable security, leaving you with maximum distance to the next.

He could see the Normans now, on the next rise. Battle at last. Now to send them back where they came from in little pieces. The Normans had moved on THEM – they were the ones who had to hold their ground. But it would be the same; soon they would hurl them back into the sea. Men were shouting now – a cry ran up and down the ranks. Now there was an enemy to see they could shout themselves hoarse.

Now, he was shouting too. He could see them moving forward, a wavering line. He lifted his sword of Offa's father – that had his magic – brought it down on his shield with a resounding clang and roared. At last, through the noise came another sound – the shouts of the enemy. The line of bowmen, the mounted men behind them. It would be now – he held his shield high. The rain of deadly points, thudding in shields, glancing off iron. There were screams in the shouting.

Now, the horsemen. His grip tightened on sword-hilt; his muscles tautened. His mind was strangely calm, fulfilled at last with fierce, cold jubilation. The pounding of hooves, the air thickening with rocks, arrows, spears. The blood coursing hot through his veins with the thrill of the battle.

Crampons bite into the dirt, crunching the baked, crinkly surface. Axes shatter it and raise the dust. I advance, gingerly, up the easy slope of the mud cone. My mind is of the not quite all there, part of it having fled, vaguely disbelieving it all. My axe dangling from its leash, a free hand searches for rock holds on the left-hand sandstone wall. It pulls a few pieces off. I scan, hopefully, for some weakness to place a piton or nut, anything that might conceivably relieve my anxieties on the hideous looming pile above. Nothing, definitely nothing at all. I have to move up on the clay cone where it steepens and narrows. Axes, slotted deep and too easily, are pulled on slowly, steadily, and suspiciously. Feet are kicked in over and over again to form something I can trust as a foothold. Will the axes hold me if the footholds crumble? Can I keep in balance on the footholds if the axes come away?

The clay cone runs out into a thin sandstone barrier before the less steep continuation of the gully. I need to get high enough to reach over the top of it. The clay starts to come away in lumps – much dust, destruction and shifting of feet. I can't get high enough. Can I kick a lump out of the sandstone wall to create a hold to bridge out onto? Can I torque my axe into the crack that has appeared in the bare sandstone at the back of the corner? Dust, heat, and frustration. Maybe a move rightwards onto the broken rock rib. Can I stretch across? A hold? Maybe a hold. A side-pull. My foot up there? Too scary. Nothing to stop a fall. The crack! Furious clearing away of the mud, and a Friend slammed into the crack. It seems to hold. Will it just crumble the rock? It's real close. Maybe it's OK. Maybe I can get my foot up there, pull carefully on the hold, delicately sidle into balance up on the rib…

A fantasy world of success at last, above that real world of the horrible crumbling groove. Now, this I can call easy, after a pause to rest and give thanks, a mud slope to kick and hack steadily upwards. Eyes scouring the rocks, the cracks, the shattered recesses, for somewhere to place protection – anything for some kind of security in this world of crumbling instability. No, that's behind a block that will come away – this is best a crack, well back in the wall, and an ice-screw hammered in like a piton, hammered in all the way if possible. The mud slope goes rightwards, and above are two possible escape routes of shattered rocks leading onto the big easy sloping ledge – the line of the monster traverse. I take the middle line. Hours are passing, but in total absorption, I do not notice. Sweating in the heat, a dry, parched throat. Curse that sun! Can I pull on this block? Miracle of miracles – a hand jam. A Friend placement as well. Now, a steep move up, try this way, try that. Will the Friend hold in *those* rocks? The heat and a dry, dry throat. That was the cause of it – the dry throat and the dusty heat. That was what beat me *that* time when I abseiled off, cautiously, from a peg.

But I came back, this time going for the chimney/corner further right. It might be too steep at the top. I traverse in. The best! A crack deep in the back of the corner. An ice screw hammered in – a block shatters away. Try again in a different place – hammer it and keep hammering till it will go no further. The corner looming steep above – too steep. The clay running out, breaking away again. Nowhere to place the axe. Everything crumbles when I try to place my feet. Out there! Bridge out there. A sunny ray of hope in the tunnel of fear. Back against one wall, foot out on the other.

Ice-pick searching for purchase in the crumbly, shattered mudstone of the sloping shelf above the corner. Thrashing at it violently, over and over, deeper and deeper, fragments pouring down, until it feels as if it's bitten on something. Then, the moment it becomes possible, the prospect of success crystallising from a distant fantasy into a real thing. Dare I? Pull!

All over, I hack and grovel up the easy clay slope to find a belay. Victor comes up with disconcerting ease and speed. His is the last short pitch, up the final steep sandstone band. There is an obvious corner, but Victor is out on the alarming arête. I am following in the gathering gloom, scraps of turf for placements, mounting disbelief and a tightening rope. I am hauled out on top. Success! Not to mention survival.

Not so far away, meanwhile, give or take 900 years, a large number of men are hacking each other apart with sharp pieces of metal. Now, on the face of it, this would appear to be a fairly bizarre and stupid thing to be doing. Yet none of them thought this at the time, and a battle, after all, is one of the more spectacular and less forgettable manifestations of the perverse and mysterious springs of human behaviour.

They had chased them, the fools, broken ranks and ran after them with cries of victory. Then the horsemen turned and cut them down... He had seen Edgar killed. Svein was dead, and Edmund hobbled, a gash in his foot where he had pulled out the arrow. He held up his shield, heavy in his hand. They were coming again. Showers of stinging iron – the devil's rain. Men were falling, the accursed horsemen! He was falling down in a lust of fury, flying hooves and chain-mail – falling. A crash like thunder – he was lost. The smell of the soil and the sharpness of the iron that would find him - but which never came. Staggering into rebirth, back into the shield wall with the housecarls who had saved him. They had them running again now. He stepped over another body, nearly severed in two. It was Edmund's. In all the clamour, a cry went up with a new note of panic and despair. The man beside him understood it: 'Harold is slain.' Two men beside him were slipping away. One of them threw down his shield, and they ran. The shield wall was breaking; the fugitives were shouting, drowning fear and shame in their cry: 'The King is dead!' Torches flickered in the night as they looted the bodies. Corpses lay naked, bloody and mutilated. The best men of England lay there, dead with their King.

The beer was flowing freely in The Globe as Lobby's photos were passed around a group of hard-core North Londoners. Lobby – actually Phil Butler, and so-called from an incident in Scotland involving a bivouac with lobster pots had been the special photographer on the last stage of our monster traverse. We had called the route Reasons to be Fearful, an adaptation of the title of one of Ian Dury's records and particularly apt, we thought, since that record came in Part One, Two, Three, etc. Clearly, such a momentous event as as the completion of the traverse had required recording, and Phil Butler, 'Smog Capers' correspondent, was just the man to do it. One of his photos even reached the pages of the climbing magazine *High*, spreading our fame to the furthest corners of the land. (sic)!

This evening, Victor and I are recounting our adventures in versions usually much to the disadvantage of the other. But, all in all, with the increasing alcohol-to-blood ratio, there is a certain creeping satisfaction as we bask in the – if not glory – perhaps at least fleeting, localised notoriety of our somewhat eccentric enterprise.

'You bastard! I hate you! It's all your fault!' Victor is leading off, on the final stage of the traverse. Three feet later, in a milder tone: 'Do you mind if I put in a little protection here?'

'No, surely, that seems a good idea to me.'

Pause.

'Umm, I think I'm going to need another screw here.'

'OK then.'

Pause.

'I'm feeling really bad today. I'll have to put another screw in...'

'Jesus! You'll end up virtually aiding it, and it'll take us forever; can't you just fucking well go for it?'

This is the encouragement Victor needs, and he's soon slipping along nicely in the usual speedy Victor way. The trouble with a traverse with doubtful protection like this is that you need to trust your partner. Victor, I trust not at all – except not to fall off. This is what matters. As for myself, I am not so confident, especially on one long runout with the mud/stone steepening, becoming ever more brittle and shattered, the placements only materialising after frantic excavation, and the screws ever more doubtful, the potential pendulum ever more spectacular...

A few moments of less than total certainty, that adrenalin seeping into the system again, and then easy ground – grass, boulders, and guano. Now, all that remains is for us to find a way off through the upper sandstone band, and there's some urgency in the situation because of the rapidly gathering gloom. Soon, I am seconding this final pitch with some curious bridging and chimneying manoeuvres, Victor's torch dazzling me by way of encouragement. The axe swung into excellent placement quality mud, and no serious obstacles between me, the chip shop, the pub, and talking about this little adventure for evermore.

On level ground, and our man from 'Smog Capers' is on the spot. An arm over Victor's shoulder for the victorious pose in the dazzle of the flash. Two conquering heroes immortalised. Gathering our hardware, we clank off through the trees, leaving the cliff – our 'battlefield' – behind, silent and deserted in the darkness. Seagulls fly, silhouetted in the orange glow of the town, and a fox is scavenging on the foreshore. Ghosts fill the gloom – the ghosts of battles past, drifting with the sea breeze over a battlefield with no corpses.

There is either action or no action. Making a move in climbing, or in life, can be scary. Venturing into the unknown, where the outcome is uncertain, is the essence of adventure.

Letting loose and going for it can be a liberating experience. Unshackle the chains that might be holding us back and launch into the unknown.

This is where life truly begins.

Tim Emmett

Dorset

He descended and came to a small basin of sea enclosed by the cliffs. Troy's nature freshened within him; he thought he would rest and bathe here before going farther. He undressed and plunged in. Inside the cove the water was uninteresting to a swimmer, being smooth as a pond, and to get a little of the ocean swell, Troy presently swam between the two projecting spurs of rock which formed the pillars of Hercules to this miniature Mediterranean. Unfortunately for Troy a current unknown to him existed outside, which, unimportant to craft of any burden, was awkward for a swimmer who might be taken in it unawares. Troy found himself carried to the left and then round in a swoop out to sea.

Far From the Madding Crowd (1874) by Thomas Hardy

Loch Ness has its Monster, The Minch has the Blue Men, Cornwall has mermaids, and Ireland has the Merrow. What does Dorset have? A mer-chicken. I'm not joking. This mythical sea monster has been claimed to have been sighted off Weymouth and Portland every so often since its first appearance in November 1457, which was described in *Holinshed's Chronicles* published in 1577:

'In the month of November, in the Ile of Portland not farre from the towne of Weimouth, was seene a cocke coming out of the sea, having a great crest upon his head, and a great red beard, and legs of halfe a yard long. He stood on the water and crowed foure times, and everie time turned him about, and beckoned with his head, toward the north, the south and the west, and was of colour like fesant, and when he had crowed three times, he vanished awaie.'

Mind you, that's not the only massive cock to have been observed in Dorset. There is the Cerne Abbas giant, for one, not to mention Mikey Robertson's commando climbing exploits.

Dorset's Jurassic Coast, stretching from East Devon to Studland Bay, has over 90 miles of coastline with long stretches of cliffs of variable quality rock. Swanage and Portland, in particular, provide excellent venues for traversing and deep water soloing.

Dorset is the most popular venue for deep water soloing in the UK. The combination of cliff height, water depth, and water temperature makes for the most conducive conditions in the UK for this pursuit. DWS in Dorset was preceded by trad climbing, sea-level traversing and sport climbing, but the earliest recorded ascent of the Dorset cliffs from the sea occurred long before then.

In a blizzard, on the sixth of January 1786, the Bengal-bound *Halsewell* foundered in a violent gale on the cliffs below East Man hillside at Winspit. Two crewmen, the cook and the quartermaster, scaled the cliff and raised help. According to Rodney Legg in *Dorset Life*:

'At two in the morning, the *Halsewell* East Indiaman, 758-tons burthen, commanded by Captain Richard Pierce, bound for Bengal, was lost in the rocks between Seacombe and Winspit quarries in this parish. Never did happen so complete a wreck. The ship long before daybreak was shattered to pieces, and a very small part of her cargo saved. It proved fatal to 168 persons, among whom were the captain, two of his daughters, two nieces and three other young ladies. Eighty-two men, by the exertion and humanity of the inhabitants and neighbouring quarries, at the imminent hazard of their own lives, were saved... 27 men found refuge on what is now known as the Halsewell Rock.'

Facing page: Jonathan 'Woody' Woods on Amazonia, F6b+ S1, at Connor Cove. Photo Mike Robertson.

The plight of the immediate survivors was described in *Remarkable Shipwrecks*: 'At length, after three hours of the keenest misery, the day broke on them, but far from bringing with it the expected relief, it served only to discover to them the horror of their situation... They were completely engulfed in the cavern and overhung by the cliff, nor was any part of the wreck remaining to indicate their probable place of refuge. Below, no boat could live to search them out, and had it been possible to acquaint those who were willing to assist them with their exact situation, they were at a loss to conceive how any ropes could be conveyed into the cavern to facilitate their escape. The only method that afforded any prospect of success was to creep along the outer side to its outer extremity, to turn the corner on a ledge scarcely as broad as a man's hand, and climb up the almost perpendicular precipices... In this desperate attempt, some succeeded, while others, trembling with terror, and exhausted with bodily and mental fatigue, lost their precarious footing and perished.'

The vicinity of the wreck, today, has recorded rock climbs and intrigued to discover the grade of the route that the *Halsewell* survivors were forced to climb, I asked one of the first ascensionists of these routes and longstanding Dorset pioneer, Scott Titt:

'My view is that the cliff line has changed a lot since the *Halsewell* came to grief. The wreck is described as by a cave, but there is no cave at the wreck site. I think that there has been quite a lot of erosion here; the wreck site is a bit offshore, not right up against the cliff. The rate of erosion can be seen by the loss of some of the routes to the right of Lost Souls. It is also likely that the top of the cliff was quarried long after the disaster, so a very different landscape now as existed then.'

Prior to the 1950s, the Dorset cliffs were mostly known for quarrying, smuggling and shipwrecks rather than as a venue for rock climbing. However, that all changed when climbers based in the south of England, in search of local venues beyond Southern Sandstone, realised that limestone was successfully being climbed further north and that there was a large area of untouched limestone cliffs around 100 miles from London. What differentiated the cliffs at Swanage from those in the Avon Gorge and Peak District was that traversing above the sea was a worthy objective in itself.

Climbers soon realised that they had to approach these cliffs equipped technically and mentally to deal with the demands of not only the rock but also the sea: lassos, pendules, Tyroleans, and swims along with the standard climbing techniques were necessarily added to the repertoire of the competent sea cliff climber.

In the late-50s, two separate groups explored the area. One was from Southampton University, and their leading light was civil engineering student Barrie Annette. According to John Cleare:

'He moved through the limestone scene like a comet. Experience gained at Swanage from his often solo explorations put him in a strong position when he moved to the more traditional limestone climbing grounds... At Swanage, he was often forced to climb solo through lack of powerful companions. It took a long time to live down an epic which occurred during one of his early attempts at the obvious line, which was to become Marmolata. He was benighted, possibly intentionally, at the cliff bottom, a place reached easily only by abseil, and settled down to a comfortable and well-equipped bivouac. His failure to return for a date led to questions at the police station. Eventually, Barrie was forcibly rescued by the Swanage lifeboat after a night-search operation. The first of many subsequently similar questions were raised in the local paper about mad-fool climbers, though how were they to know that he was as much in control of the situation when sitting on a ledge above the waves as the lecture theatre.'

The other group belonged to no particular club but were all alpinists of considerable experience who lived in the London or Guildford areas: John Cleare, Gunn Clark, Peter Bell, Tony Smythe and their friends. In 1959, Barrie Annette's *Limestone Climbs on the Dorset Coast* was first published by the Southampton University Mountaineering Club, which covered developments at Swanage and elsewhere. The potential for climbing at Swanage was outlined further by John Cleare in an article in the *Alpine Journal* in 1961. Cleare had been introduced to climbing at school by one of his teachers and remembered the impressive cliffs at Swanage from early family holidays there. He went back when he was at Art College to investigate. In his article, Cleare described a climbing area not far from London, with unexplored cliffs which, although only 100ft high, compared favourably with the Dolomites.

In 1963, Rusty Baillie and John Cleare's lateral exploration of the cliffs between Tilly Whim and Subluminal produced the first modern-style sea-level traverse in Britain. Traverse of the Gods covered a distance of only 250 yards on the map but resulted in a route involving over 2,000ft of climbing and scrambling, various rope moves, and a short swim through an unclimbed area where the cliff bottomed in deep water.

Above: Peter Gillman, belayed by Dave Condict, in action on the Traverse of the Gods. Facing page: Peter Gillman on the first pitch of Traverse of the Gods. Photos John Cleare, 9th October 1966.

Göttertraversieren by Peter Gillman

While we are spinning and singing,

whereon stretch we the rope?

Götterdämmerung by Richard Wagner

Unlike the sandstone outcrops nearer London, the Swanage cliffs have a definite feeling of the wild. Most of the cliffs in their broad westward sweep are visible from Swanage, and on a clear day you can see boats far out in the English Channel. On a rough day in winter, spray bursts over the top of the cliff, and climbing is out of the question. When climbing, you must take tides into account, and storms can blow up quickly. Two climbers drowned one January when a big wave dragged them into the sea. Access is difficult, and escape in some places impossible. Often the weather over Swanage is fine although it may be raining 10 miles inland. On a good day, the sea and the sky give the climbing here a feeling of spaciousness and freedom.

Layton Kor, Bev Clark and I had a mad scramble among boulders, dodging waves to reach the foot of a climb. We remained dry until the last dash of all, when, to the others' delight, I was stranded yards from safety with a wave bearing down on me. It was so big that I crouched behind the nearest boulder instead of on top of it so that I wouldn't be taken out to sea when the wave bounced back off the cliff. Fortunately, the day was hot, and I steamed dry during the climb.

Directly below the lighthouse is a 1,500ft stretch of cliff with only one route up it, and that can only be reached by climbing down it first. From the sea there is an obvious traverse line that is just at the point reached by a high, high tide. The ledge has been scoured out of the cliff by the sea and is extensively overhung. The cliff itself is pierced by zawns and swings in and out dramatically.

In 1963, John Cleare, Rusty Baillie and Ian Martin came home from the Alps where Rusty had just made the second British ascent of the Eiger Nordwand with Dougal Haston. The cliff below the lighthouse was then totally unknown ground – no one had even made a reconnaissance by boat. The three reached sea level east of the lighthouse, climbed westwards for seven hours, and were finally stopped by a wide zawn which they called the Black Zawn from the colour of the rock. Fortunately, it was no later in the year than September, so Rusty took his clothes off, tied on a 150ft rope, and swam across. They rigged up a rope pulley and hauled their gear across, John following in the sea. Ian Martin had left the climb at the one possible escape route halfway along, a chimney now known as Scotsman's Retreat.

John and Rusty named the climb partly in Rusty's honour and partly because, as John said, 'It was a hell of a traverse.' They reckoned it had 3,000ft of climbing and was XS and A2. I became a journalist in August 1964 and started climbing in May 1965. As I enjoyed climbing, I naturally enjoyed writing about it. In November, John Cleare and I went to Switzerland to do a background feature for the coming attempt on the Eiger Nordwand Direct. In February and March, I reported the climb for the *Daily Telegraph*.

'Now that you're writing about climbing, we'd better get you up some climbs,' said John when I came home, 'We'll start with Traverse of the Gods. It's the longest serious climb in Southern England.' John has always taken an avuncular interest in my climbing, and his eyes lit up as he described the pendulum, the Tyrolean traverse, the irreversible bridging move across a zawn and, finally, the swimming pitch. To add to the enticements, he declared he would photograph every inch of the climb. It was to be another Cleare climbing extravaganza, with death-defying ropework and using all conceivable items of modern American equipment for the sake of some passing glory in the *CCJ*. We were to spend the whole summer on the climb, finally completing it on December 12th and ending with what can only be described as an epic due largely to my own inexperience.

In the summer, Dougal Haston, Mick Burke and Layton Kor all said they wanted to do the climb. The first party actually to arrive on the cliff top comprised John, myself, Royal Robbins and Tom Patey. Tom was nursing the hand he had impaled on an ice axe in the Alps and was to provide musical accompaniment only. But as we walked past the way down Subluminal Ledge, a girl ran up to us: 'There's someone shouting down there.' At sea level, we found a boy unconscious, half in the water. His partner was supporting his head. He had been climbing a difficult crack with no protection and had fallen 30ft. It was the end of our day's climbing. We eventually evacuated him by rubber dinghy and then the Swanage lifeboat.

We came down again a month later. This time John and I had with us Roy Smith, just home from the British Andes Expedition, and Dave Condict, who climbed Rondoy with the 1963 LSE Andes expedition. Roy had with him Barry Cliff, but Barry didn't think he would climb as he was still nursing the back that he broke in a flying accident in 1964. We arrived at the eastern end of the cliff at one o'clock. John looked like an overdecorated Christmas tree laden with his hardware and cameras. The London Sub-Aqua Club were sunning themselves on the rocks and seeing us uncoil our ropes, some came across. 'Rescued a climber from the sea here two weeks ago,' said one, 'More dead than alive when we got him out.' We thanked him for his kind thoughts.

The first obstacle was a vertical wall undercut by the sea. Halfway along the wall was a nose which jutted out above a zawn. Roy, who was leading, had to climb free to the nose, fix a rope there, and then continue to a wide ledge on the far side. The rest of us were to pendulum across the zawn. Roy, climbing in a vest, white trousers which looked like longjohns, and boots, appeared contemptuous of the problem. The traverse line followed very thin cracks high on the wall. Climbing delicately, he reached the nose quickly and then continued to the ledge. Dave said he would prefer to climb free, too, but John instructed us to pendulum for the sake of the photographs. The traverse line to the hanging rope was just above the waves; Dave never really had much chance.

John told him to tie a loop in the rope for a foot and to attach a hero loop with a Prusik knot to the rope for a hand. Dave did so and swung. But with his foot in the loop, he could not jump for the ledge and went straight into the water up to his waist to cheers from the divers. He untangled himself from the rope and ignominiously climbed on the ledge. He found a place in the sun and took off his masters and socks to dry them out. John went next, using only the hero loop for his hand and just caught the ledge with his feet at full stretch.

'OK, Pete, you come next,' John shouted. I was annoyed already because before even trying the first of John's rope manoeuvres, I had been soaked by a large wave as I stood innocently on the first stance. I advanced along the base of the wall. The situation was like a whirlpool, drawing you further and further into its clutches so that any retreat became more and more impossible.

At the edge of the zawn, you had to lean out and clutch the rope. From that moment you were committed. The holds were precarious, forcing you to turn to the greater insecurity of the rope itself. I stopped thinking about the problem, grabbed for the rope, and swung. For a moment, as the ledge came near, I thought I had made it, but my flailing foot just failed to get a hold. Roy, on the ledge, held me firm on the climbing rope, pulled me in the necessary couple of inches, and I grounded gratefully. Barry decided not to follow on account of his back.

Chuckling happily, John continued ahead with Roy, climbing unroped. The ledge now ran round the back of two cavernous zawns: 'The sort of place you expect to find a dead sailor washed up, lashed to a spar,' said John. Water dripped from the roof, and the rock was black and wet. Dave and I roped up, and I led round the first cave onto the nose separating it from the next. Dave followed, and I led again into the back of the second cave and out again. The handholds on the ledge were good, and there were usually good footholds below it, although they were sometimes wet. There followed a third cave.

We came out into the sunlight on a wide ledge at the back of which was an overhanging crack. The traverse line petered out here, and we had to go up the crack and strike along higher up. A fine bridging move on small holds took you off the ground, and then by placing your left foot up near your ear, you were on top of a large block – a mild VS move that gave a peculiarly strong feeling of achievement after the continual sideways scuttling of the traversing thus far. Having gained height, we now had to lose it again. We walked along a ledge for several yards and then traversed gently down a slab to below the original line we were on and perched on a nose.

Roy now had to lead up a crack in a corner, hand traverse along the top of the left-hand side of the corner, and slide on to yet another nose, offering a very constricted stance which, to compound our difficulties, was covered with guano. Roy climbed the pitch successfully and ensconced himself away from the mess, with his legs dangling over the edge. I followed. The crack offered narrow finger jams, but I followed a natural tendency to back up the left-hand wall of the corner. At the top, I found out why this was the wrong thing to do: I now had my back to the holds I was to use next. I turned around precariously and launched myself off on the hand traverse. When I arrived at the stance, I found that there was only two feet of space between it and the roof that overhang it.

'You've got to sort of back in head-first,' said Roy, 'There are some holds on the roof.' I did as I was told and slithered onto the stance through the biggest and most nauseous pool on the ledge. 'Smells a bit, doesn't it, Pete?' said Roy. I sat in that pool for the next two hours. There was no room for anyone else on the ledge, or so we thought, and Roy set off on the next pitch.

Facing page: Peter Gillman on the first pitch
of Traverse of the Gods. Photo John Cleare.

Above: the first pitch of Traverse of the Gods involves some rope work to cross a small zawn. View from Tilly Whim Caves. Roy Smith leads, Dave Condict belays.

Below: John Cleare in action in the Black Zawn towards the western end of Traverse of the Gods. Photos John Cleare, October 1966.

The obstacle was another zawn, about 20ft across, with wave-swept rocks at the bottom. The idea was that Roy should perform a diagonal abseil onto a block halfway down the back of the zawn, do an unpleasant stomach traverse along a guano ledge running along the opposite wall, and thus gain a substantial ledge on the far side. There were two ropes to manage – the climbing rope and the abseil rope – as well as the one I had climbed on. Roy set off on the abseil but came back after a long time, saying that the rope was jamming. He thought he would try a hand traverse along the continuation of the ledge we were on and climb down the back of the zawn onto the block. Eventually, he came back from that line too. 'You'll have to do it, John,' he called.

John came round and joined us on the stance, a heap of tangled ropes and bodies. I moved away from the scene of action while John and Roy sorted things out. Roy muttered to John, 'I thought I was going to come off then.'

'What was that, Roy?' I asked.

'Nothing, Pete,' said Roy.

It was nearly half an hour before John was ready to set off, did the abseil and remained on the block while Roy followed him and led through. John wanted to be in a good position to photograph the next piece of circus tomfoolery. Roy had by now reached the far nose and was securing the end of the rope he had climbed on. Dave and I, squashed onto our stance, passed the other end through a rusty ring-piton, coiled it, and threw it across the zawn to Roy.

'OK, Pete, that's it, across you come.'

'You're joking,' I told him. 'On what?' I refused to contemplate a Tyrolean traverse – whatever it eventually turned out to be – until the rope was fixed to a newer-looking piton. Everything had to be undone, and Dave fixed the rope to another piton around the corner of the nose.

'Right, Pete, it's quite easy. Just hang on to the rope with your hands and hang your feet around it as well.' John gave his instructions from the block at the back of the zawn, his vantage point for taking pictures of terrified climbers silhouetted against the sky. Dave belayed me, and I set off. At first, it was easy – the rope sloped downwards towards Roy's ledge. But as I passed the halfway mark, the rope sagged more and more, and so the last few feet were uphill. To make things more difficult for myself, I had wrapped my feet too tightly around the rope, and this counteracted the pull of my hands. Roy strained a hand out. I managed to hunch myself forward into a semi-sitting position and grabbed his hand, and he pulled me in. John wound his film on happily.

'You rotten swine,' I yelled at him, 'You... '

'Great action shot, Pete,' said John.

Dave, on the wrong side of the zawn, admitted he wasn't looking forward to the acrobatics but came across nonetheless. Roy and I had managed to tighten the rope, and gaining the ledge was rather easier for him. We now tried to pull the rope across. It had, of course, stuck. It was six o'clock, and it would be dark in under an hour. There were still three to four hours of climbing, and Dave and I had to return to London that night. John and Roy were staying overnight, and John said they would come back for the rope the next morning. We were now just below Scotsman's Pinnacle. Two escape routes lay on either side of this – one, Scotsman's Retreat, on the far end of a Hard Severe pitch, the other, Rhodesian's Retreat, on our side of the pitch. The two routes meet halfway up the cliff. We took Rhodesian's Retreat. The Swanage rock lived up to its reputation. John hurled down a few boulders as he went up, and Dave, who came last, was narrowly missed by one that the rope dislodged.

The months passed. We made plans to return to complete the climb on several occasions, but they were always thwarted by the pressures of work or bad weather forecasts. 'We can't let it run into next year,' said John. At the end of the first week in December, the forecasts were still consistently bad. But on Saturday December 10th, a partial clearing was forecast, and we arranged to meet at Swanage the next morning. I travelled down with Dave, and John brought Ian Howell, fit and enthusiastic from the Andes and Yosemite, and his friend Adrian Lang. We went down Scotsman's Retreat at 11 o'clock. The sea was calm, but the sky in the west was dark.

By coming down the wrong chimney, we had missed out a pitch, so Ian led across it, followed by myself and Dave, and we then climbed it back again. It was a pitch typical of Swanage – a deep recess for your hands and a sloping ledge for your feet. John said that on the first ascent, Rusty had so disliked the look of it that he had climbed down onto the ledge below it, run across dodging the waves, and climbed up at the other end. He had been put off by a formidable bulge halfway along, but as it turned out, it was reasonably easy to balance past this. John stood below us on Rusty's ledge to take his pictures. Since our last visit he had got married, and his wife had kicked £800 worth of Nikon cameras over an 800ft cliff in Sutherland. He now had the latest waterproof Nikon, designed – unlike our krabs, pegs, and peg hammers – to withstand the corrosive action of salt. We now approached what John said was the hardest pitch on the whole climb – the one that he and Rusty had avoided at first with an artificial pitch of A2.

The barrier, as usual, was a zawn. Ian climbed down to a block at the back of the zawn, followed by Adrian and myself – the principle being to pack as many people onto the stances as we could to provide front and back belays wherever possible. John looked into the back of the zawn and laughed – we were crushed together like rush-hour passengers on the London Underground. The hard pitch followed. As usual, there was a recess for hand jams but this time, literally, nothing for your feet. Two-thirds of the way along the pitch was a corner, which would provide rest.

Ian climbed up out of the back of the zawn to the traverse line and placed a runner high up. He stuffed his arm into the recess, pulled his legs up to almost the same level, and monkeyed his way leftwards, pausing in the corner and then pushing on to the nose that formed the next stance. Once again, its height was severely limited by an overhang. Adrian went next, tackling the pitch differently: he jammed as far out along the crack as he could and then swung on his arms, just reaching the first foothold. John arrived in the back of the zawn, told me to wait until he had his camera ready, and then sent me off.

'It's easy for the first move,' he said, with his usual bonhomie. I had to bridge across the corner to unclip from the runner. Adrian's method seemed to be the better one. The jams were poor, and the rugosities all sloped the wrong way. But brooding on the problem only wasted strength, so I swung. My left foot hooked on to the first foothold, and I gained enough purchase to bring my arms across. Another move, and I was in the corner. It wasn't very restful as there were no footholds there either.

'You've done it once you're in the corner,' lied John. The overhang above the stance prevented me from pulling straight over, and I had to move out around to the point of the nose. Ian looked blandly at my hands as they scrabbled for a hold among the coils of rope and his feet. I failed to find one, sank back, and then heaved over the edge. Dave chuntered around next, finding the same difficulty in finishing, while Ian and Adrian continued ahead. The next pitch was a wider zawn which looked easier than it actually was. The holds were good, but awkwardly spaced, and the shelving ledge that provided footholds was wet from the water trickling down the cliff wall. The pitch was probably just VS. It ended on yet another nose, after which came a longer pitch of some 60ft.

When I arrived, Ian and Adrian were already at the far end. I waited as Dave and John came round. The pitch consisted of three typical recesses and should have been much easier climbing than I made of it. You can practically walk into the first recess, but I mistakenly traversed with my hands where my feet should've been. I baulked the last move long enough for John to tell one of his interminable jokes about the Pope. I decided to go just as he came to the punchline, which I missed, thus inflicting the whole joke on the party again at the other end of the pitch. The end was in sight from the next stance. An easy ledge led to a narrow zawn just beyond which was a rope hanging down the cliff face. It had been drizzling earlier, but we had been sheltered by the overhangs. Although it was dry now, the sky was darkening. The tide was coming in. Fortunately, I didn't realise that it would be another four hours before I was at the top of the cliff. John explained that we had now come to the irreversible move.

'Alvarez had to be hauled across, but he's short. Ha ha!' he said. The zawn was about five feet wide. 'You just fall over until your hands can reach the opposite wall, then you bridge across, move your hands down and jam them into the crack, bring your back foot over, monkey round into balance, and then hand traverse to the end,' he said proudly. 'It's very simple,' he added, with the confidence of a man who knew he was big enough to make the initial move. I only arrived in time to see Ian flexing his drained hand muscles at the far end of the pitch. Adrian fell across the gap, found he had his feet wrong, came back, and then bridged across at full stretch. He pulled across successfully, extended himself on the hand traverse, and then flexed his muscles too. Now my turn. I didn't so much fall across to the opposite wall as make a reluctant flop. I found myself staring at the waves that were tumbling into the back of the zawn 20ft below. I brought my foot across. It was a good 12 inches short of the foothold just below the crack where I was to jam my hands. I came back. 'It won't reach, John,' I said.

'Step a bit lower down this side,' said John. I did and strained across, thigh muscles aching. 'Put the back of your heel on that nick just below where your foot is now,' said John. I did so, an act of utter faith. Poised above the sea, my stomach level with my chin, my left foot pawed for the hold. One lunge, two, and then it touched.

'I've got it, John,' I shouted, 'I've got it.' My moment of glory lasted perhaps half a second. There was a baffling hiatus until I found myself five feet below the traverse line, suspended above the sea by the ropes. For God's sake, don't pull me back, I thought. But Ian and Adrian won the tug of war, and I got hold of the rock on the other side of the zawn. I climbed up to the traverse line, did the hand traverse, and pulled over onto the ledge. 'I hope you got that,' I shouted back to John.

'I was too busy hanging on to you,' John shouted.

'Perhaps it's just as well,' I said.

Dave arrived next. 'I thought if Gillman can reach it, I'll be all right,' he said (he is four inches taller than me). 'But it was at full stretch.'

John, last across, was delighted with the way things had gone. It was 3:30pm, and we had come to the end of the climb. In summer, you follow a ledge system for about 25ft until it arrives at water level, strip down, and then swim across the Black Zawn. The first man takes a rope to haul everybody's gear across. The way up the other side is easy. But this was December. Ian said he'd do the swims so that John could take a picture, but we managed to talk him out of it. The way out, as Dougal Haston says in *Eiger Direct,* was up.

John produced from his rucksack his pair of magic jumars, a souvenir of the Red Wall television extravaganza. Only Ian had used them before. I had never used Hiebler clamps or Prusik knots before, either. I assumed it was like standing in slings – all right once you got used to it. The rope hung down from an overhang above us. John went first, steadily but rather slowly. It was about 70ft to the top of the cliff. The rope was laid, not as easy for jumaring on as Perlon, and the clamps did not appear to slide up the rope very easily. John rested a number of times and negotiated the overhang by reaching up over it, clipping a Hiebler onto the rope, and pulling up on that, he disappeared.

By this point, it was half dark and had begun to rain. The wind was blowing up, and waves were sweeping powerfully into the Black Zawn to our left. Ian gave me a lesson in the essentials of jumaring – weight on one sling while you push up on the other clamp. I had the usual total confidence that anything an experienced climber told me to do would work. I made progress at first, although it was very slow, and I was obviously using too much arm strength. The rope swung about, and the clamps seemed very hard to push up. Then my foot came out of one of the slings, and it was difficult to tie it on again. The rain was heavy now, and the rocks glistened in the light of the headtorch shone from below. The waves were lashing against the cliff. I had swung away from our ledge and was now dangling free above the irreversible zawn. It would be tedious to detail the way I gradually lost control of the situation. My feet started to splay outwards, and I danced on the rope as if it was being shaken violently from above. I got a foothold on the rock face and stood there only just in balance. I had a sling between the top jumar and my waist which was meant to allow me to rest, but it was too loose. I knew I had to get off the rock face, and in trying to get back into position on the rope, I flipped upside down.

I had only shouted 'Help!' twice before in my life, both times when I thought I was drowning. Once was in heavy seas during a regatta off the Isle of Wight, and the other when I got mixed up with the breakers at Sunset Beach in Hawaii. I knew the moment I shouted it this time that it was a pointless thing to say, as I was the only person who could get myself out of trouble. There should now follow the story of how I fought with myself, forcing myself to balance upright in the slings, painfully relearning the technique of jumaring, and pushing myself inexorably up the rope to safety. I pulled myself upright and looked up. The top 25ft of the cliff were smooth, with no holds for rests anywhere. I knew that once I had left the ledge I was now on, I was deeply committed. I had used up a great deal of strength already. The others on the ledge below were yelling encouragement, whatever they may have been thinking. John was invisible above. I did a simple sum in my head. Surely four of them could pull me up? And that is why the story that should follow does not. I shouted up to John: 'Let me down on the safety rope.'

The three below shouted: 'If you come down, you'll only have to get back up again.'

'I'll swim,' I shouted back.

'You can't,' they shouted, and they were quite right. The waves were now pouring in like tanks.

John did not seem to have heard me. 'Let me down on the safety rope,' I yelled into the wind.

His voice drifted back: 'I have.'

I found that I was still clipped into a jumar with a sling. I hauled myself up on it three times before I got enough weight off it to release the krab. I had just time to think, this shouldn't be happening, as I plummeted down, and then I felt the beautiful deceleration of the rope. I was dangling over the irreversible zawn – John must have paid out the safety rope in response to my demands until there were 20ft of slack. The others pulled me onto the ledge with the back rope, and I sprawled there apologising.

Ian at once took control. He climbed up to where I had left the jumars – I had thought of bringing them down with me but decided not to in case I dropped them. Ian came back down with them and sent Adrian off. He made very good progress. As he climbed, I asked Ian: 'Do you think four of you would be able to pull me up?'

'I don't know, Pete,' he said rather disconcertingly.

I looked at the white-streaked waves pounding at the cliff below us. 'We should have a couple of hours before the tide reaches us, shouldn't we ?'

'I don't know,' said Ian again. I wished he wouldn't keep saying that. As Adrian reached the top, he gave a thumbs-up sign in the light of the headtorch we were shining from below. Now, it was Dave's turn. He went very slowly at first and then came back to sort out his slings and start again. He set off, and this time, his progress was slow but steady. He eventually disappeared over the top. Ian told me that if they were going to try to pull me up, I was to try to help them by climbing on the jumar rope. I told Ian I would be able to climb on the cliff until those last 25ft. 'Good luck,' he said as he set off.

'Good luck to you,' I replied, though aware I was not in a strong position to say it. I shone the torch on him, sometimes swinging it down to look at the sea. Occasional waves were now throwing spray up onto the ledge where I stood. Ian disappeared over the edge of the cliff, and I was alone. I started thinking about my children and told myself to stop being morbid. I kept looking at the rope to see if it was being pulled in. At least six times, I decided to give up climbing.

After about 20 minutes, the rope started to snake upwards. There was a tug at my waist, and I started off up the wall facing me. I had made only three moves when the pull of the rope swung me off the holds and round over the zawn again, where I dangled free. But the steady pulls from above continued, and I managed to kick up on holds on the wall. About two-thirds of the way up, the pulling stopped. I managed to stand precariously on a ledge, the rope tugging so hard at my waist that my head began to swim. I gripped the rope and tried to walk up the wall. But in no time, I was over the top, gasping for breath like a stranded fish. Facing me were John, Dave and Ian, who were pulling on the rope. Further up the cliff Adrian was acting as brakeman. I didn't know whether to make jokes or appear contrite. There was a fierce wind and driving rain, but the others seemed cheerful enough. 'Thank you very much,' I said.

'That's OK,' said John. Ian said he had enjoyed himself, as he'd never been on an epic in England before. Sodden, we stumbled to the cars. To complete our day, both of them had to be pushed out of the mud.

Above: John Cleare in action crossing the Black Zawn during Traverse of the Gods.

Below: Tyrolean traverse on Traverse of the Gods. Photos John Cleare, October 1966.

The Cormorant by John Alcock

I Suppose a Cormorant's Out of the Question Then?

The name of an E5 route on Pabbay

I was in trouble. For the second time I tried to pull myself out of the water; for the second time the waves sucked me back. It was as if a consortium of octopuses had grabbed hold of my legs and ripped my torn hands from the barnacle-encrusted rock. I had used the angry swell to propel my drenched body onto the cliff, but now it was tearing me back into its frothy chaos. I was tiring rapidly, and there was no chance of help. It wasn't meant to be this way.

When he rang me up a few days earlier, Gordon Jenkin had made it sound like fun. Gordon had spotted that two successive Swanage guidebooks said that no one had completed a one-day traverse of the entire cliff. Named the Cormorant, the route comprised five miles of climbing, scrambling, boulder-hopping and swimming from Tilly Whim Caves to the headland at St Aldhem's. The cliffs were lined with horizontal cracks which invited sideways movement. By staying close to sea level, we hoped waves had smashed away the loose rock which often plagued the crag's upper sections. The guidebook said that the individual sections 'can provide good sport on a hot summer's day'. They had been done in stages but apparently never in a single push. It sounded like more than a worthwhile challenge.

Gordon had assembled a strong but disparate crew. Jenkin himself was the elder statesman and our respected captain. A former author of the Swanage guidebook and new route pioneer, we thought that we could rely on him for local knowledge. Gordon was famed for his attention to detail. He was full of generous advice and could tell you the exact rack needed for a Yosemite big wall or – to the inch – the length of a particular dyno. You could spot Gordon on a far-off climb by his bare chest and bright bandana. He was to develop a mission to cover every obscure bit of South West rock with well-bolted climbs for the masses to enjoy.

The fat-free, prematurely white-haired, Frank Thompson was one of my closest friends. I admired him as a man of stern principle. Although once known as Frank the Milkman because of his work for a dairy, he'd actually been strictly vegan (and also teetotal) for decades. We'd done many hard climbs together. Once, after completing the North Face of the Piz Badile, we'd got back to the hut late to find the guardian could only offer us a meat soup. Frank went hungry rather than pollute his body with animal-derived food. To my ongoing surprise, he tolerated my wicked drinking and eating habits, while I loved his loyalty, kindness and regular outraged rants about anything from inconsiderate drivers to the ineptitude of the England football team.

The friendly, tousle-haired Sue Hazel was simply a legend, but in another extreme sport. Back in the 1970s, Sue was one of the first British women to skateboard and went on to compete around the world. If there were any hard landings on the traverse, she was most likely to survive them. She was no slouch at climbing either, being universally identified as 'Strong Sue'. Her only weakness was Raynaud's Syndrome – a condition which made cold-weather climbing impossible. Sue was clearly highly motivated as she'd bought a new wetsuit for the day.

The team was completed by one of the country's best young climbers, Ben Bransby. Ben was a loveable scoundrel, a mischief maker and one of my favourite climbing partners. He could be relied on to sneak into my rucksack to steal my sandwiches or reverse all the carabiners on my quickdraws for a laugh. On one occasion, Ben lassoed the gear stick of my car from the back seat, and we lurched to Pembroke with him in control. He'd been thrown off the British climbing team for wearing a Viking helmet during a competition and falling off when it fell over his eyes. We'd

Facing page: View westwards along the Boulder Ruckle towards the prominent Marmolata Buttress feature from the far western end of the Subluminal Area. Photo John Cleare. Right top and middle: John Alcock and Frank Thompson on the day of The Cormorant. Photos Alcock collection. Right bottom: Sue Hazel topping out in Dorset. Photo Hazel collection.

spent much of his teens together battling our way up Littlejohn and Fowler classic South West climbs before Ben got too good for an ageing crock like me.

After fuelling up at a cafe in the picturesque town of Swanage, we headed for the cliffs. Passing Tilly Whim Caves brought back memories of new routing with the eccentrics' eccentric: Crispin Waddy. On the drive south from Bristol, we'd stopped at Crispin's empty family home. Typically, Crispin didn't have keys. So he'd climbed a drainpipe, squeezed through a skylight, and raided his parents' fridge for supplies.

On arrival at Swanage, we scaled the fence barring entrance to the cave and spent a memorable night gazing out at the stars before he dragged me up ferocious new lines the next day.

But today was not a day for new routing. In fact, it arguably wasn't a day for going near the cliffs at all. The sky was a depressing monochromatic grey. The air was an off-putting, enveloping damp. And the sea greeted us with unwelcome pounding violence. It's relatively rare for waves to batter the south coast, let alone in August, but today the ocean was glowering and malevolent. It was as if the gods had installed a no-entry sign at the top of the cliff. Our leader was decisive.

'It's too dangerous. We'll start at Subluminal instead.'

Sue and Ben immediately turned around, but to my surprise and concern Frank ignored Gordon and said:

'I'll give it a go.'

Frank likes Gordon and is a good friend, but he doesn't like being told what to do. I was faced with a dilemma. Gordon was undoubtedly being sensible, but if we didn't start at Tilly Whim then the dream of a complete one-day traverse would be over. Moreover Frank was at that time my lodger, and it seemed disloyal to let him die on his own.

Reluctantly, I swung down after Frank heading for the 2,000ft Traverse of the Gods, which the guidebook warned was 'a serious outing and should not be undertaken lightly'. Keeping to a high line made the climbing harder and more intimidating, but staying low increased the chance of being washed off. It was a lonely quest, hemmed in by much harder climbs above and the churning sea beneath. There was no chance of being heard by our teammates, let alone being rescued.

Of course, being without ropes, we missed out various abseils and other guidebook-recommended cheating manoeuvres. There were some wild, committing swings around bulges, some desperate over-gripping on wet breaks, and occasionally even some fun climbing. Given the seemingly endless task ahead, it became a blur of continuous movement, mainly racing to keep up with Frank. When the climbing got too hard, short swims circumvented the difficulties. Finally, we reached the ominously named Black Zawn and a compulsory 75ft swim. The swell seemed to have grown, and I watched Frank take two attempts to get out the other side.

When I jumped in the water, the cold immediately seeped into my wetsuit as the waves smashed over my head. I was not as good a swimmer as Frank, and rock boots hampered my kicks as I pulled through the maelstrom. It was hard to see where to go and even harder to breathe without swallowing seawater.

I reached the exit point and treaded water while waiting for a wave to throw me onto the rocks. I felt the surge behind me and let myself be thrust upwards. My hands grabbed the sharp rocks. My feet fumbled for purchase beneath the water. In an instant, I was torn from safety and dragged back into the waves, gasping for breath. A second attempt proved another strength-sapping failure. By now, I was getting somewhat desperate. Bashed and swirled around as if trapped in a giant spin dryer, I knew that Frank couldn't help me and that I only had a limited number of goes left. A third wave pummelled me back onto the cliff. This time, I managed to lurch a tiny bit higher and resist the downward suction. A final thankful heave and I flopped into a ledge above the water line.

When I got my breath and composure back, a series of short scrambles and boulder-hops led to Subluminal and a potential easy escape up Pedestal Crack. However, a welcome reunion with Gordon, Ben and Sue strengthened our resolve. In calm conditions, a series of non-tidal ledges along Subluminal, Cattle Trough and the Ruckle provide safe and easy access. They are built for sunbathing and relaxing, in contrast to the many scary challenges above. Today, however, the ledges were wave-swept and skiddy. Moving along them was like fighting our way through a drunken mosh-pit, with random breakers threatening to sweep us off our feet.

As Frank, Ben and Sue surged ahead, I noticed that Gordon was falling behind. Before setting out, our esteemed leader had declared that the expedition was basically a caving trip above ground. As a result he wore an orange caving oversuit and even carried a tackle bag with 'essential supplies'. By contrast, the rest of us had decided that it was a climbing trip interspersed by swimming so wore wetsuits and rock boots with energy bars stuffed down our fronts. Not only were we lighter and faster, but when a short zawn forced us all to swim, the limitations of Gordon's kit became apparent. I turned round to see Gordon floundering, more under than above the waves.

'I've got negative buoyancy,' he shouted with admirable understatement. I watched with useless concern as Gordon battled back to his launch point. It was hard to hear him above the waves, but I gathered that he was heading back, so I rushed on to rejoin the others who had rather carelessly abandoned our captain to his fate.

I was getting increasingly cold and exhausted. The dash along the Ruckle had felt like an endless episode of some sadistic reality TV show in which the producers let loose a chaos of waves to try to knock us off the frictionless ledges. At one point, Sue watched as a wave picked me up and slammed me against the cliff.

The swim from the Promenade to Fisherman's Ledge proved the final straw for me. An easy escape route enticed above while continuing would involve a committing long swim across a choppy Connor Cove. Aware that hypothermia, or cramp, could prove problematic, to say the least, I called it a day and headed up a Severe called Helix to find the cliff-top crew consisting of Ben's Mum, Jean, and their dog Winston. Frank, Ben and Sue were left to gallantly soldier on. They had to swim far enough out to avoid being smashed against the cliffs but not so far that a hidden current could wash them away.

By Blackers Ledge all three were getting cold, and Frank decided that he'd had enough. Sue and Ben thought they could stave off hypothermia if they kept moving. Short swims, boulder-hopping and a rock traverse saw them to Cormorant and then Guillemot Ledge where the sun finally broke through. At Guillemot they came across two parties on Tudor Rose – the only other climbers we saw all day. Sue shouted out:

'I know that we're all mad.'

One climber replied, 'Well, I think you're slightly madder'.

From the end of Guillemot Ledge, our remaining heroes faced an extended swim to reach the sanctuary of Dancing Ledge. After entering the water, they soon realised that they were now battling the tide. If they stopped, they would go backwards. If they tried to sprint, their energy would be dangerously sapped. Sue later wrote that it was a matter of swimming steadily and persistently towards the waiting ledge. She also figured that the currents would be strongest near the point, so she and Ben angled their swim further out from shore to the concern of a watching Gordon.

Eventually, they felt the comfort of rocks beneath their feet, and as Sue wrote in her diary, 'It all ended happily ever after.' Gordon was a relieved leader. Frank, Jean and I found them a short time later. Sue and Ben kicked back in the sun and made the sensible decision that it would be foolish to continue given the state of the sea and unknown, more serious territory ahead. In Sue's words, 'What fun we had!'

A few weeks later, Frank and I went back, intent on revenge. Frank had a secret weapon: a pair of bright yellow rubber washing-up gloves to protect his skin from the rough rock. With the aid of Marigolds, he surely couldn't fail.

Conditions could not have been more different. The sea was benign, smooth and flat as the Ruckle and shimmering a welcoming smile. Progress was fast and joyful. The Traverse of the Gods was as fun as climbing can be, a constant variety of athletic moves broken only by the occasional cooling dip.

We ran along the dry, baking Ruckle with our wetsuits pulled down to our waists to avoid overheating. Our sodden rock boots left a splodgy trail along the white

Above: View eastwards from the summit of Marmolata Buttress along the Boulder Ruckle towards Anvil Point and its lighthouse. Below: Advice for climbers. Photos John Cleare.

Above: Looking west across Dancing Ledge. Below: Abseiling to the foot of the Swanage Boulder Ruckle cliffs in November 1970. Facing page: A climber making an abseil descent of Subluminal Cliff at Anvil Point. Beyond stretches the extensive Boulder Ruckle cliff. Photos John Cleare.

limestone. I was relaxed enough to look up at the fearsome cliffs rearing more than 100ft above, lined with memories of past epics. I recognised the line of Cutlass, an E5 on which Crispin had taken numerous sweeping falls onto microwires while making the first ascent. When I came to second, I was so gripped that I jumped into space and pulled up one of our ropes hand-over-hand rather than trust my weight to the friable rock.

At least today we wouldn't have to crawl on all fours up one of Swanage's notorious loose, unprotected top-outs to the dubious sanctuary of a rusting belay stake.

In just two hours and 10 minutes, we reached the noisy sociability of Dancing Ledge. It felt bizarre to be mingling with day trippers and sport climbers as we sat back and relaxed in the burning late August sunshine.

My then partner Lorne greeted us with welcome extra food supplies. After just over an hour's rest, we forced ourselves back onto the traverse. The climbing was good but strenuous, sometimes above water, sometimes half in, half out. Hand traversing breaks a couple of feet above the waves felt very different with my legs supported by the ocean – it was climbing without consequences. Whenever it got too hard, or we drifted too high, we could just jump off and enjoy a refreshing dip in the placid ocean. The trip was worth it for the scenery alone. We swam and clambered through high, cool caves filled with water of the deepest green. Impenetrable rock walls leered overhead, confining us to our sea-level romp.

Beyond Winspit, the cliffs shrank to be replaced by boulder-hopping of the slipperiest kind. Concentration was essential to avoid falling among what felt like giant balls of ice. Suddenly a foot popped, loading my right shoulder. I felt a sharp stab of pain. I knew instantly that it was torn and I would be out of climbing for a while, but I was enjoying myself too much to be upset. Fatigue was now setting in. I felt I'd used as much energy as on an Alpine North Face. As we approached St Aldhem's Head, dense wind-flattened scrub lined the slope above. At one point, a startled muntjac deer leapt out in front of me. Finally, just over six hours after setting out, Frank and I wrestled up through the scrub to the headland. It was done.

Postscript: A few weeks later (and with me out injured), Frank set off on an even more committing coasteering adventure, a solo traverse of the whole of the military firing range at Range West in Pembroke. He was armed with his own new secret weapon – a pair of short fins to strap over his rock boots for the long swims. Naturally, Frank hadn't sought permission from the army for his solitary invasion and his successful trip was made tougher by soldiers ordering him to get back in the water whenever they spotted him come ashore.

Swanage Freak by Kevin Howett

Not that it matters, but most of what follows is true.

William Goldman, *Butch Cassidy and the Sundance Kid* screenplay

The Exeter University Club rarely visited Swanage, but on one trip I made with Paul Newman (no, not *the* Paul Newman) in the summer of 1979, we got more than we bargained for. It was a stunning day, with no wind, the sky a lazy blue and a flat calm sea. It was shorts and T-shirt weather. Paul was in first ascent mode, and I was simply along for the ride. Neither of us had a guidebook, and at the time, I had no idea which part of the coast we were on, although I thought we were near Fisherman's Ledge.

I still can't marry my memories to the descriptions in any modern guide, but I remember descending easily enough to a series of ledges and traversing them for some distance. The ledges shrank as the wall above us steepened. Paul kept looking upwards for a suitable challenge, dismissing each possible line and moving on until he stopped below a sizeable roof way above us.

We sorted the ropes, and I sat comfortably on a small but nicely situated ledge about a metre and a half above the sea, looking out to sailing boats trying in vain to sail. It was all very 'summer holiday'. Paul started up the wall looking very confident, and despite the steepness, it seemed easy enough for him. We swapped some inappropriate banter; we were both in good spirits and enjoying a relaxed afternoon. It was rather hot on the ledge, and I was beginning to relax a little too much, leaning my back on the wall to face out to sea.

I was spending too much time watching seagulls and boats and only occasionally glancing upwards to see how he was getting on. The ropes ran out steadily until he was about 15m up beneath the roof. He was arranging protection, ready to launch through it, when I noticed a slight swell for the first time.

Paul had just got himself established above the roof and was now finding it considerably harder as he searched for some good protection, clearly beginning to get pumped. Still reclining, my back to the wall, I strained my neck to watch him, realising that if he was going to come off at any point, this might be it. At that moment, the sea simply and quietly rose to the level of the ledge and rather sheepishly caressed my legs before slithering back off. Startled, I tried to figure out what had happened as any normal wave tended to splash everywhere. Watching the horizon, I noticed that the swell was getting higher and was heading towards me, this time as a definite wave.

It took a few seconds to remember that I wasn't belayed to anything, but instinct kicked in, and I quickly rammed a Friend into a nearby crack. I realised this wave was going to have a bit more attitude as the sea level under the ledge subsided by a good two metres or so, and the wave hit the ledge and spewed all over me and some way up the wall, this time with more force.

I felt it sting my face then retreat back to whence it came, leaving me rather drenched and considerably colder as rivulets of salt poured through my shorts. I looked up at Paul, who had missed the whole episode, being too engrossed in finding gear before tiredness would decide what was going to happen and called to him, pointing out to sea.

As I turned back, I saw the horizon rise substantially higher. The sea below me shrank down by four metres. I braced myself for the worst. The bulk of the wave hit me full-on and forced me into the wall. It lifted me into the air where I was halted momentarily by the belay runner; then I was spun around with my legs above my head at one point. I couldn't see anything and couldn't hear anything. Then the cold hulk of water fell back down to deposit me on the ledge as it returned to the deep and stole the very air around me as if it had created a vacuum. I was gulping for breath when I realised there was a dead weight on the ropes.

There was a moment of confusion, then concern, wondering what had happened to Paul. Amazingly, I had held onto the belay plate, and Paul was hanging in space not far above me, also wet. The spray from the impact had got him. This and the fact that I had probably pulled the rope taught while being thrown around by the wave had pulled him off the rock. I lowered him to the ledge.

We sat for a short while to see what else was in the offing, but the swell slowly subsided and vanished, the azure sea recovering to its former benign self. The sun continued to beat down on us, and as everything rapidly dried out, it was as if nothing had happened. But we had had enough and pulled the ropes and walked the now steaming ledges back out. We could give no explanation why this had happened, and we realised it could have gone horribly wrong, especially if I hadn't placed that Friend and managed to clip a sling into it in time. On the up-side, I had experienced a G1, a G2 and a G3 glubbing, all in one day. That was something to be proud of when you are 21 years old and searching for thrills.

Dorset in the 80s was ground zero for the deep water solo revolution that later swept the UK. Gordon Jenkin's 1986 Climbers' Club guidebook to Swanage piqued interest with a sensational picture of Nick Buckley soloing the Conger on the cover. The beans were spilt in an article in the May 1995 issue of *Climber* magazine by the Cook brothers. The DWS guidebook to Dorset, *Into the Blue*, came out the following year, but what really set the fuse was the inclusion of a teaser for Richard Heap and Mark Turnbull's earlier 1996 film *Under the Sky and Above the Sea* at the end of their iconic *Hard Grit* which brought then obscure DWS in Dorset to the attention of a large climbing audience in 1998.

Le Premier Psicobloc by Nick Buckley

I started climbing on Sheffield grit about 1972; not quite the washing-line scenario, but close. I was actually more interested in caving then and didn't climb much until I'd been at college in Portsmouth for a couple of years. The Pompey crew were drunken dirtbags mostly, but we managed to climb everywhere from Scotland to Cornwall.

Swanage was our nearest crag, however, and so we ended up there more often than not. By this time, I'd failed my degree and grown some muscles courtesy of my job as a plasterer and fancied myself as a bit of a hotshot (in retrospect, I was a not-very-big fish in a very small pond). We vied with the Swanage locals and the Bristol boys for new routes, soloing the sea traverses while prospecting for new routes and annoying the rangers by dossing all over the country park. Kev Turner and I led Freeborn Man in 1979 as a trad route. I think we originally gave it E3 6a. The name comes from an Ewan McColl song. I didn't solo it until well into the 80s, and Pete Oxley may well have done it before me.

The guidebook cover photo of the Conger came about as I was in Swanage sometime in 1983 with my then-girlfriend, who is now my wife, Roz. She isn't a climber, but I persuaded her to do Helix anyway, and we paddled in the pools at Fisherman's Ledge. My mate Angus Dalgliesh appeared in a canoe. I gave him my camera and told him I'd solo Conger if he'd take some photos. I hadn't done it before; it was an onsight solo. We didn't think about the 'deep water' aspect at the time as DWS wasn't a thing then, but I was soloing a lot then on the mountain crags as well as sea cliffs, so I didn't really worry about it. Obviously, I knew I wouldn't die if I fell off.

We moved north a couple of years later and mostly climbed in Derbyshire, so we missed all the DWS explosion, only really learning of it from lads who worked for me in rope access. The totally undeserved title 'Le premier psicobloc' wasn't coined until about 20 years later.

Right above: The cover of the 1986 CC guide to Swanage. Right middle: Pete Oxley on Freeborn Man, F6c S1. Photo Oxley collection. Right lower: Grant Farquhar on Freeborn Man. Photo Jethro Kiernan. Facing page: Nick Buckley on The Conger, F6b S1. Photos Angus Dalgliesh.

...and Captain Blood's Cavern by Jonathan 'Joff' Cook

There remained the sea, which is free to all and particularly alluring to those who feel themselves at war with humanity.

Captain Blood by Rafael Sabatini

'We should go to Fisherman's Ledge today,' Damian said one sunny Sunday morning in 1989. 'The guide says there are loads of roof routes we can go for there.' When we arrived above Connor Cove, there were already rucksacks at the top, but no one to be seen. We had to figure out how to descend to the ledge. The most logical option was the abseil down the arête of Helix, which put the roofs of Fisherman's Ledge on the right and the architectural delight of the Conger cave on the left. As we dropped in, to our amazement, there were four unroped people on the route, which we quickly identified as the Conger; after all, it was on the front cover of the guidebook. One of them, a tall, blond, fit-looking bloke, was resting to the left of the crux chimney shouting encouragement to his buddies. We found out later that this was Crispin Waddy.

We watched as two of them crossed the crux to join him. One peeled off, spinning feet first into the sea 30ft below. He was fine – came up laughing even. Well, that was different. He swam over, climbed onto the ledge next to us and shot up Helix, leaving little puddles along the way. We stuck to our plan and climbed the Ritz (or something) in our usual dodgy style, but Damian was particularly struck by what he had seen and announced that he was going to try and solo the Conger there and then. Again, we watched as, despite his fledgling talent, he struggled to decipher the starting chimney and awkward corners. But soon enough, he was chest-flat to the streaked hanging block that must be rounded for the crux. Sure enough, he rounded it with his knees up under his chin but was maybe too contorted to span his left leg across the apex of the zawn and, in the attempt, relinquished his grip, dropping cleanly into the sea. He loved it and vowed to be back as soon as his (only pair of) boots dried. And the next time we joined him, but it was Dom's and my turn to take the fall while Damian cruised it. Eventually, we joined his success, and all three of us leapt off the 50ft cliff-top to celebrate. That's how it can start. Three idiots with a sense of adventure, feeling the thrill of nature's challenge and the benevolent safety net of the sea. Well... that's how it started for us.

The next challenge was set by Mike Ford. It was 1990, and he told Damian and me that Freeborn Man had been soloed just a few times, and he aimed to be the next. Being a policeman, he had a strategy for all his plans, and in this case, it involved setting up an abseil rope and practising it a few times on a Petzl shunt. It was all new to us, but he showed us the mechanics of it, and I set off to buy a shunt so that we, too, might get in on the glory.

On a perfect summer Sunday afternoon, Damian and I took my brother's Yamaha 125 across the Poole chain ferry and rode over the Purbeck hills to park up at Durlston Castle. Halfway along the one-mile walk-in, we saw Mike with a couple of friends walking towards us, beaming solidly. He had sent it and was off to the pub to celebrate. That was the Dorset way: tick something good and stop for the day to enjoy the moment. It was all the extra incentive we needed. Despite knowing he was a better climber, and had prepared rigorously, we now had the belief.

I went first, struggling with the unfamiliarity of using the shunt, as it jammed and had to be unlocked every time I even lightly loaded it on the furred and ancient dynamic rope. It was not a good dry run, and I figured I was still a long way from having the confidence to try soloing it. After all, as far as we knew, no one had yet safely tested the fall. It looked like the drop from below the top crux would give a flight line awfully close to the cheese-grater slab. Perched at the top, we went to exchange the shunt, but Damian fumbled it. Down the rock it bounced and, with a splash, disappeared into 35ft of opaque, green sea.

'Looks like we're soloing it then,' declared Damian.

Above: Team shot early 90s. Back row left to right: Pete Oxley, Steve Taylor, Mark Williams, Mike Robertson, Joff and Damian Cook. Front row left to right: Jan Rostron, Jane Wylie, Carol Robertson and Martine Huismann. Facing page top: Into the Blue CC guide to DWS in Dorset. Facing page middle: Liam Cook on Privateer, F7b+ S2. Facing page bottom: Jacob Cook on Fathoms, F6b S1. Below: Mark Williams on Hornier than Thou, F7b+ S0, Stair Hole,1998, as Dominic Cook chalks it up with a sponge on a pole. Photos Joff Cook.

I watched from the base, ready to mount a rescue, as he cranked through the slab into the pocket sequence to arrive at the rest below the blank slab crux. I was willing him not to fall, but not exactly wanting him to succeed because that meant I would definitely have to try next. As he smeared up onto his right toe and nailed the most relieving rock-over in Dorset, I was delighted for him, but I knew I could not baulk at it now. All I can remember then is him calling beta and encouragement as I followed his sequence move by move. I could barely believe it when we both stood together, dry at the top of the line. The sun was out, and the day was fantastic for more climbing, but the moment was so special we agreed to pack our stuff and go catch up with Mike. The ride home, smiling and laughing and smiling – those moments never fade.

Jon Biddle persuaded us to try Fathoms. It was the natural successor to the Conger and a real crowd pleaser since after you arrive at the biggest jug in Dorset, at 35ft up, you have to commit to the tenuous, flaky crux, and that saw most suitors airborne. Damian nailed it first try; Dom and I fell. So the next time, Damian abseiled over to take photos of us flailing.

'You know, I think there's a line coming across to the left here that's not in the book. We should try it.'

So we came back for it on a sunny mid-week evening, and after down-climbing the beautiful Funky Nomadic Tribes, Damian set off first, traversing on the fault line ever further left, past the starting ledge of Fathoms and into new territory. The most puzzling bit of the route is how to come up into the start of the arcing corner from where the fault line comes to an end and the cavern starts. After a few ineffective attempts, he cracked it and was now moving up and left on the undercut rail to its end at a modest rest before the bulging second crux.

Now anyone else with a first ascent firmly in their sights of an absolutely beautiful line, destined to become a classic, would've focused on themselves, shaken out and committed to nailing the final tricky move to the adulation of the crowd, but not our Damo. He shouted back from the shake out, 'Joff – you've got to get on it, man! It's unbelievable!' And he hung there, calling out the beta that allowed me to get straight into the first crux and hit the undercut rail.

'Superb. Let's finish this.' I leaned out, shaking out and watching as he tackled the final bulge – a bulge neither of us had thought to drop down and clean before attempting the route. As he finger-swept the grit off the holds, he got more and more tired until he finally pumped out completely, dropping 40ft into the water.

'Go on… go for it,' he called from the sea, and so I did, and with Damo looking on, I pulled through the second crux and took the first ascent of a three-star route that by all rights should have been his. He named it …and Captain Blood's Cavern after a ride at Paulton's theme park and the TV advert song. In those days, the routes

in the Swanage guidebook read east to west, so this line would be the last in the chapter.

My brother Dom had a physics degree from the University of Exeter, and he said that this homemade 'base jump' would work. Who were we to argue? Two cut-off pieces of broom handle, six lengths of paracord, and a PVC all-weather car cover. Once it was bundled up and ready to go, it needed someone to throw it out behind you as you jumped from the cliff, holding the handles. We thought it would be daft to just drag it behind you since it might get snagged. Dom looked to Aidan, the older, wiser brother, to perform that duty.

'3, 2, 1, go!' Dom Jumped; Aidan cast the bundled PVC parachute out behind, and everyone held their breath. A woosh of crinkled plastic burst into a wide canopy overhead, and he smoothly descended gently into the sea. Now everyone wanted to try it.

Great inventors typically want to see just how far they can push the concept, so it was with Dom. The 50ft drop into the deep waters of Connor Cove was no longer enough. He set his sights on the 80ft ledge at the western edge of the Promenade crag above the infamous route Benny. I climbed to the bottom and had planned to take a photo or two of this huge 'base jump'. As he left the cliff, the parachute filled with air a little too quickly, causing him to pendulum back towards the cliff; he was descending inches away from the cliff face. In the horror of the moment, I did not press the shutter button, just held my breath that he might not catch the cliff or the rock near the base. It's a testament to how well the 'chute worked that although Dom landed in shallow water, he was still relatively unscathed. He sprained and strained his back and shoulder but broke nothing except perhaps his confidence to repeat that particular feat.

I was psyched for DWS and couldn't get enough, so even when there was no one available to head to the crag with me, I decided to go alone. I did that quite a few times, actually, but the one that stands out was a solo venture to Connor Cove around 1992. The wind was up, and the sea was very choppy, so I thought I'd stick to familiar ground. After repeating a couple of classics in the eastern edge of the cove, I figured Fathoms would be a sure bet and abseiled onto the Funky ledge. Harness off and clipped to the ab rope, I waved to a couple of sea kayakers about 100m out and started swinging along the fault line, doing my best to stay dry as the waves burst upwards from the cliff base. As I started to ascend, I gave the sea state no real thought since I had total confidence.

The last real move is a mantle onto a ledge a few feet below the cliff top, and that's where I went off-script. I was in the air. Why was I in the air? – I should be topping out. A block of limestone, the shape of a medium-sized TV set, was next to me in the air, having pulled free, and was dropping adjacent to me, and some of the way down, I was facing it, and then it was behind me. Oh, shit!

Facing page: Damian Cook on …and Captain Blood's Cavern, F6c S1, Connor Cove, 1990. Photo Joff Cook.

A feet-first entry into the sea was a hell of a relief, almost as much as the absence of an impact from the block of stone against me as I pierced deep below the surface. Once I had my head back above the water and re-oriented myself, I swam for the ledge and the inevitable several-attempt exit as the waves teased. You think you're out? No, you're sucked back in. Finally out of the sea and onto the ledge, I heard, 'Hey, are you OK?' The kayakers had seen the whole thing and, due to the sea swell, had not seen me surface. So they'd raced across to look for me.

'Thanks. Yeah. I'm OK now.' I used to solo alone. But I wouldn't let my boys do it; it's been the undoing of the best.

Snorkelling out through the caves at Stair Hole one summer day sent shivers right through me and chilled me to the core. God, I'm skinny, but I had no idea how much water lay below Animal Magnetism. Were we talking touch-down or full bouldery impact if I took the ride? There was only one way to find out. Finning back and forth, looking up at the route through misted glass and down into the kelp. Where would I land? Is there enough water? Damn, it's too cold to keep this up. I just have to hope it's OK.

Pete had put the route up and bolted it generously. We had followed and believed it solo-able, but no one had tested that yet. Sitting in the late afternoon sun until the tremors abated and wondering what would happen, I waited for the innate command: Time to go!

Damian waited on the slab, just in case, and provided the usual psyche. I couldn't miss and executed the sequence I knew so well in my head to the huge resting jug that Richie White would one day rip painfully free. Only the exit groove to realise, and thanks to friction, or psyche, or just being skinny, that too was now in the bag. God, this feels so good!

Jon Biddle was the first one I ever heard call it 'deep water soloing'. He had climbed with Crispin and done the classics in Connor Cove at the beginning of the 90s, then set about adding some brilliant lines at Connor and Stair Hole like Jellied, Anarchy Stampede and Meet Herbert the Turbot.

'I'm thinking of writing an article for the mags all about deep water soloing,' he said as we rested between routes above Connor Cove. Whether he got the name from Crispin or had made it up on the spot, I have no idea. A year or so later, he still hadn't got around to it, so Damian and I wrote it instead.

It's a little bit strange, I think: if we used to have a sport climbing or trad project, and your mates knew you were the one who saw it first, there is an unspoken rule that says: hands off until it's had the first ascent. But with DWS, that never really seems to have been the case. If you failed, your mate would know you'd be off back to the beach or gear spot to dry off, and they had the perfect window to get in there and try their hand.

So it was with The Despicable Terrier, F7a S0, at Stair Hole. I failed once, so Mark Williams, who was always impatient to burn off his mates, took advantage of me being vacant for a few minutes and nabbed it first. And good on him, I say. It's not like I was in the way, dogging it on a rope. Of course, it is only fair that I got to name it, and so I named it after him. The same rule applied when The Thieving Gypsy, also at Stair Hole East, was first climbed.

That was always a big part of the fun: A team of keen individuals jockeying for position to get the first send, such as when Pete beat a team of three or four others to the first ascent of the Laws Traverse. He backed himself to swing in along the improbable-looking roof, and the others hesitated. His was the glory.

When it came to Hornier Than Thou, the super-low footless alternative to Horny Lil' Devil, Damian and I had stiff competition. Damian had spotted the line, but we were trying it alongside many of the Dorset crew and even Dave Henderson, fresh from mega exploits in Devon. Until someone cracks the sequence, it's crank and splash and go again. When the battle drags on to the cooling end-of-summer days, special tactics would emerge. Big towels and a change of clothes, as close to the route as you could. Down jackets and Thermos flasks at the ready. Liam Cook was so keen to nail The Brotherhood at Stair Hole that he continued his campaign into September, shivering uncontrollably on the slab below under a silver survival blanket between attempts. It's so shallow underneath that he could only attempt it for the two or three days of the highest spring tides, regardless of how cold or blowy it was, and it sure was.

Perhaps my most despicable steal was from Damian on Gates of Greyskull. Mikey Robertson, Damo and I all wanted it, but only after Damian had gone first and shown us the splash-down did not occur via impact on the projecting slab of rock. Once that was established, it was game on. The three of us took turns to push to the final dynamic crux: slap the right flake, slap the left flake, slap the finishing slot... except that we couldn't.

Damian was closest and would have been first, but he was in love, and as we neared the prize, he missed the vital last couple of trips, opting to follow his heart after the Hungarian hottie who had caught his eye at the language school where he worked for the summer.

More decent folk than Mike and I might have waited for him... one for all and all for one and all that... but, nah. We got ourselves down there, leant over the lip of the finish, and with a big Magic Marker drew cross-hairs to exactly where the blind finishing slot could be located. It worked a treat, and the two of us had it nailed the same day. All is fair in love and war. Damo took the love, but Mike and I won that war.

A Trad and a Sport Climber by Damian Cook

A trad and a sport climber were once embroiled in a fierce debate as to which was the purest form of climbing. The argument raged on until they sought the advice of a visiting guru from a distant land. The guru, being a clever sort, avoided the question altogether and answered:

'Free soloing is the purest form of climbing. You eliminate the rope; you're up there like a bird or a lizard. It's about as free as you can be.'

Well, that told them.

But as the guru wandered away, he started to ponder on the wisdom of his words. Something about his answer did not ring true, and it occurred to him that the solo climber was not entirely free since he must either pre-work the route to be sure of success or carry with him the fear of falling to injury or death. If these could be eliminated, while some consequence of a fall were maintained, then the climber might truly be free. Then he thought, bollocks to all that philosophising, and off he went deep water soloing with his mates.

Footnote: The guru quote is directly from John Bachar in Masters of Stone – the video everyone was watching back then.

Where Worlds Meet by Damian Cook

Deep green... deep, deep blue. So close but a world apart. Just a few seconds on the other side. I'm never quite prepared for it but always so grateful, so relieved, when once again I'm welcomed into the Green Room. A brief moment of bizarre calm. A shimmering glass ceiling, the endless blue walls, the bubbles of my impact and exhaled breath. The exaggerated movements slowed, timeless, as if captured on canvas. Failure is rarely so sweet, so rewarding.

The perspective is startling: I've fallen through a gateway to another world. I resisted for so long. I held on with my last strength, and when that wasn't enough, I expected no more. But here it is, just one step away from the cliff, the cool, forgiving blue. Already my body glides upwards towards the light, where my lungs can rest. I know they'll be laughing and cheering. They know the feeling. It will be one of them next... and then my turn again.

My head and shoulders tear through the silver seal, back to the sunlight, the air, and her. Towering over me, her pockets and edges which ultimately led me here promising, 'next time'. Yes, next time it will be different. Exhilaration and relief subside a little, and that hint of disappointment that drives a perfectionist cools the water. I swim quickly to the side where I stand in the drying sun. Dripping, smiling, shaking more from adrenaline than from the cold, I look back towards the other world. If I weren't so obsessed, I'd like to spend more time there, overcoming my fear and learning the secrets of the deep.

Left top: Jacob Cook on Hornier Than Thou, F7b+ S0, Stair Hole. Left middle: Liam Cook off Adrenochrome, F8a S1, Stair Hole. Photos Joff Cook. Left bottom: Joff Cook on Spittle and Spume, F6a S0, Cave Hole. Photo Mike Robertson.

Above: Damian Cook on Crab Party, F6c S2, Cave Hole. Photo Mike Robertson.
Below: Damian Cook on Crime Wave, F6b S1, Connor Cove. Photo Joff Cook.

Above and below: Crispin Waddy attempting Swordfishtrombones, F7a+ S2, as a DWS prior to leading it with ropes. Photos: Waddy collection. Facing page: Crispin Waddy. Photo Grant Farquhar.

The Waddy Show by Crispin Waddy

Get ready and it's rock and roll

Showaddywaddy...

Hey Rock and Roll by Showaddywaddy

I started climbing in Dorset, where I'm from, around 17 years of age, during a year off, which I spent hitching to Swanage to climb easy trad routes, mostly HVSs. I then went to the University of Bristol to study Physics and climbed mostly in the Avon Gorge. In the early days of our explorations of Swanage, carless as we were, we had no choice but to hitch to the crag – a time-consuming but generally interesting experience. As a result of that, and the fact that there were no campsites very close to the crags, our tight budgets and overheavy rucksacks, we never brought a tent, preferring to make use of the various quarries and caves for shelter. This wasn't as easy as it might sound due to the illegality of it. The procedure, at the Tilly Whim end, was to gather firewood from the woods bordering the path and lob it all over the edge at a certain point, then backtrack to a vicious wrought-iron gate guarding a tunnel entrance. This was stumbled down, headtorchless (for fear of being spotted by the, maybe imaginary, warden) into a large man-made cave. After a while, this came to feel like an important, ritualised start to a climbing trip, and the rules were strictly adhered to. Considering the huge fires we used to light, the no-torches part seems particularly absurd.

The ceiling of the quarried-out caves is a limestone bedding plane held up by occasional, somewhat rickety, pillars of rock. It looks like a nearly completed abstract jigsaw with the missing pieces conveniently laid out directly below the holes in massive chunks or fittingly, perhaps, like sarcophagi. Frighteningly easy to solve, mentally. But of course, we felt indestructible and slept blithely through to the sparkling mornings, breakfasting in style in the crisp, bright mornings on a ledge overlooking the sea, a short walk from the climbing.

In the summer of '85, Andy Ford and I spent a few days here, picking off a few excellent new and established routes in the Boulder Ruckle before poor weather forced a retreat. In our downtime, we used an *Encyclopedia Britannica* (what's that?... consult Google) for inspiration for some route names in a part of the Ruckle with a Greek legend – Jason and the Argonauts – theme. Hence Symplegades, The Clashing Cliffs, which he had to row between before they crushed his ship, and Dragons' Teeth, which were sown by one of his enemies to produce fully armed warriors to attack his men. Apparently.

A few days later saw us loading up with several days' worth of supplies and, that night, walking the coast path to a similar (though fortunately more stable seeming) quarry that was better situated for climbing at the western end of the crags. At one point, stumbling through the twilight, I caught up with Andy, who had lost his footing and was trapped under the ludicrous weight of his sack like a beetle on its back. He was helpless under the load (we used to carry so much stuff around everywhere), and I was helpless with laughter. It was a good thing I was behind, or he'd have been there for hours, ever more feebly shouting into the night.

Topmast Quarry is even better appointed than Tilly Whim: a beautiful, secluded ledge with a grassy lawn that overlooks the sea. This was near or above a series of undeveloped crags, and we planned to rectify that. In the morning, we settled into the routine we had established the previous week. The top of a lot of the Swanage cliffs comprises somewhat unstable rubble and grass. It nearly always relents in angle, and it's generally fairly easy, though a little scary, to negotiate – typically, it's reminiscent of the top of some of Millstone, though at worst it can be far more extreme – but it's nice to try to clear it off, especially before abseiling over it. If suitably armed with a crowbar, you can produce some satisfying trundles, which may have been part of the point, in fact. Anyway, we both disappeared over the edge to while away a little time spotting possibilities. Eventually, I found myself at sea level at one end of a long, arched sea cave with several entrances. There was an interesting, though

desperate-looking, roof crack running perpendicular to the sea. The whole venue was bathed in a surreal green light welling up from below as the sun shone around and under the pillars supporting it all, which appeared to be narrower underwater. The crack led to an easier-looking horizontal chimney leading to a slab hanging in the roof, which curved away out of sight at the far end.

A little later that afternoon, the tide had dropped sufficiently to access the start of the roof. I made no attempt to free climb this. The first 40ft or so, aided, took a fair amount of time, especially with the basic equipment we had in those days. By the time I was able to start the free climbable section, Andy was already having to rearrange belays and ropes to avoid the rising water. And when I had re-emerged into the sunshine, it was impossible for him to get to the start dry. We left the ropes in place for the night, and Andy prusiked out. I was expecting to swim back and do the same.

Looking around, I realised that nearby was what looked like an easy and pleasant line and set off up it. It was above deep water, and it didn't seem too crazy to solo. I was always inquisitive and just set off. I could see the top and just thought I'd see what happened. I'd already jumped off the top of Conger, so I wasn't worried about falling in. The route wandered about in an interesting, exposed manner towards a deep hanging corner that led to a ledge and gave a good view of the left side of an appealing hanging arête, the edge of a huge block of limestone hanging out over the sea. That looked worth investigating too.

That first solo turned out to be a very enjoyable VS and ended up being called Aquamarine. I think it was the first deep water solo I ever did, though I didn't really appreciate the full potential and usefulness of the method at the time. It's sadly rare that such easy routes are above deep water. The arête also turned out to be an excellent route, Zircon (F6a+ S0): another little gem, though maybe not on a par with the very best ones at Connor Cove half a mile away, most of which were still unclimbed.

It had an interesting, unexpected historical twist. The crux is during the initial leftwards traverse along the lip of the overhang, with nothing but the sea beneath. CRimPs, EdGes, unDercuts, smeArs. The normal stuff of everyday climbing. But also, almost for the first time, a beautiful sense of freedom, making challenging moves unencumbered by gear and fear – exciting but not dangerous. I had never really been particularly drawn to soloing, finding that, for me, the very real possibility of death in the event of an error or even bad luck (even though, just by choosing to be there, you've brought that on yourself really), didn't result in easy flow, but more typically staccato movement: a refusal to trust friction as I would otherwise.

Once the arête is gained, it becomes easier due to the presence of some huge holds. Close up, it is evident that these are chipped – carved would be more accurate given their size – but weathered enough to blend in beautifully with the rock. They must have been made by the quarrymen for some purpose, and somehow it

Crispin Waddy on A Drop in the Ocean, F7b S0. Photo Richard Pollard.

Crispin Waddy on Tidy Tackle
Campaign, F6a S0, Pembroke.
Photo Mike Hutton.

seems to me that the skills these people had are evident in the quality of this work, but maybe I'm overthinking it. Presumably, there had been a crane on the ledge, similar to the one still present above Octopus Weed on Portland, and a barge would have nosed in below the overhang to receive the quarried blocks. The highly-prized stone ultimately ending up in a grand London house or somewhere similar. Those jugs must have been cut to enable some sort of access or control of the load.

Considering that they had an authentic, practical purpose – there was evidently a fair bit of work required to make them, and, of course, they were part of a quarry anyway – gives these huge crafted buckets a genuine legitimacy that enhanced the route for me, as well as making the final exposed hanging arête enjoyably casual. This contrasts with the unimaginative hackings on some gritstone slabs, for instance, at Cratcliffe, which somehow evoke in me a feeling of Merchant Ivory sepia-toned picnics, with butlers serving tea and cakes to the gentry on a jolly from Chatsworth, while the menfolk, unrestricted by whalebone corsets, show off their physical prowess on crude non-boulder problems. Maybe that's unfair, though. Maybe it was just 'local youths messing about'. You could argue that little has changed really...

The following day, we had to get up quite early for the low tide. By now, it was obvious that this wasn't a particularly convenient route. It ideally needed low tide and calm sea for both the start and the finish, so to give ourselves a chance, we tried to get in when the tide was still dropping. So much for our leisurely breakfasts. We abbed into our respective ends, then there was a long delay, muffled shouting, and occasionally a small amount of rope would become free until, finally, both ropes started to move. It turned out that they had become trapped underwater by the high tide in the night. It was only by repeatedly plunging head first amongst the boulders, feeling for the tangles and sorting them out, that Andy had managed to free them. Then he had waded to the start and seconded the line, by which time he was freezing and opted to jumar out to warm up. So that was Sea of Tranquility. I only have vague impressions, now nearly 40 years on, but I have the feeling that it could be above water and solo-able. If so, I suspect it would be very hard, but it could be a brilliant, atmospheric route.

Andy graduated as a vet a few years later but tragically died, with his girlfriend, falling on the descent from Mount Aspiring in New Zealand during their travels before starting work. How illusory that youthful feeling of indestructibility proved to be. He was one of the most uncomplicatedly decent and generous people I ever knew, and I miss him still.

More than a year later, Toby Foord-Kelcey, Phil Windall, and I were back, this time focusing on Fisherman's Ledge. I had been wondering what potential there was on the long, low wall to the west of The Conger cave. The rock here is mostly very sound, and little had been done. An obvious starting point for getting to know the area was the ledge at the start of The Rise and Dear Demise of the Funky Nomadic Tribes: a Thin Lizzy song and Kevin Farrell route from seven years earlier. It was perhaps named due to the highly appropriate first line: 'Out of sight, do it.' From the starting ledge, a juggy break led off leftwards. I thought I'd have a look at what there was along there. Rude not to, really. Toby was getting set up for his lead. Part way along, an easy rising overlap leads up into bulges. This goes easily on good undercuts and footholds until a perfect round porthole appears at its end. This feels like it should be an appropriate finish, like a full stop, but it's not quite as perfectly well-punctuated as that. A sense of urgency, emanating from your forearms, takes over if you hang around too long. You're quite high now and facing the crux. On the right are a few less-good holds, though fortunately, the angle leans back a little. If you commit, you'll be fine either way.

Soon, I rejoined Toby on the ledge where he was still sorting the rack. I think he was a little surprised to see me coming down the rope. The whole business had only taken a few minutes, and an excellent new route, Fathoms (F6b S1), came to be. It was such a simple, effective way of racking up the routes, and I was hooked. That day I ended up doing eight new routes, two on ropes with Toby and the rest were solos. Some were superb, and some less so (several have now felt the call of the sea), but from that day on, I really started to actively seek them out. Until then, these relatively low cliffs, especially those with the sea directly below, had generally been neglected in favour of their more impressive neighbours, as the hassle of access and setting up belays for such short routes presumably didn't seem justified. But these attributes were the very reason that it worked so well. There were plenty of excellent routes to be had, and it was the start of a whole new phase.

I'm not a strong swimmer. DWS, for me, was a way to do new routes really quickly. No one had bothered with them because they generally need a hanging belay, and they are short. I tried to DWS the first ascent of Swordfishtrombones but was worried about hitting the slab if I fell. I used to carry stones in my chalk bag to drop and see where they landed. I ended up climbing it on a rope in 1987 at E5 6b. Andy Donson later made the first DWS ascent in 1995 at F7a+ S2.

A few years ago, I was surfing at Sagres in Portugal and decided to do some DWS. I measured the cliff with my rope: it was 100ft high. I was on my own; there was a proper big swell and no place to get out for miles. I hung my rope into the water with a jumar attached and climbed with a harness on. To my surprise, the Portuguese fishermen chucking their fishing rods over the edge and hauling fish up simply ignored my antics. I didn't fall off.

I've done a few new routes at Swanage in the past few years. I was top-roping Great Shark Hunt and noticed a jug in the arête to the right, which I thought would make a great route. This became A Drop in the Ocean. The Vanishing was the route that got away from me, but I added a safer start to it, and a variation to Conger called Eel Hook which used a jug that has now gone.

Freeborn Man by Toby Foord-Kelcey

This route was first climbed when E4 was a top grade, and before sticky rubber and DWS was even a twinkle in anyone's eye. On the first ascent Nick was unable to top rope it so reasoned that a lead might give him the incentive to hang on. I made an early repeat of this route a long time ago as a roped lead, and it was quite epic to fix the gear. I have also soloed it, and it is far easier that way. Try leading it, and you will find out why it's E4.

Chris Flewitt

For a couple of years in the mid-80s, I toyed ineffectually with being a 'full-time climber'. After a promising start in Nepal, Australia and one or two other exotic places, this lifestyle choice gradually lapsed into an extended stagnation in Bristol, interspersing long periods of doing not much with poorly funded trips to nowhere in particular. A few other climbers in Bristol shared this life, one of whom, Crispin Waddy, would carve a successful trajectory from it into the future, whilst the rest of us eventually slid into office enslavement or other variants of normality.

Sometime in late summer 1986, Crispin, already revealing glimpses of higher ambition, encouraged three of us – Phil Windall, Jamie Ayres and I – to join him in Swanage, where he'd set up home in the Tilly Whim caves. I have various fragments of memory from the week or two we spent there: evenings nursing a single pint for hours in the gloomy Durlston Castle. Stumbling later down the long dark cave-access tunnel within which our host banned the use of a torch. Sleeping fitfully after failing to find a flat place to sleep; waking up to diffused but dazzling sunlight burning through the sea fog each morning.

We climbed for a few days in the Boulder Ruckle, then were led by Crispin over to Connor Cove. At that time, there were almost no routes recorded there, probably because of the low height of the cliffs and limited number of belay ledges. Crispin rigged abseils optimistically in a couple of spots, and I held his ropes on some first ascents around E3/4 on what was later named the Funky Wall.

One day, I was down on a ledge sorting ropes whilst Crispin eyed possible lines around us. Then I noticed he was gone and – more puzzlingly – was not tied in. He traversed a long way leftwards above overhangs, vanished around a corner and eventually reappeared on the cliff top. Reunited later, he talked through the physics of what he'd done: the cliff was steep and not that tall, the water was deep, and there'd been no real danger. I listened but didn't comprehend. The route was Fathoms; years later I realised it was one of the earliest first ascents in Britain made intentionally in DWS style.

A day later we were below the Conger and made a convoy solo. Crispin insisted that this was the normal way to do the climb, a statement that would only cease to be an exaggeration a decade later. Most people are elated after their first deep water solo. Lacking any peer group endorsement that it wasn't madness, I set off scared, had a moment of relief after sketching across the hanging chimney crux, then reverted to terror, not helped by Crispin and Phil's laughter, as I topped out on friable shards of crud.

As I get older, I notice that the world is split into people who grasp opportunity whenever it appears in front of them, whilst others examine opportunity, succumb to caution and say 'no thanks'. Handed a privileged chance to get involved in British DWS pioneering from its inception, I turned my back and didn't solo above water again until 2002.

Left: Toby Foord-Kelcey on The Conger, F6b
S1, in 1986. Photo Foord-Kelcey collection.
Facing page: Grant Farquhar on Freeborn
Man, F6c S1. Photo Jethro Kiernan.

Above: Andy Long on The Swinging Nineties, F7b S1. Below: 2003 DWS fest. Photos Mike Robertson.

DWS did not take off in Britain until the mid-90s but then became fashionable very quickly. Notably, in the late 1990s/early 2000s, it was popularised through a series of DWS festivals, mostly held at Swanage, which combined crowds of swimwear-clad climbers attempting DWS routes with rave-style after-parties – a conceptually radical evolution for British climbing at the time.

I attended precisely none of these events, considering myself even then to be too old and boring. However, I eventually became afflicted with a degree of FOMO, not helped by the irony that I had tasted DWSing a decade before most of the organisers of these events, let alone the paying punters. In August 2003, Dan Donovan and I paid a visit to Berry Head to climb Rainbow Bridge and Magical Mystery Tour. The former was considered quite hard at the time and had not been documented in a guidebook as a DWS.

We had some beta from the web forums but little else. The early parts of the route were just above the sea, but the crux takes you high above the waves into a pumpy technical groove. I was absolutely blown away by the intensity of the experience. Being with Dan, there was an inevitable 'safety-third' aspect to the day, resulting in us soloing two sections of the route that have boulders, rather than deep water, underneath and are usually bypassed.

At the end of the route, after about 300m of climbing, there is a giant wedge-shaped ramp rising out of the sea on which to recover and sunbathe. Somehow I had not fallen anywhere on the route, and as it was a very hot day, I felt perversely cheated. Already awash with adrenaline, it seemed absolutely appropriate to hike to the very top of the ramp and take a blind running jump off the cliff beside the ramp without any thought to fall distance or the adequacy of the landing. Thankfully, it was 'only' 15m or so to the water, and the waves were soft.

In 2004, I was still living in Oxford, renting a small house by the Thames close to the centre of town. My 'employment' that year was acting as finance director/cat-herder to an ill-disciplined software startup attempting to exploit a clever idea in financial econometrics from a local academic. This sucked up my time to a greater extent than I anticipated without generating any income. Meanwhile my son, Leo, was getting older and deserved more attention. Consequently, I was not climbing as often as I would have liked. Furthermore, Oxford is about as far from decent outdoor climbing as it's possible to achieve in the UK. Swanage was one of the least-distant options, well connected by fast roads for most of the 180km drive from Oxford, so I made several day trips there.

By then, Freeborn Man was on my radar as a route I should try. This may be Swanage's best-known route; certainly it has the most YouTube videos. The route tackles a 15m overhanging face directly above the sea, topped by a short slab. Nick Buckley led the first ascent in the late-70s and graded it E4, but more recent ascents were in DWS style, and the route was graded F6c. The 1990s Swanage DWS guidebook, *Into the Blue*, suggested that soloing Freeborn Man for the first time is 'a major life event comparable to driving your first car or burning down your first public building'. The problem was that conditions there were hard to get right. It needed sunshine to be dry but also a calm-ish sea to be able to access the bottom of the route safely.

In June, a friend, Roger, and I hiked to Connor Cove for 'a look'. We soloed the easier Troubled Waters next to Freeborn Man but contrived various excuses not to try its harder neighbour. Throughout the summer, the route stayed front of my mind, but opportunities to try it did not arise. On the last day of July, I was scheduled to spend the weekend at home, as we were setting off for a long family holiday in Ireland a few days later. However, I could not help checking the south coast forecast: it was perfect. Frustrated, a mad plan popped into my head. Central to my reasoning was the rationale that everyone, even a non-bread-winner at the weekend, is entitled to a 'lunch break'...

After yelling a 'back soon' lie through the front door, I jumped into my car with climbing shoes and chalk bag. My diary does not document the start time, but I recall it was mid-morning. Average speeds in the 'fast lane' of British motorways are 130–140 kmh, so a 90-minute drive time was just about feasible and, I believe, achieved. Unfortunately, the approach hike to Connor Cove is a further 3km or so. Somehow, I found the energy to run most of this.

At the cliff top, I decided I had about an hour available to climb, at the very most. However, the diary records that I allowed myself 'a few zen moments' before warm-up laps on Troubled Waters, then another pause 'taking everything in' before scrambling back down to the shoreline. An absolute cardinal rule of DWS is to never do it alone in case you fall awkwardly into the water and are unable to swim due to injuries such as collapsed lungs and a broken back or get hit on the head by loose rock. Unfortunately, though it was the weekend and conditions were ideal, there was no one else in the area. I put that detail out of my mind.

As I remember it, the first few metres of Freeborn Man are steep but fairly straightforward up to a committing traverse on small pockets and undercuts into the crux bulge. There is a strong sense of summit-or-fly at this point as the prior moves feel irreversible. The exit onto the slab is where most people fall off. I remember rocking up optimistically and finding a small crimp that I knew I was definitely strong enough to hold. Then, pulling into balance on the slab with a huge release of tension.

For an encore, I decided to climb the Conger as well, touching those holds again for the first time in 18 years. The diary records that it was 'a long expedition and not trivial'. But with all the traffic on the route since 1986, it definitely felt more solid. After that, I just had to reverse the run and the drive. The diary claims that I arrived back at the Oxford house at 3:30pm. It does not record whether anyone had noticed that I had gone out.

Swordfishtrombones by Andy Donson

Nineteen Eighty-Nine was a great year to be a young and devoted climber in Bristol – midweek Avon Gorge was teeming with climbers pre-UnderCover Rock indoor climbing, and the hordes would cram four to a car and head to the greater ranges at the weekends. Usually, this would mean Pembroke, with the slightly crumblier and definitely less crowded Swanage a second best. Crispin Waddy, a Bristol regular, provided a reliable source of inspiration for top routes for the latter, having been born and bred on the seacliffs of Swanage. The weekend would start on Friday with drinks at the Durlston Castle Pub, then descending to the convenient Tilly Whim caves to bivi (sleeps 100) – all totally legitimate at the time. After a weekend usually spent on steep Boulder Ruckle routes, Crispin would invariably end up convincing us to go to Connor Cove to try something called deep water soloing, which seemed to be unique to this one venue at the time.

The Conger was the main event and would be soloed on loop, passing under Nick Buckley's imposing Freeborn Man, the main rarely-repeated challenge here that, after a few nervous attempts, seemed beyond us. Even more mind-bending was Crispin's own project Swordfishtrombones that launched out through the huge roof above the Conger traverse. Crispin told the tale of how he had spectacularly failed to climb this as a deep water solo, where Steve Taylor captured him on film in mid-air sporting a bandaged leg from a previous mishap. After this, he resorted to a more conventional roped lead to complete the first ascent of this brilliant and improbable line.

Toward the end of that long summer of climbing, with my dole days nearing an end and a proper job looming grimly in the near future, we ended up once again at Connor Cove on a sunny Sunday afternoon. For a change, the opening moves on Freeborn Man felt less desperate and anxiety-inducing than before, and I was soon pulling the lip. With this surprise confidence boost and nothing else to do, I quickly made my way back down to Swordfishtrombones.

Below the big roof, some committing steepness with fists led to a back and foot chimney rest and time to recover and face the consequences of my whim. The roof here is offset, forming a short hanging groove that crosses the roof 40ft over the sea. Going onsight on a 6b that had already ejected Crispin was likely to end the same way. Stretching out right from the chimney, a surprisingly good side-pull materialized, then a perfectly situated foothold at the bottom edge of the hanging groove.

With that, I was away and crabbing urgently sideways across the roof to the relief of juggier ground at the lip. Soloing Swordfishtrombones was a surreal bookend to the climbing that year. All too soon, I was stuck working in London and hobbling around on crutches after trying to solo on Southern Sandstone, ironically slipping off a route called Banana.

Right: Tim Emmett on Swordfishtrombones,
F7a+ S2. Photo Andreas Hechenberger.

Stairhole Euphoria by Sue Hazel

'You will probably never be able to climb properly again.'

The surgeon told me. The available evidence and my age profile (I was in my late 50s) possibly clouded his opinion, giving a false impression of my inner resolve. As I sadly wended my way back through the hospital corridors towards the entrance, I found it hard to process what I'd just been told. What an assumption for the consultant to make? Well, if he wasn't keen on trying to reconstruct my thumb, or at least sound a bit more positive, I'd find someone who was.

Every once in a while during our climbing lives, we hit an abrupt setback, a slap in the face that alters everything and threatens to take our climbing away from us. For me, this most recent instance (I've had a lot of practice) was a severe thumb injury from a hand shoved out to protect a skateboard slam. The pain consumed me for a few days while I employed a regime of self-administered first aid before the realisation that this was probably not just a sprained hand. It would take another three months of continued doctors, hospital and physio visits to get an accurate diagnosis on what I'd actually done to it. My thumb sat at a really weird angle; the muscle had already atrophied, and it was useless for all but the simplest tasks, with pain on exertion of the smallest force.

Once in possession of an MRI image, the full extent of the damage became apparent: two separate torn ligaments. The initial surgical specialist I saw tried to break the news gently that this was a serious injury and it was likely that my left thumb would never be able to function normally again. I told him I was a committed rock climber and would not be shy of effort to facilitate my recovery in any way possible. He looked at the range of medical images in front of him and considered his own evaluation of my hand. The prognosis didn't look good: two operations, a year apart, to repair the damage; if, in fact, it would be a good idea to attempt the second risky procedure at all?

Fast forward another month to the recovery ward of a different hospital post-operation. The brilliant hand surgeon who finally put both ligaments of my thumb back together seemed pretty pleased with himself as I came around from the anaesthetic but left me in no doubt that from then on, it was all about the rehab. I was delighted, of course, and was absolutely committed to getting back as much function as I could, always going a little overboard with what the physio had prescribed. Progress seemed slower than I expected; I think the medical staff knew exactly what was ahead of me even if I didn't, but that's how it is for ligament surgery.

It seemed natural to set myself a goal during the rehab; Horny Lil' Devil became that goal. Just to get to the midway rest would feel amazing, and it would be an absolute milestone in my recovery. Six months after the operation, my knees slid into that double knee bar again, followed by a sketchy finish to the route. I sent my surgeon and physio a copy of the picture as a very grateful thank you for saving my thumb. It was not an altogether bad outcome after hearing the initial dismal prognosis.

Left: Sue Hazel on Crime Wave, F6b S1, at Connor Cove
in 2005. Photo Mark Stephenson. Above and facing page:
Sue Hazel on Horny Lil' Devil, F7a S0. Photos Dan Brown.

A Vanishing Summer by Mike Roberston

I was an E2 climber when I first set eyes on Freeborn Man at Conner Cove. But it was so damned inspiring. I got well past the crux on my first try and then got befuddled on that weird slab move at the top. Of course, I was undeterred; the fall was genial, the climbing was just so good, and when you've done every move but the very last one… Freeborn was my first E4, a mega-classic. The 90s gave us Cave Hole, then almost completely undeveloped. 1994 onwards was a time of such balls-out fun down there, and I loved taking newbies down there for the frolics and the guaranteed air time. One of the highlights for me was dropping my car keys in the drink and borrowing a mask to find them. A small circle of crabs had formed around my car keys, and I laughed so much I had to surface. And then I looked up; the huge roof crack of Crab Party was thus discovered.

The Vanishing, F7a+ S1, is a hell of a beast. Maybe my finest hour, in new DWS route terms. It was such a lucky find for me. I did it on my own mid-week. I could have called the route 'Hiding In Plain Sight', given the number of folk who'd previously guessed at its existence. Lurking right under and behind the mega-classic the Conger at Conner Cove, it'd been swum under a whole host of times and no doubt imagined by every south coast activist. But imagination needs a true outlet, and in the summer of 2001, I was the first to take a look at the line that, due to its wholly subterranean nature, could never be top-roped or inspected. I took a low tide with me on the first try, thinking I could do a lengthy, low traverse into the back of the cave. It gave way to me at about F6c with plenty of barnacle action and seaweed. From a tiny ledge in the rear of the huge cave, I pounced with trepidation up the steep lower wall (S1 for the ledge – take a spotter) and found my way onto an enormous tilted 'sofa' feature (this seats two; take snacks).

From there, the true nature of the line becomes clear. You're now at the start of a horizontal chimney, which gradually flares and smooths out until it reaches sunlight. The way forward is unclear; even the proposed whereabouts of your feet and head are uncertain. But the best thing of all, when you're trying for a first ascent of a yet ungraded or established route? Water, water, water. All deep, close by, safe as houses all the way. So I got into it, grasping, squeezing and body bridging, to finally, with a shout, slip from the glorious hidden pocket that would be my salvation. Splash. Darn.

Take two. A week later, swells and tide have stolen my traverse. But I'm prepared; I load my clothes, rock shoes and chalk bag into a dry bag, along with a small towel. A brief naked swim, and I'm rubbing myself down on that tiny ledge. Go, go. Big move, do it, move up onto the sofa. Chill, Mike. Relax. Look out, down and along that imposing and immaculate grey flaring chimney. Remember what to do? Of course not! But I will have my way. I launch out with a nervousness that denotes the importance of the moment. Into the final section, ass in, turn, twist, reach… And I have the pocket. A few more moves, and I'm into the post-crux section of The Conger. It's done. I look down between my feet at a piece of rock that I first set eyes upon some 10 years previously – why did I wait so long? The Vanishing was thus born – possibly my finest contribution to the incredible cliffs of Dorset.

That year, 2001, was also my best S3 summer. S3 is a headspace – it's a zone; a yearning to push it just a little bit further, to test the absolute boundaries of the genre. And so it was for me that summer. I'd somehow managed to compile a short list of S3 deep water solos for that season. Well, once it's written down…

First on my list was a previous sport route of mine, over at Dancing Ledge. A sheet of overhanging limestone that went by the name of White Rave at F7a+. So I abbed in and had my first good look at 'that' boulder in the water. The scenario: don't fall outwards (like we always have a choice). So off I went, and, yes, it felt very high, with the upper crux coming in at around 60ft, a high and steep rockover. And, no, I never quite got that boulder out of my head. Yet it was exhilarating; fun, fun, fun.

Left: Mike Robertson on the first ascent of Richard Ger, E4 5c, in Mongolia. Photo Grant Farquhar. Facing page: Mikey on the first solo ascent of Rocket USA, F7b S3, Smokey Hole. Photo Steve Taylor.

Facing page: Mike Robertson on the first ascent of Last Great Innocent, F7b S2/3, in the Black Zawn, Swanage in 1997. Right: Mike Robertson on Swinging Nineties, F7b S1, at Cave Hole in 1995. Photos Joff Cook. Left: *Deep Water*, Mikey's outstanding 2007 Rockfax guide to deep water soloing in the UK and elsewhere around the world.

Next up was a true monster, an old traditional E6 6b/ F7b at Smokey Hole. Concocted by Pete Oxley back in 1991, Rocket USA was not one to receive regular traffic. But a beastie it was, soaring up a huge and bulging upper groove. An inspiring line, but with three issues: 1. Very high indeed. 2. A flat boulder a metre or so below the high tide mark. And 3. Some odd fins of rock in the steep upper groove, which, were they to snap, would probably propel your unwilling body directly to the aforementioned flat-topped boulder.

So I took all this into account, cleaned the heck out of it, and took Barry Clarke and Steve Taylor down there. Barry was wetsuited up for rescue; Steve was the nominated cameraman, having taken a sickie just for the occasion. I can only say it went basically to plan, but as I pulled onto the top slab, I felt close to tears. This one had taken it out of me, and the hard moves on the protruding fins had scared the bejesus out of me. In my own words, in *Deep Water*: 'I wouldn't do it again for a thousand pounds.' But, what a stupendous route; VERY stressful. My hardest DWS and unrepeated.

Not one to be deterred by such fickle things as plain old fear, I next turned my attention to the back of the world-famous Durdle Door arch. The main event there was Pete Oxley's sport route Riding To Babylon, F7a+. It rose from the sea at a constant 35 degrees and featured a crux that was, quite literally, at the very top, 90ft from the briny. The water depth was adequate, at about 10ft, but the top crux was alarming, with a certainty that I'd be pretty pumped when I arrived there. No photographer in tow that day, but Gavin Symonds provided the backup for a rescue down below.

Game on. The lower stretch out of the way, I arrived at the precarious rest at the halfway mark. From there, it was all or nothing; the top headwall bulged and jostled above me. I pressed on. The plan was to try and get a reasonable rest at about 70ft, but, in reality, this sort of failed miserably, leaving me in a position of cataclysmic uncertainty. Go or retreat? Neither is a good option, old bean. So, in true brains-out Robertson fashion, I just went for it, balls-out. I arrived at the top crux, a series of open-handed slopers, utterly pumped out of my box. Four mighty slaps later, I grovelled over the top, utterly exhausted. The best-laid plans of mice and men? You bet. I didn't climb again for over two months. But what a summer. Riding To Babylon was very stressful indeed, but the absolute icing on the cake. I did actually stop climbing for two months that very day. It was finally repeated as a solo fairly recently, which is great news and, maybe, proves that I wasn't totally mad in 2001.

Above the Blue by Jerome Mowat

Durdle Door is a grand, sweeping arch of rock that curls around from the mainland, one foot planted firmly into the water, a knife-blade ridge running along its spine. The feature itself could have been lifted from a fantasy novel, the name picked from a nursery rhyme. Quite simply, this jewel of the Jurassic coast begs to be scaled by the vertically minded. Until now, soloing Riding to Babylon had been a pipe dream. My DWS partner-in-crime, Liam Cook, would often discuss it when stuck in the gym, indulging in escapism. Liam is not only of strong climbing pedigree but also one of the best partners one could hope for: motivated, bold, adventurous, encouraging and supportive. Our excursions with him are rarely dull. It was with Liam that I went for a recce of the great arch.

We woke up early to make full use of the high spring tide that was due to peak at 10am. The mission was dysfunctional from the start: A lack of gear, knowledge and familiarity with the terrain, making a straightforward approach pretty outrageous. At one point, after the third failed abseil down the wrong section, I turned to Liam and was about to suggest sacking it off. The day hung in the balance. Was it an omen? Was it not our time? 'How have we not died yet?' I said instead, half-jokingly. With no abseil rope and no slings or carabiners, Liam made an exposed scramble down shattered, overhanging rock. The muffled confirmation a few minutes later was music to my ears.

By the time we got to the bottom, it was nearly high tide, and I set up the rope with renewed speed. There was no chalk or other evidence of the route having been done recently. I wondered with Liam when the last time it had received attention. I enjoy climbing neglected routes. It feels like one is honouring the past, awakening a giant from slumber for just a short while, then letting it rest until the next suitors arrive in some years' time. The climbing itself is sheer joy. A corner at the start leads to a no-hands rest at eight metres before questing out over the arch on a sculpted ramp line. A few metres from the top, the jugs run out, leaving one to puzzle out a crux sequence on flat edges that bar progress to the top. On neighbouring Portland, the more accessible and highly popular sport climbing area, it would be an instant classic.

I was at the top of Durdle Door, having led it clean. Back at the base, Liam asked if I was going to go for the solo. I could feel my mind whirring away with all sorts of internal checks and balances I wasn't party to; it felt like a slight humming. 'Yeah, why not?' I surprised myself with the answer. The unconscious decision-making was nuanced and complex, but the tangible result was a set of reasonable conclusions: conditions were good, the climbing felt steady, and the tide was still high. Essentially, I could think of no good reason why I shouldn't.

Oh, yes – just the water depth. Joff, Liam's uncle, had arrived in a kayak to spot. He offered to snorkel the drop zone to check for depth. He swam around, face down. 'It's about three metres here,' he said over the easier first half of the route. He swam out a bit and dived down. 'And really deep here,' he said, over the top, more important half. 'At least four metres.' We had different ideas of the definition of really deep, but it would do. Stepping out from the ledge – unhindered by rope, released from fear and doubt – was a relief. A phrase popped into my head as I climbed: the unbearable lightness of being. It turned into a mantra. I had no idea what it meant, but it felt right, just like the climbing. Before long, I was shaking out on the jugs below the final sequence. Thoughtlessly, I pulled on the small, flat holds and mantled out the final shelf. There was joy – and closure.

Deep water soloing is all about stripping things back to the essentials: the sound of the sea, the texture of the rock, the feeling of breath, and of life. These are the moments that I seek: to do something so meaningless and yet meaningful, both arbitrary and profound.

Left: Jerome Mowat on top of Durdle Door. Facing page: Jerome Mowat on Riding to Babylon, F7a+ S3. Photos Liam Cook.

Dorset

Head Above Water by Martin Crocker

Martin Crocker describes the first ascent of Borstal Bash, F7b S3, Portland, in 2000.

Below: Pete Oxley on One Cool Vibe and This is the Life, both F6c S3. These two routes at Cave Hole were the first DWS on Portland in 1989. Photos Oxley collection.

Facing page: Jonathan Woods on Crab Party, F6c S2 at Cave Hole, Portland. Photo Mike Robertson.

It's that moment again: the only time you'll ever experience anxiety on a 'deep' water solo. Poised 25ft up beneath irreversible, overhanging rock. A shallow sea laps benignly below, concealing hidden dangers. It's that stressful moment of prevarication before you commit yourself for good. No ropes or gear to get you back in contact with the rock, no one to drag you out of the sea and no second chance. This is absolutely the last opportunity to quit.

So let's get back to being a responsible citizen for a while, and here we are again, hammering to work in the car, 10 miles into the 23-mile journey. Back and forth, day after day, forth and back. Forty minutes of rush-hour hell alongside every other poor sod who dreams of being somewhere else. Day in, day out, year after year. And for what? To steal moments of precious time before we're reduced to another eight hours in the office, pinned down. Maybe I'm being extreme, but has anyone else noticed how afternoons drag? Got to shut it all out. Switch on the radio; there's a tune that we like. So maybe we're too old for dance music, but the stimulus to move is still too powerful to ignore: slamming bass, jazz chords, and – of course – the rhythm. Time to slip out of our cars, out of our shells, and maybe just a little bit out of our minds, but it's difficult to break out dancing while driving at 70mph.

There's no other option but to dream of an awesome undercut and overhung line of weakness through the roof of a sea cave, too remote and wave-washed even for the hardiest Portland bolters. The route's got to be hard, maybe F7b. And the water's shallow, maybe only two to three feet in places. But what a line. I'd saved it up from last year, waiting for the perfect moment. And now it's arrived. I abseil to the small ledge just above the reach of the waves on the opposite side of the cave. Time for those last-minute system checks. Glasses secure and not likely to disintegrate; my daughter's armbands clipped either side of my harness to reduce depth of immersion in these shallow waters – like hell! Jumars dipping into the sea to save me from the half-mile swim out. It all helps to complete the delusion of safety that I need to get me through this.

Time to belly-flop along the narrowing ledge. The ledge runs out, so I start to hand-traverse left towards the back of the cave. It's new, unseen territory. The rock gets wetter and wetter, darker and darker. It's hard to see anything, let alone get a grip. I reach a jug. I'm only four metres up, but the sea is a mere skim over boulders of smooth, algae-coated rock. 'I must reduce my height!' I shout at myself, frustration echoing around the cave. But the only way to get to the dry boulders at the back of the cave is to jump. I think this little adventure is going pear-shaped as I struggle to get a boot off with one hand while hanging onto the jug with the other.

Somehow, the boot comes free, and I sling it to the back of the cave to keep it dry just as my elbow tendons say: no more! I half-fall and half-jump, splash-landing on the boulders. 'Owww!' Not perfect, but not quite a twisted ankle. One boot is wet. (And I do so like to be in control.) I scuttle into the back of the cave to dry out, massage my ankle and ego, and reflect for a while.

I have fond memories of climbing in Gower when I was 15. We'd rope up on a few Jeremy Talbot routes, but the real fun would begin when the tide came in. Leaving our heavyweight clutter of hawser-laid ropes, Hexes and steel karabiners on the top, we would solo aimlessly above the sea, oblivious to any consideration of danger. There was something alluring about the sparkle of sun on the soft, convulsing waters, mesmerised by the rhythms of the swell. Wavelets would reach for the soles of our boots as we scraped and bounced across the ragged, barnacle-encrusted rock. We'd engineer collisions with the spray from breakers and find any excuse just to throw ourselves in, so irresistible was the lure of the sea.

We were, of course, the first climbers to invent 'deep-water soloing' – along with all the other climbers who had been doing the same thing since climbing began. Naturally, soloing above the sea is now fully systemised: there are ethics, styles, grades, stunts, as well as the recorded routes that you can take or leave. There's no right or wrong, or better or worse way of doing it. Some of the south-coasters like partying, while others, like Crispin Waddy, undertake solitary explorations for the

full fix. That's the beauty of it. You do what you choose to do in your circumstances. At the end of the day, we all still want to be alive, buzzed up or not.

I return to the cave roof and spot the line of chalked-up pockets that lead to an enormous block jammed in the ceiling of the cave. The crux appeared to be the sequence to get to the block, but I managed to swing in to top-rope this beforehand. Below the crux, there seemed to be a slight pit between boulders in the sea that might provide three, four, or even five feet of water should I fall. Falling won't come into it, of course, but you become programmed to go through this façade of systematic risk appraisal in order to delude yourself that you're not going to die. Doubt is the killer.

Back on the rock again, but the wind and swell have increased, meaning I have to traverse steeper rock at a higher level to gain entry to a corner. Swell can be your worst enemy above shallow water, not because of the risk of being hit but because of the troughs between waves that can effectively reduce water depth. Fall at the wrong time into a trough, and you can hit boulders that were previously submerged. (At least what remains of you can then be cleaned up by the following swell. Now that's environmentally friendly. Best not to fall at all. That's my motto.)

I swing up into the corner, but it's soaking wet. Not from spray but from condensation, the bane of the deep water soloist, especially on these overhanging routes at Portland. High humidity – yuk. Water condensing out of mild and moist sea air onto cold, impervious black chert rock spells *slippery and loss of control*. This is a serious and unplanned impediment, and I wonder whether the chert-pocket-dependent traverse will be viable. At times like this, it's hard not to get wound up. You can wait months for the right complement of conditions – tide, water depth, wind, humidity, water temperature and still get it wrong. Shit happens.

Squelching up to to the roofs, soloing above rock, my heart pounds stronger and stronger in the usual psycho-physiological run-up to the moment of commitment. A narrow wall sandwiched between huge roofs runs rightwards to the first of the chalked-up pockets 10ft

away. I'm rarely so hesitant, but normally things are not so unexpected. I stop for a while, hoping that something within can deliver me from the tedium of reason and reservation. And while I debate how the risks of being spiritually demoralised if I retreat are probably greater than the threat to physical health if I carry on, something marginal presses the on-switch. This is it.

I reach the pockets and get pulled into the momentum and sheer joy of the rhythm. It's more like dance than mountaineering. You expand into your own micro-environment, head back, flowing and powerful, conceived mainly as a wild dream but now a simple reality. From pocket to pocket, high above the water, until that awfully long, but familiar, reach above the lip of the cave gains the block. Exulting as I heel-hook the block, my naked spine exposed to the expressionless boulders waiting just beneath the two-faced surface of the water. Curiously, there is no fear; there never is any fear. (But then, there is no finer master of delusion than the sea.) Jonathan Cook once said that this game 'is an expression of freedom'. It is, but more than anything, it is the freedom from the impasse of self-satisfying restraint and 'calculated risks' that you leave behind the door of commitment. Here, you enter a new world of dangerous dance, choreographed spontaneously by rock and sea.

With the cave roof behind, it's easier ground now, but it's high, steep and a little loose: time to seek help from rational man. The shakes of success are starting, and two minutes later, I top out with the customary grunting and strutting around in a sweat of adrenaline, just like coming off stage after a really good gig. That was the fix.

But it doesn't end there. Like every genuine addiction, you need another, and another, and another – chain soloing, perhaps. Thank God for low tide. Six or seven solos later, more and more boulders have emerged, and seaweed floats on the surface. The fishermen are packing their bags as the sea level drops, squeezing the players out of the picture. But you've sensed already that statistically, you've gone as far as the probability of not dying will allow that day. Addicted or not, you know it's the right time to quit. Responsible, rational man awaits.

Mark of the Beast by Pete Oxley

Potters Bar, where I grew up, was a grey nothingness. I felt isolated at school, with not a lot of friends. But, I discovered a passion for the outdoors, which I experienced during holidays with my parents to Cornwall and other places. In some ways, climbing feels like it was pre-determined. I was searching for a sport that would fulfil me internally, such as orienteering and cycling rather than team sports. Around 1979, I saw Ron Fawcett on TV; it transformed my life. Fawcett and Pete Livesey on the *Rock Athlete* programme clicked with me; that was what I wanted to do more than anything in the world. A lot of sports are not very interesting, but climbing is like a combination of dance and mental approaches. It has the fear factor, and it clicked with me. I wanted to be a 'rock athlete' from that point on.

My parents made the brilliant decision to move to Dorset when I was 10 years old. I started climbing, in 1980, when I was 14. At the time, there was nothing happening in Dorset: no climbing walls, no training, no bolts. Nothing, except an interesting history of trad climbing; Swanage is a fearsome place to climb with super-intimidating cliffs.

Climbing, for me, was a rebellion. I didn't enjoy school, and I was in a lonely place. No one else seemed to be into the outdoors. I couldn't join the local climbing club because I was 14, and you had to be 16 for their insurance. So, I asked myself: what did Ron do? Answer: he went climbing eight hours a day. With no one influencing me and no rules, I could be as creative as I wanted to be. So I chipped holds on bridges at Poole and Wimborne and spent the summer traversing on them. I created a route that was about E5 6b with a homemade peg. That was my first ever lead. At that time, the highest grade in Dorset was about E3.

When I left school, all I wanted to do was go climbing. I was set to go to Sheffield Uni to do Geology, but I was taken with what was in front of me, and I didn't see the point in going to Sheffield to be just another climber in the Peak. So, fueled by an inner passion, I made the decision to be a full-time climber in Dorset at 18 years of age. In those days, climbing was an underground movement in the south with characters like Martin Crocker, Ken Palmer, and Mark Edwards. The Cook brothers, Crispin Waddy and Nick Buckley, in particular, were very inspiring characters to me. We lived through great times when new ideas were forged in empty updrafts of sea spume in Dorset.

Music was deeply motivating to me from punk and bands like Joy Division and Helmet. It was a mirror to climbing. I've always been an independent, iconoclastic type of climber, so I was an indie-band type of guy. I was pumped up: adrenalised. I climbed five or six days a week for 10 years on the Enterprise Allowance in my Purple Capri. The southwest has always been overlooked by the climbing media. There were not a lot of figures to look to down south. Martin Crocker was highly active with abundant raw skill. I was blending the idea of athleticism and training with a creative approach to raise standards and take the grades further. Only a few people were doing this, and we were up against a wall of trad ethics in the south. But it was a cool time.

Seeing the 1980s French sport routes blew my mind. I was there when Jerry Moffat onsighted Chimpanzodrome. There was this skinny guy sitting there in baseball socks. The south was cut off from the media up north. It's one of the reasons I left the country. I was a full-time climber for 15 years, climbing five or six days a week on the dole's enterprise allowance from 1985 onwards. Geoff Birtles and people like that were highly critical. There were also access issues with the Wells family, who owned Lulworth, resulting in meetings with landowners and the BMC. There was Martin Crocker and Crispin and me. I did a monumental amount of shunting on my own, and so did Martin. I was the first person to build a cellar in the south for power training. Martin was a machine at around E5. I had a serious accident in Cheddar on one of Martin's routes – I took a 35ft decker and had a big argument with him. That's when I decided to bolt routes properly.

I made the second ascent of Freeborn Man shortly after Nick Buckley did the first ascent. I thought it was quite a serious route at E4. At that time, the routes in Dorset were heading away from the cracks. That is a blank wall, I thought; where the hell are

Above and facing page: Pete Oxley on the first solo ascent of Lulworth Arms Treaty, F7b SX. Photo Oxley collection: 'My first solo ascent of Lulworth Arms Treaty at Stair Hole in 1992. This went at E7 6b due to the serious start. It's probably never been soloed since, as it's a half solo / half DWS hybrid route – another style invented that never took off. This shot is interesting as it shows the trad history, sport history and modern DWS – trad thread from 1987, sport U bolt and DWS towel. I love this juxtaposition of styles living together.'

Below: First Dorset Rockfax by Pete Oxley: 'The image that started it all on the cover. This was Mark of the Beast first ascent on trad at E7 6b in 1987. Quite a story. With the fifth edition about to be published, it has become one of the top places to climb and DWS in the UK.'

Tom Livingstone on Mark of the Beast, F7c S2. Photo Mike Hutton. Facing page: Pete Oxley on the first DWS of Mark of the Beast. Photo Oxley collection.

the cracks? It looked to be another level; it was the first 6a move in Dorset. Nick was a guru – he dragged Dorset up to E3 and E4. I was looking up to these mythical Portsmouth figures. Even to see them in the pub was scary. I thought: what's next, Nick?

When Nick soloed The Conger in 1983, DWS wasn't on the radar. I climbed Troubled Waters and The Great Shark Hunt, F7a S1, deliberately as deep water solos in 1985 and soloed Freeborn Man around this time, but the first person to take a systematic DWS approach to routes in Dorset was Crispin from 1986 onwards.

By 1987, the move had been made in Dorset to the fully athletic style of Euro climbing away from the flat walls and cracks. This was really what I was after, what I brought to the table after Crocker/Waddy and before them Buckley/Littlejohn/Monks. It was like a changing of the guard. Training started in Dorset at a sports centre on a vertical wall – that rocketed standards for me. I brought in the notion that really hard routes could be soloed as DWS objectives in their own right, and Mark of the Beast was a massive breakthrough in level and psyche for me.

I first came across Stair Hole in 1982/3 – I was doing a lot of sea-level traversing. I was looking for something steep that would emulate what Ron and the climbers in France were doing. Stair Hole is the microcosm of the progression of British Climbing. It traced the history of development in Dorset. It started off as a backwater. There were a few VSs. Mark of the Beast started off as a trad route. The name came because it's a gothic crag, I gave it E6 6b, and it's 60ft long: 666. I did it first go and then did a victory hang on the lip, which came off, and I had to lead it again. It was around the same time as I did Roaring Boys, which was the first E7 at Swanage.

In 1993 I deep water soloed Horny Lil' Devil, the name of which came from an Ice Cube track, and Mark of the Beast – for me, it was a training route for Laughing Arthur (which, to my mind, is the greatest route in Dorset; I regret giving permission for others to retrobolt it). My approach was different. I was into onsighting, but I had a cut off. I wanted to get stuff done. I would inspect the E6s to make sure I got them. I rarely fell off. I only ever fell off DWS about three times in my life. I took it in a professional light. I wasn't in the onsight camp. I liked the idea of projects, but I wouldn't want them to drag on.

I was good mates with Mike Robertson. After a wild party in London with my niece, we ended up driving around South London at 3am for four hours. We started inventing this mystical character called Herman Borg and riffed on his adventures. The car had a fan belt problem. How is Herman Borg going to deal with this fan belt problem? And it went on from there. That's where the name of the route, Herman Borg's Basic Pulley Slippage, came from.

Mark of the Beast marks a critical change in mindsets that occurred over a few years (in the late 80s and early 90s). I started as a trad climber, but the exposure to training and 'Euro-style' led everyone to consider how we should now approach the mega overhanging lines that were gagging to be climbed, the blankness beckoning beyond the cracks and grooves of previous times. Mark of the Beast blew a hole in the preconceptions of what could be climbed in the south, mirroring the north, but in complete isolation for me at the time.

I think Mark of the Beast is the best F7c in the UK. My route Adrenochrome, F8a S1, recently had its first probable flash DWS ascent from Buster Martin. Also, I heard that Cailean Harker declared it to be the best route in the country. I would quite possibly agree – these routes are two UK ultra-classics with nothing else like them other than heading out to Europe's limestone meccas. King Lines.

It's hard to believe that it has been over 30 years since I did these routes. Now ancient history, I feel like a Jurassic fossil from the past embedded like one of the prehistoric tree-bowl holds down at Stair Hole itself. If we ever go back to live in Dorset when we retire, then it is at the back of my mind that I could do the 40th-year anniversary send of Mark of the Beast. Who knows, and probably who cares? But I have that thought stashed in the back of my psyche closet.

For Whom The Swell Tolls by Dave Pickford

There is nothing else than now. There is neither yesterday, certainly, nor is there any tomorrow.

Ernest Hemingway, *For Whom the Bell Tolls*

When deep water soloing really kicked off in the UK in the late 90s, I was spending most of my school holidays near the Dorset coast, where I could climb as often as I was able to. In August 1998, aged 17 years, I was staring wide-eyed at the 45-degree overhanging wall above Stair Hole's East Cave at Lulworth, Dorset, with a mixture of trepidation and excitement. Pete Oxley's mega-classic Mark of the Beast takes the challenge of this awesome wall head-on.

It's a perfect line up a unique piece of rock, and one of the best deep water solos in the UK. I'd climbed quite a few routes of the grade before, but sport climbing is different: up there, I'd be on my own. I set off before I got second thoughts, climbing quickly through the first section – big pulls on big holds through super-steep terrain. I shake out for a few minutes at the half-height rest, where you can relax before the upper crux. After a while, I went for it. The upper part of the route is defined by a series of increasingly powerful moves, the very last of which is the crux. There were a few shouts of encouragement from my friends on the sea-level ledges below as I climbed higher. I thought the top was in reach - it looked so close. But then, the famous move barred its dragon's teeth: from a couple of shallow pockets, I spied what looked like a hold out left, made a wild lunge for it, and hit something that wasn't a hold at all. Boom! Splashdown.

Two seconds later, I find myself swimming in the clear, cool water of the English Channel, my friends cheering the fact that I got to the last move on my first try. It was a revelation and the beginning of a lifetime's fascination with this style of climbing. I realised I could climb a route that was close to my limit without a rope, free as a gull flying across the waves. I was hooked. A little later that summer, I went back to the East Cave and sent Mark of the Beast one evening as a north-easterly wind blew out through the cave under me, sketching strange shapes across the wide, blue water.

The revelation of deep water soloing transformed my experience of climbing. From that summer onwards, I continued to push into the twilight zone of my own physical limits and, simultaneously, the psychological height limit at which I was prepared to fall into the sea. The process deepened my powerful addiction to climbing and at the same time, extended my passion for the sea that had been instilled at an early age whilst sailing with my dad.

Over the years that followed, I continued to push the edge of the envelope high above deep water. Some climbs were wild, some were foolish, and others were textbook cases of just getting away with it. In 1999, I made the second ascent of For Whom The Swell Tolls, F7b+ S1, another hard Oxley masterpiece at Connor Cove on the Swanage cliffs. The climbing is sustained and technical; the crux, as with Mark of the Beast, is right at the top. It's much higher than the Lulworth routes; the hardest move could be 60ft above the sea at low tide. That's a long way to fall into water; from this height, you have to get your entry right to avoid injury.

I went down to the crag alone one evening at the end of July. The clifftop grass blowing in the wind had been burnt to pale ochre by the summer heat. I threw a rope down the top section to brush away the thick coating of dusty lichen that accumulates on these routes after the winter rains. I was truly glad I'd done so. The holds were completely slick with a fine dust, and the route would have been unclimbable without cleaning.

Left: Pete Oxley on the historic first DWS of Mark of the Beast, F7c S2. Photos Oxley collection. Facing page top: Tom Livingstone on Mark of the Beast. Photo Mike Hutton. Facing page bottom: Pete Oxley on the first DWS ascent of Privateer, F7b+ S1, in 1998: 'Just one of the best hard solos in Dorset. It became part of a challenge I called the Devil's Triptych at the ledge: E666. Three x E6 solo in one day. The other two were Herman Borg and For Whom the Swell Tolls. Nobody has repeated this yet.'

Once the holds were brushed and chalked, I jumared up, moved my rope across so it fell on the ledge where the climb begins, abseiled down, took off my harness, and set forth into the unknown. I swung out quickly along the hanging rail that approaches the edge of that vast sea-level cavern to the west. Inside the cavern, the swell boomed and surged. I reached the point where the route blasts up towards that dauntingly smooth, grey headwall.

I took stock here for a minute or two before pressing on; an oystercatcher flew low over the sea to my right, its shrill call breaking the silence. I used it as a starting pistol; it's showtime. The headwall was suddenly directly above me. I pulled strongly into the complex series of moves that form the crux. A sequence of layaways and crimps led to the final move: a big lunge for a jug below the final ledge. I could see it, but it looked miles away. I briefly looked down. The sea glints far below, like a strange beast sleeping beneath the fastness of the wall, its belly rising and falling with the motion of the swell. It's waiting, it seems. Just waiting. It's got time, after all.

But I don't.

Suddenly, my left hand exploded off the intermediate sloper and caught the jug as my left foot popped off a tiny smear. Both feet cut loose, and I grabbed the jug with both hands. I'd done it. What a route. I was breathing hard and shaking with adrenaline. Sometimes it feels like it shouldn't be possible to climb solo in this style, but it is possible – and that's the thing that makes deep water soloing so addictive.

Marked by the Beast by Grant Farquhar

He [the beast of the sea] forced everyone, small and great, rich and poor, free and slave, to receive a mark on his right hand or on his forehead so that no one could buy or sell unless he had the mark, or the name of the beast, or the number of his name.

Revelation 13:16-17

The first route I ever deep water soloed was Animal Magnetism. Not being comfortable with the concept at that point, I climbed it more like a solo rather than a DWS and overgripped every hold, resulting in a massive pump by the time I got to the top.

The first time I saw Mark of the Beast was at the first DWS fest at Dorset in 1998. For a few years, this became a highly entertaining and well-attended annual event. At that first one, there was an informal and open 'competition' to make the first ground up solo ascent of the route, which was 'won' by Leo Houlding. I was too intimidated by the crowds and the good climbers to even give it a go as I envisioned myself falling off right at the bottom, and the initial bit is above a ledge, so there was the prospect of a very public humiliation.

The following year, I had done a bit more DWS and was a lot more comfortable with it, so I gave the route a go. There were still a lot of people spectating, but it had less of a gladiatorial atmosphere. To my surprise, I made it all the way to the crux at the final moves, which features two poor sloping diagonal crimps – a marked contrast to the good, but spaced, holds below. Just before those crimps, there is a really deep one-finger pocket, which actually feels like a jug, so I was going up and down trying the moves from that before eventually committing and falling. The following day, I had a few more goes but still couldn't unlock the sequence using those crimps. On my last go, the beast showed her teeth and bit my bollocks.

Falling off the crux, yet again, I ended up doing a massive belly flop, and I'm convinced that the first point of contact was the most tender parts of my anatomy – that's what it felt like anyway. This was followed immediately afterwards by my chin, which slammed my jaw painfully shut. Now, from 60ft landing on your bollocks is extremely painful, I can tell you. But, later that day, driving back home, the pain down below was being eclipsed by a growing pain in my lower jaw which increased to a screaming intensity by the time we reached home in North Wales around 2am. I didn't sleep that night.

The following morning, I request an emergency appointment at the dentist. 'Is it an emergency?' the receptionist asks. I'm not sure. 'Are you in pain?' 'YES!' 'You will need a root canal on that tooth,' says the dentist, and when he puts in the anaesthetic, I fall asleep in the chair. The root canal didn't work, and the dentist later had to remove the molar. I've still got the gap in my dentition as the mark of that day when I was bitten by the beast.

A few years later, it was another DWS fest at Connor Cove, with the usual circus-like atmosphere in the balmy English summer with the top of the cliff crowded with climbers and spectators and a steady stream of climbers on the classics like The Conger and Freeborn Man. There was a boat on hand with a dedicated captain who could take us into the back of the cave and save swimming in.

Sometime in the preceding few years, Mikey Robertson had discovered and climbed The Vanishing which takes the fault line in the apex of the cave all the way from inside to finally join The Conger. Adam Wainwright and I availed ourselves of the boatman. I joined the line of climbers soloing The Vanishing. Meanwhile, Adam was popping out from new lines on the left of the cave. A tricky move low down on The Vanishing guarded the steepness above, which fortunately had lots of rests in spectacular positions before popping out, pumped, in the chimney of The Conger.

Left top: Adrian Baxter on Mark of the Beast. Left middle and bottom: Martin Atkinson on and off Mark of the Beast. Photos Grant Farquhar. Facing page: Leo Houlding making the first ground up DWS of Mark of the Beast in 1998. Photo Joff Cook.

The Vanishing by Stu Bradbury

'You have to come to Swanage and do The Vanishing, Stu, you'll love it. It's right up your street; one of the best DWS routes I have done. We need to sort out a trip.'

Rich Pollard was waxing lyrical. How can you refuse such enthusiasm? It wasn't the first time he had mentioned it, and it was obvious he wasn't gonna let it lie. The following weekend was shaping up to be settled with a high-pressure window; conditions were looking good for a day-trip dash.

Amazingly, a whole bunch of us converged to share two cars, and after an early start, blasted off up the A30 and turned east for an appointment at Swanage with Fisherman's Ledge. The day was stunning. Blue water lapped the base, contrasting with the steep white limestone. Light danced, diamond-like, off the gentle movement of the ocean, creating an updraft of familiar coastal smells and the sky reeling to the sound of gulls.

Access into the depth of the cave for our main objective meant getting wet (Grrr)... Pulling myself onto the cramped stance under the roof, I was lifted by the swell only to be clawed back and kissed by razor-sharp barnacles which shredded skin. Looking up into the darkness of the cave, the line of The Vanishing looked improbable; it was hard to see where it went. The start was greasy in just the wrong places, the rocky ledge below spelling out the price of failure. I watched silhouettes of Rich and Luke disappearing up into the darkness to get stuffed into the apex by way of some pretty interesting contortions.

Pulling off the ledge, I rocked over onto the cramped balancy start. The traverse across the initial lower wall gains an overhanging arête via a powerful move, followed by a dynamic layback-type pull, to find enough momentum to swing around to its right and gain a standing position. From there, I got a good view of Luke pulling off some wacky shapes to reach the dead end at the back of the cave.

Meanwhile, Rich had managed to concertina his leggy frame into a small niche in the roof and was pasted into the very ceiling and milking the rest. Looking seaward from there, a glare of light pierced the cave along the hanging V in the roof, highlighting the line and showing the way to freedom. The climbing is wild but full of surprises: 3D bridging, back and footing, and the odd knee bar keeps you held in place whilst grappling blindly uncovered hidden jugs and powerful undercuts, which kept you moving between wedged rests (I even found some jams). What a find – Mike Robertson must have been buzzing when he bagged this beast.

Even while you're climbing, you can't believe the situation you're climbing into. I was starting to appreciate why Rich had been so keen to share the experience. I was totally engrossed to the point that I wasn't even thinking about the height or the potential fall into the drink. I nestled into the niche in the roof for a rest and was treated to a bird's eye view of Maca putting in the effort in the darkness of the cave behind me.

Grunts of effort echoed through the cave, followed by whoops of joy as rests were reached or cries for beta as gravity played its hand. But as always there is payback, and you soon realize those glorious jugs have funnelled you into a tight spot. The way is barred, and the cramped position offers a claustrophobic rest, but it's tough to leave...

By the time you've figured it out and committed to dropping your arse out into space via the hanging undercuts you're in the wide hanging flare on the other side with no way back. It's too late; the jugs are gone, and nature has played her hand. Just when you thought freedom was yours, she's spat out the crux, and the breeze carries her laughter on the updraft. Now, for the first time, you are aware of the sea clawing below; one false move or release of pressure on this smooth rock, and a beautiful onsight will be over. The consolation prize of a saltwater enema will be yours. Hang in there; lateral thinking and whoops of encouragement wins the day. The finishing jugs accept defeat and allow passage to the vertical world. The spell is broken, and the only thing that's vanishing is the strength of the autumn sun as it slowly sinks in the west.

Above and below: Andy Steinberg on The Vanishing, F7a+ S1. Photos Crispin Waddy.

Fear and Loathing in Lulworth Cove by Richard Bingham

He said it would make me higher than I'd ever been in my life... pure adrenochrome – or maybe just a fresh adrenalin gland to chew on.

Fear and Loathing in Las Vegas by Hunter S. Thompson

Above: Richard Bingham in 1999. Opposite page near: Richard Bingham on Adrenochrome, F8a S1. Photos Paul Barton.

Opposite page far: Dave Pickford on the second ascent of Windowchrome in 2008. This links Window of Opportunity into Adrenochrome at F8a+ S1. Photos Pickford collection.

If you ignore playing around in trees, my first memory of climbing was as a child in the Outside shop in Hathersage – they had a stone wall, which I presume was for testing climbing shoes. While my parents browsed the store, I climbed up various routes in my trainers, which was fun, but I didn't realise that rock climbing as a sport existed back then. Several years later, I have my dad's friend, Jon, to thank for taking us to The Foundry in Sheffield and introducing us to the world of indoor climbing. Dad and I improved quite rapidly, and it was not long before we started exploring the local crags – our first outdoor route being an HVS on Rivelin Needle.

In hindsight, we probably overestimated our ability, as our second route was a scary multi-pitch on Plum Buttress. My first climbing in Dorset, and my first deep water solo, was also with my dad. We visited Portland, Lulworth and Swanage, ticking the classics. I remember scoping out Mark of the Beast and Adrenochrome, but it would be several years before I had the ability to climb them.

Adrenochrome is a chemical formed by the reaction between adrenalin and oxygen – both needed in plentiful supply for the aptly named central line up the steep side of Stair Hole at Lulworth. I knew that it could be soloed safely, with just the mild sting in the tail of hitting the cold water if you failed. The moves are brilliant and varied. You go from dynos and footless swinging on jugs, to crimps, then pockets and slopers – it has everything. I had practised it on a rope the year before, then returned in July 1999 with it firmly in my sights. After some more roped practice, I was ready for the solo.

Conditions are critical at Lulworth, and I had struggled previously because of moisture on the holds at the crux. I was lucky on the day of my ascent; it was cloudy, quite cold and with a steady breeze to dry the rock. I only had one pair of shoes and a single chalk bag, so I was fully committed to one attempt. I got through the two dynos and rested on the jugs before moving right into the crux. There is a hard move into a pocket for the right hand, and once I got this, I knew it was on. All my previous falls into water had been from below this height, so the adrenaline really kicked in. I was shouting, 'COME ON!' between every move; I have never been so pumped, either before or since.

At the time, I think it was the hardest DWS around. Personally, I saw it as a significant achievement as the route had been a goal of mine for a number of years, but I didn't see it as a breakthrough in the grand scheme of things. Living in Sheffield, I was constantly impressed by the local legends completing super-hard test pieces on the gritstone (above rocks, not water!). Looking back now, it was clearly the highlight of my climbing career.

A few years later, at Christmas 2004, I suffered a sudden, serious illness – Acute Disseminated Encephalomyelitis – causing swelling of the brain and spinal cord. After 10 days in intensive care with life support, I started a slow recovery but was paralysed from the waist down. My residual arm strength and bouldering skills came in useful, as I quickly learnt how to transfer myself from bed to wheelchair, helping me to get independent again.

But learning to walk again took several months, and it was a full year before I could run again. I'd like to thank the support I received at this difficult time from the climbing community; I still have a little box with all the messages of support that I received – thank you, everyone. My climbing ability never fully returned, and now my energy goes into cycling and long-distance running. I still climb, but only in the summer holidays when I can be seen enjoying sport routes up to about F6c+.

High on Adrenochrome by Jacob Cook

Stair Hole has been the ultimate natural playground for me growing up. Its cliffs, arches, caves, tunnels and hidden beaches have been the source of many adventures, often involving rock-scrambling, swimming, snorkelling, kayaking, cave exploring, make-shift rope swings and fossil hunting.

Perhaps my earliest memory there was my eighth birthday when my family had arranged a Pirates of the Caribbean-themed party for me. This consisted of a series of challenges set in place to find pieces of a wooden treasure map hidden around Stair Hole. Each piece of the map we found would lead us to the next until, ultimately, all the pieces fitted together like a jigsaw puzzle to reveal the location of the treasure. This we found was buried in one of Stair Hole's caves.

During my teens, when I was training board diving twice a week, my main hobby was cliff jumping and diving, and the area provided many rocky platforms with deep water beneath for me to enjoy this.

It wasn't until I was about 18 that I actually started climbing and deep water soloing there. I suppose it was inevitable I would at least try it, as my dad and cousin Liam were keen climbers. From the beginning, I was fortunate to receive lots of support from them: driving me to the crag, offering their sage advice and beta, belaying me on projects and most importantly, taking epic climbing photos, making me look better than I actually was.

My brother, Lucas, was also a great help. His hobby is snorkelling for spider crabs. I have often found his presence swimming in the sea beneath me reassuring, as he would check the water depth, ward off kayakers from paddling beneath me, and be ready to pull me out of the water if anything went badly wrong.

I have always had a particular anxiety around deep water soloing, exclusive to other forms of climbing. I think it is the height and exposure – combined with the consequences of falling in. Being quite skinny, I would only have so many splash-downs in me before it would take too long for me to warm up again, and I would have to call it a day. However, for the same reason, I love this form of climbing, as it discourages half-hearted attempts, forcing you to fully commit.

Perhaps it is because of this adrenaline-fuelled anxiety that, to me, climbing at Stair Hole often seemed like an arena. Somewhere to test your physical and mental resolve against your opponents (the routes) on the main stage (the rock face) while spectators could sit around the auditorium shouting encouragement or heckling, often both being equally effective as sources of motivation.

Left: Jacob Cook at Long Quarry Point, Torbay. Photo Joff Cook. Facing page: Jacob Cook on Adrenochrome, F8a S1. Photo Liam Cook.

Jacob Cook off Adrenochrome.
Facing page: Jacob Cook on
Adrenochrome. Photos Joff Cook.

If Stair Hole is a combat arena, then Adrenochrome is its reigning champion. Running up the centre of the east cave, its moves are so dynamic, technical and theatrical that it almost seems unnatural, as if someone had designed and sculpted it into the cliff to be climbed. With overhanging dynos, heel hooks, technical crimpy sections and a pumpy finish, it truly is the whole package, and it became my dream route.

Although for years I considered it well above my ability, I would still throw myself at its spectacular dyno, which is rather low down on the route, as it was just too fun to resist. After a couple of seasons of being capable of this, I could consistently get further, up to and sometimes even past the first crux. It wasn't until I had sent Mark of the Beast in 2022 that I considered dedicating more time to it. Even then, I had to play on the route several times on a rope before I truly considered it possible.

This was during the early summer of 2023. I had just finished my degree and booked flights in just over a month to go travelling for the rest of the year. This left me with an unrealistically short window and meant I would need consistently good conditions and a good team to pump up the psyche.

Unbelievably, though, both materialised; a constant stream of drying winds blew up out of the mystic east and carried with them to the crag some strong locals in the form of Marwan, the ultra-strong and mega-supportive Croatian, and JJ, Dorset's version of Jason Mamoa, who just like Aquaman relentlessly cleans the ocean of undesirable elements.

With these advantages, I made decent progress on the route over the next couple of weeks but rarely made it past the second crux. As I got closer to my travel departure date, the climbing conditions became more hit-and-miss, and I accepted that I had simply run out of time. So, I resolved to just focus on having fun with the route without any real expectations.

As my departure date approached, the 10-day forecast was not good, except for one day – the last Thursday of June. We went again – to play – no pressure. I surprised myself on my first attempt that day. I made it past the second crux and somehow battled through to slap the top jug before powering out and taking the ride to the 'green room'.

Although I was disappointed I had not topped out, I was still satisfied with the effort. I decided to rest and have a second go – for the sake of the training, if nothing else. I had one pair of dry shoes left, one dry chalk bag, and genuinely no expectations to make it as far as I did before, as I thought I had spent all my energy on the first attempt.

But this time, to my disbelief, I found myself boxed out of my mind but pulling over the lip and sending the route. The team below whooped, then chanted for me to do a victory jump into the sea. This I relished. Oddly, it was the first cliff jump I had done at Stair Hole in years. In a sense, things had come full circle. The joy of flight. The beauty of it all – crystallised into one ecstatic moment.

The intrinsic merits of a sea cliff cannot be translated in terms of a mountain crag... the sea was ignored at first, then its presence was acknowledged as a nuisance, and finally, climbers were forced to come to terms with it... new skills had to be perfected to move along the waterline mindful entirely of the sea and its tidal variations.

John Cleare

South Devon

You step blind round a corner of sheer rock and move carefully down into the vast, dank mouth of the cave. It seems as big as a cathedral: a black, thundering dome, like a lunatic's skull, water boiling along its floor, birds flitting in the dark air.

Al Alvarez, Moonraker chapter in *Hard Rock* by Ken Wilson

Sea Traversing – The Truth by John Fowler

Sea traversing and climbing on sea cliffs form an important part of Devonshire climbing and will no doubt become increasingly important in the future. Critics will say that sea traversing lacks the element of danger or exposure that a good rock climb supplies. To my mind, this is one of its advantages. It allows the average climber to fester without actually doing nothing and also without the appearance of festering. The lack of exposure allows the relative beginner to attempt hard moves that he would normally be unprepared to attempt, even on the end of a top rope. It also gives the 'permanent second' a greater feeling of equality as he is participating more equally in the climbing risks.

When climbing a long way above the ground, the average climber is unable to push himself to the limit and, therefore, rarely knows where this limit is. When sea traversing, even the inexperienced climber can do just this and is often agreeably surprised as to how far he can go on greasy, sloping footholds. In this respect, however, the exposure is considerably greater than on a boulder problem, and failure is greeted by cheerful cries of derision from all those who remain dry. The moves cannot be undertaken in the reckless manner normally seen on boulder problems. One other notable advantage is that being near the sea and subjected to frequent inundation, the standard of a particular sea traverse does not seem to be affected greatly by wet weather. It may not seem harder in the wet than in the dry, even when a rock climb of similar technical difficulty is rendered virtually impossible.

This results in an all-weather climb somewhere between a boulder problem and a true up-and-down route suitable for all grades of climbers. The qualifications for a sea traverse of which the hardest moves may be technically VS or HVS are, in fact, fairly basic: a change of clothes and a little ingenuity. To this could be added a little technique and an ability to swim. The result, one should hope, is that the climbers with the most technique and ingenuity get less wet than the others. Those who cannot swim and have no technique or ingenuity are in for trouble.

One exasperating factor of sea traversing is the amazing variability in standard from one section to the next. One minute the party is strung out on an easy section of rock, and the next, the leader is brought to an abrupt halt by a seemingly impossible wall. This is where ingenuity comes to the fore. Tactics depend on the temperature of the sea and the time available; hence the time of year is crucial. The quickest method is to swim, and in summer this is quite acceptable. But in winter, when time is at a premium, one is unfortunately less inclined to do this. If swimming is unavoidable, the problem then arises as to whether it is preferable to strip off or swim with one's clothes on. In winter the temptation is to leave them on, but if the traverse is at all long, it's advisable to keep them dry.

Facing page:
Dave Thomas soloing Flaming Drambuie, E6 6b, on Sanctuary Wall, South Devon. Technically he is soloing above deep water, but it's probably not most people's idea of DWS. At this point on the route he is facing a fall of 40m into the sea: 'More shallow grave than deep water.' Photo Mike Annesley.

The now well-tried Tyrolean traverse is the most satisfactory method of getting a party across an impossible section with only one member having to swim. One intrepid member, usually the one in the most hurry, swims across the hard section with both ends of the rope and tensions these up by means of Prusik loops. The centre of the rope is looped over a spike or through a tape or peg, which someone volunteers to leave behind.

The rope can be crossed by either lying on top of it with one leg hooked on the rope or by hanging beneath it, preferably in a sit-sling, and attached to the rope by a karabiner. If the second method is used, it is important to be attached by a second karabiner to one of the ropes only, as otherwise, the climber will be completely unattached to the rope should the thread or peg give way. In theory, as in abseiling, the rope can be pulled through when everyone has crossed and the traverse resumed. In practice, the rope will jam, and somebody has to swim back to free it. Another snag occurs if the Tyrolean is close to the sea and is awkward to tension. The weight of a climber is then sufficient to stretch the rope until the climber touches the water. The lowest part of the climber, usually his backside, is then submerged. Nasty!

If nobody feels like swimming (it is surprising how many excuses are produced in January), the party is reduced to a series of pathetic attempts to lasso spikes, throw nuts on the end of the rope into cracks and, finally, to throwing a loop of rope over to the other side in the hope that it will snag on something. This last technique is surprisingly successful and has saved several January swims. There is, of course, a moment of high tension when the first hero, or sucker, (whichever is the case) attempts to cross the rope. Once the first man is over, the rope can be fixed properly, and the rest of the party can cross in relative safety.

If the hard section is short, it may be crossed by a series of sling moves or by some cunningly devised pendulum. This involves flicking a loop of rope over a spike and penduluming across to a ledge or good handhold. The perils are, however, numerous. The spike may look adequate from below but may be no more than the point of a sloping ledge. As soon as weight is placed on the rope, it will roll off the spike, dropping the climber in the water. If this manoeuvre is to be attempted, it is important to reach your objective with the first swing. There is nothing more pathetic than swinging gently backwards and forwards, watching the rope gradually roll off the spike.

Frigging of this nature, and Tyroleans in particular, are time-consuming and often irreversible. Having completed such a section, the suspense mounts as, at any moment, one may turn a corner and find something worse still. Often it's impossible to escape upwards, and the whole party must take to the water.

The climber's changing attitude to the water is also interesting. Initially, one avoids it entirely, keeping scrupulously dry, but as time progresses and one is caught by a wave or is forced to use a foothold just below the surface, the standard deteriorates. Gradually, one is prepared to use lower and lower holds until you are prepared to go in up to the knees or deeper to make a not particularly hard move a little easier. The act is justified by saying that if the tide was a little further out, the hold would be uncovered anyway. This underlines another exasperating factor in sea traversing. A desperate section can be transformed into a walk by a falling tide, even in the time it takes a party to cross a particular pitch. The converse can also apply, and it is therefore important to choose the correct state of the tide for your traverse.

The action of the sea is such that cliffs that are unsuitable for rock climbing at higher levels are often perfectly sound close to the water. Not all such rock is good, however, and it sometimes produces weird, wafer-thin handholds, especially in limestone. These cannot fully be trusted, but the climber sometimes has no choice but to use them. A well-known local climber at full stretch on a particularly hard move was heard to say, 'I think I've cracked it.' This was followed by a loud crack upon which he disappeared from view in a cloud of foam to surface a few seconds later in the middle of a series of tape slings which were floating away in all directions. This also underlines the need to keep all gear tied onto something, as not all climbing gear floats. This includes climbers, as they don't float, either, when loaded down with gear.

Sea traversing certainly includes the unexpected, and as in true rock climbing there is always doubt as to the final outcome, especially during exploration. The party can find itself in some truly tremendous situations. For instance, penetrating deep into the cliff in huge caves or traversing around the sides of a huge zawn where upward escape is utterly out of the question. A rough sea can also make the simplest traverse a major epic, and a sense of timing becomes of the greatest importance. It is amusing to watch a friend wait for five minutes to run across some boulders that are alternately covered and uncovered by the waves and then walk straight into the biggest wave for half an hour. However, it is not so funny when it is your own turn.

Above all, it's essential to approach sea traversing in the right spirit. One should expect to get wet and be thankful if one doesn't. The subject can be taken more seriously by insisting that no contact is made whatsoever with the sea, but this is only for purists and hardmen.

Facing page: John Fowler, John Cleare, Sue Cross and Alison Chadwick-Onyszkiewicz discovering the truth on a Devon sea-level traverse. Photos Frank Cannings.

Sea Mountaineering by Peter Biven

One soon comes to respect the sea as a potentially hostile force. Not everyone can swim, so the sea mountaineer must learn to deal with a watery environment as well as seaweed, barnacles and wave-fretted rock. The most frequent comment on sea cliffs is that they are 'serious'. This is true – in my view one of the most serious cliffs in Britain is the Boulder Ruckle at Swanage. Safety lies at the top of a sea cliff, and rising wind and tide can add a new dimension to the usual meaning of the word 'gripped'. The second is often at risk from the waves, and several accidents have occurred where the whole party has been washed away because of an inadequate or nonexistent belay.

John Cleare, recalling an epic November night at Swanage, writes: 'The only means of escape lay along the cliff bottom, half a mile of raging white water broken here and there by huge boulders. One at a time we struggled from boulder to boulder, one man always tied and belaying the other with the rope. Laden with equipment in the pitch darkness it was painfully slow.

'Sometimes we were climbing, sometimes swimming, always pounded by the waves. Soon, my glasses were smashed. Once Barry was swept away and knocked silly, his head gashed open. After five hours, we reached the final 300ft of open water, deep and angry. It was too far for our rope and exhausted as we were, we were not prepared to swim unsecured. We crouched on the highest boulder to wait for dawn.'

Such objective dangers are normally outside the experience of the land-fast crag climber. In fact, sea mountaineering is a different game. This is the crux of the matter. Horizontal climbing, now known as traversing, has a distinctly different character and requires skills such as swimming, lassoing, a knowledge of tide and wave patterns and the ability to deal with angry fulmars and cormorants attacking from above, and sharks, jellyfish and congers from below. Climbers rarely buy tide tables, an omission which often causes embarrassment, but the AA book gives tidal constants for various points around the coast. For example, for high water at Gogarth subtract three and a half hours from high water at London Bridge, remembering to convert from Greenwich Mean to British Standard Time. Simple!

At nesting time, sea birds resent the intrusion of climbers. A determined attack by herring gulls, fulmars and razorbills can be frightening and adds yet more emphasis to the charge that sea cliffs are serious. The continual screaming of the gulls is very unnerving and gradually wears down the climber's morale. Then there is the question of the temperature of that other element involved in sea traversing: water. In summer, the problem of swimming a zawn (a Cornish word migrating steadily up the Celtic fringe for a great gash in the cliff face) is simple: strip stark-bollock-naked and put everything – cameras, watches, karabiners, socks into the tough polythene bag which, thoughtfully, will have been brought along. This performance is usually so hilarious as to rob the zawn, white and angry as it may be, of its terrors. The polythene bag is put into the rucksack, and the non-swimming climber can now play his trump card. Inflated, the sack is then put on back-to-front, and the straps tied together. Tying on to the rope, which is an imperative for any swim, the climber then hurls himself clear of the rocks and with his own personal buoyancy is virtually unsinkable. The first man over can often fix a Tyrolean traverse.

There is another aspect of sea cliffs which has been somewhat overlooked until quite recently: the sea stack. The virgin summit has long since disappeared among our home mountains – even among the Alps. To do genuine explorations as such, one needs to travel very far. But stacks are, for the most part, unclimbed summits, which were the original objectives of mountaineering. And numerous possibilities surround our coast. Climbing these may often be far from straightforward. Many are vertical and undercut at the base by the sea. Sometimes the quality of the rock is suspect, so new and ingenious tactics are devised. One stack at Ladram Bay in East Devon thwarted all efforts, even with a bow and arrow, until it was discovered that builder's nails, set in the mud seams of the sandstone provided adequate protection. On another, it was found that by the loan of a kite from a child on the beach, a corulene line could be flown over the top. A bold experiment was tried with portable footholds employing carefully selected limpets on slabs – but they stubbornly refused to cooperate. On one stack, there was no firm rock on which to fix a descent abseil, so the two climbers simultaneously roped down opposite faces – yet another example of safe expedients being used, uniquely, by sea cliff climbers. But no doubt, in the course of time, when the ethics of this sport have become less fluid, this may be considered improper.

In 1967, the television programme on the Old Man of Hoy brought awareness of the sea stack. By the end of the 60s, stacks from Land's End to far beyond John o' Groats were having rough cairns built on their bird-limed summits. Some stacks are elegant and slender, like the Elcgug Needles of Pembrokeshire or the Devil's Chimney on Lundy. Others are broad and lumpy, like battleships riding at anchor. But they all have a quality which is extraordinarily seductive, a quality of remoteness and inaccessibility which, for most climbers, is a welcome feature. Offshore islands, which have to be approached by boat like those in the Scillies, offer unlimited opportunities for traversing and the ascents of stacks and isolated pinnacles.

The rising and falling of the tides and the ever-changing background of the sea, coupled with the myriad distractions of the shoreline – caverns, creeks, crystals, flotsam and strangely-formed sea and time-worn rocks, make sea cliffs fascinating to explore and delightful to be among.

Peter Biven (1935–1976) by Barrie Biven

Pete began serious climbing at the age of 10 years on the roof of Smith's department store, Leicester. Smith's is a pretentious five-storey Victorian building, topped by a copper dome and a crumbling life-size statue of Britannia. To get there, he had led me out of our attic window, up a dusty slate roof and over a high-level traverse of rotting stonework. With an emphatic grin, he sat on Britannia's lap savouring the trespass. We were spotted, however, by a patrolling bobby who recognized us and called at our home. Our father was firmly told that although the war had just ended there were safer and more acceptable ways of celebrating.

As brothers, we had shared many sports and pastimes, but it was clear to me then, as an envious eight year old, that Pete had found an activity at which he excelled. In his school notebooks, he drew up plans to climb all the tall buildings in Leicester. He sent me on reconnaissance missions around the town in the belief that a small boy in short pants, gazing up at the aeroplanes, would not arouse suspicion. Ultimately, the Clock Tower, the War Memorial and many other buildings all yielded to his courage and skill.

We then began to explore the Leicester countryside and found a mass of small outcrops and some intimidating disused quarries. We knew little of the climbing world beyond. Pete's exceptional ability showed as he found countless thin routes under curtains of ivy and dry moss. We did not have the equipment to rope down these faces to clear vegetation and assess a route.

Above: Norwegian mountaineer Odd Eliassen fights his way across the crux Overhanging Bastion pitch of the Magical Mystery Tour. Below and facing page: Peter Biven climbing at the Old Redoubt. Photos John Cleare.

He loved to churn up a climb, gardening as he went, and it was my task to follow, brushing up the holds as I went. His joy was not so much in the finished product of his pioneering but in the act of creation. His keen eye for a route and the tension of the climb as he sought to match his mental photograph of the route were at the core of his passion for climbing. He gave the rock another dimension; he made it pliable and supple in his hands. So much so that when I followed, I seemed to be involved in another activity. It gave him affectionate enjoyment to see me struggling.

By the time Pete was 15, we had moved away from the sullen, unrelenting Leicester granite to the Derbyshire grit. We did most of the routes at Cratcliffe, Cromford and Brassington before moving north to the Peak. Each Sunday we hitched to Stanage or Froggatt or Gardom's until we met Trevor Peck in his white Rolls Royce one wet weekend. This meeting welded a team that survived for nearly 20 years until Trevor's tragic death.

For years we scabbed our knuckles and wrenched our arms, working methodically along each edge. Pete forced some new routes on the way: Congo Corner, Ferryboat Highway, Easter Rib, The Flange, and BP Super at Stanage; The Eye of Faith, Hearse Arête and Moyer's Buttress on Gardom's Edge, and numerous aid routes on Lawrencefield and Millstone. He found time to make the first ascent of High Tor prior to National Service in the RAF. Our ritual Sunday trips had ended, but he was busy putting up more routes on the Whin Sill cliffs of Northumberland.

A year later, in 1954, he was elected to the Climbers' Club and soon after, the Alpine Club. Later still, he served on the committee of the Climbers' Club. During an extended leave in the summer of 1955, we all paid a visit to Cornwall, where we were astonished to find so much glorious rock. Pete was inspired. He found new strength and boldness as the next few years were to testify. It is a story of energetic and sustained rock pioneering of the highest standard.

His routes were always exposed and daring, following obvious but previously unacceptable lines. Much of the main face of Bosigran was covered in a thick carpet of ivy and had to be gardened on the sharp end of a rope. Pete's ability at this time can be measured by anyone who cares to reverse the traverse pitch on Suicide Wall, wearing plimsols and without protection, a move he had to make on the first attempt. It was a golden era. His companions were nonchalant about new routes, and they always seemed to be on a thin stance, paying out the rope in an incessant and rhythmical stream. Pete seemed to have little need for deliberation. He used to laugh at my bent knees and limp semaphore across the rock as I attempted to force a new route.

His love of Devon and Cornwall and our friendship with Cliff Fishwick brought him to Exeter in 1961, just three years after the first ascent of Malham Cove. Trevor and I had pressed him to join us, as he was no enthusiast of artificial climbing. Finally, criticism from 'admired contemporaries' caused us all to abandon aid climbing as a central part of our climbing repertoire.

In Exeter, he managed to pursue his academic career, first as a teacher and later as the Social Education Advisor to Devon County. An interest in young people inevitably led to weekend climbing trips with young novices. Some of the talents he nurtured are among the most respected names in sea-cliff climbing: Cannings, Littlejohn, Darbyshire and Chadwick all accompanied him on a wave of exploration of the Devon and Cornish coasts. More fine routes followed with a few brief trips to the Alps. But he was never really happy there. His Alpine tantrums are legendary but have not always been understood. A childhood ear infection caused him pounding headaches and altitude sickness. In the valleys, he became moody and irritable – a temperament out of keeping with the carefree English rock expeditions.

In recent years, he had put up ferocious routes on Berry Head, Baggy Point, Lundy and numerous other cliffs in the southwest. He occasionally strayed further afield to Pembrokeshire, Anglesey and the Little Orme. He also joined the BBC team for the Old Man of Hoy climb, taking a camera to the summit. During the late-60s, he collaborated on three Climbers' Club guidebooks. These guides, his numerous articles, and his climbs have been a tremendous contribution to British rock climbing.

Pat Littlejohn adds: Throughout the 60s and early 70s, Peter exerted a strong influence on South West climbing. As well as making countless first ascents himself, many of which have become regional classics (e.g. Moonraker). He helped to mould the ideas and attitudes of many promising climbers in the Exeter area (among them Frank Cannings, Keith Darbyshire and myself) by the force of his own personality. His approach to climbing was the antithesis of the arid, gymnastic activity it has become for some people. At the core of his approach to climbing was a thirst for adventure, and the greater the variety of demands a climbing day made on his skill, ingenuity and courage, the more he enjoyed it. He particularly loved sea cliffs and was a master of coastal traversing. Sometimes this gave expeditions comparable to alpine mountaineering in length, seriousness and arduous nature.

Pete's climbing code certainly influenced the way in which many very difficult modern climbs in the South West were tackled. For Keith and I, abseil approaches to new routes were taboo if there was any possibility of approaching them aidlessly. Sometimes this involved a level of commitment bordering on foolhardy and meant weeks of waiting for the right tides and sea conditions.

At the height of his powers, this was Pete's way to maximise the elements of a climb so as to bring it nearer to the big mountain experience. It is also a means by which many British sea cliff climbs can transcend their outcrop stature.

Facing page: Peter Biven crossing the Great Cave on The Magical Mystery Tour by Tyrolean traverse. Photo John Cleare.

The Great Picket Rock by John Cleare

Ladram Bay near Exmouth boasts some spectacularly weathered 80ft-high sandstone pinnacles (with the hardest one, Bonetti Tower, being a challenging extreme), but they are all very public, easily accessible at low tide and lack the appeal and commitment of the true sea stack. Mind you, I can testify to the extreme embarrassment factor in the event of being cut off by the tide. Having to aid up a disintegrating, overhanging wall into someone's back garden to the accompaniment of a crowd of cheering holidaymakers and one irate house owner is, so far, a unique experience in my climbing career.

Mick Fowler

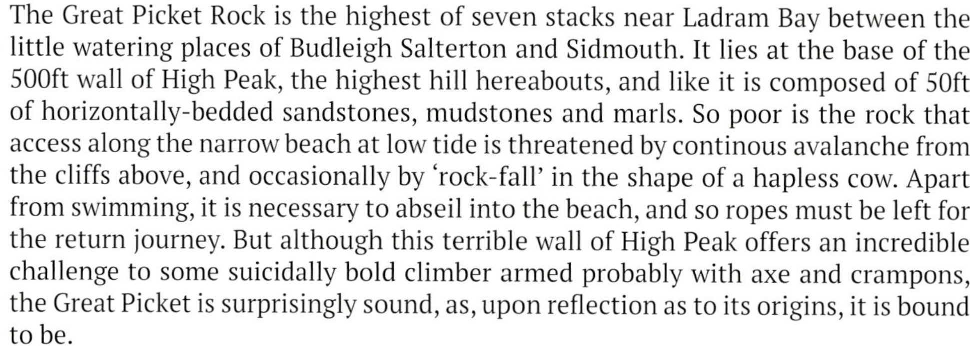

The Great Picket Rock is the highest of seven stacks near Ladram Bay between the little watering places of Budleigh Salterton and Sidmouth. It lies at the base of the 500ft wall of High Peak, the highest hill hereabouts, and like it is composed of 50ft of horizontally-bedded sandstones, mudstones and marls. So poor is the rock that access along the narrow beach at low tide is threatened by continous avalanche from the cliffs above, and occasionally by 'rock-fall' in the shape of a hapless cow. Apart from swimming, it is necessary to abseil into the beach, and so ropes must be left for the return journey. But although this terrible wall of High Peak offers an incredible challenge to some suicidally bold climber armed probably with axe and crampons, the Great Picket is surprisingly sound, as, upon reflection as to its origins, it is bound to be.

The Great Picket is only 120ft high, but we have graded it as 'very serious', a system which, like the Vallot adjectival system, takes into account the overall nature of the expedition including the escape potential. At low tide, a fingery scramble is made just above the sea around the base of the stack onto the big ledge at the bottom of the seaward arête. On all stacks this is where the rock is at its best. Above, everything appears overhanging – narrow, horizontal flutings jut out barring the way upwards, and the soft sandstone is dusty to the hand and brittle to the fingertips. The climb seems most unlikely. But on the south face, there is a slight chimney about 60ft up, below large overhangs, and a steep and wandering pitch, winding between the flutings, leads into it. The belay, a foot of angle iron hammered into a sandy fissure, is poor although the stance, bridged tightly across the chimney, is fair.

The next pitch is the crux. Bridging widely up to the fluted roofs and using purely psychological protection from a six-inch nail hammered directly into the sandstone, a swing is made onto the smooth and overhanging right wall. An airy position. Sand must now be cleared from a row of small, carved holds before they are hand-traversed out across the wall to the prow, where a leg can be swung up and an awkward mantelshelf made. A belly crawl beneath some flutings leads to a small ledge where a nest of six-inch nails gives an illusion of security. Rope drag has been a serious problem.

The third pitch continues the line of the traverse outwards towards the seaward arête where, every few feet, a gap in the flutings above allows mantelshelfing onto a higher level. This very exposed line ends right on the arête, where a final mantelshelf leads to the flat and muddy top. Scores of gulls and a few stunted sea cabbages offer little in aesthetic pleasure, but the climb has been full of character, and the route finding far from easy. The whole enterprise has seemed perhaps harder than it is; perhaps, technically, only a stout VS. But every hold is suspect, and testing it beforehand means nothing here – the rock snaps under pressure like stale biscuit.

Like certain mountains, sea stacks always offer descent problems, especially so on the Great Picket Rock, where it is impossible to place adequate pitons in the summit mud. This has led to the development of the 'seesaw' abseil. With each end of a long rope thrown down opposite faces, two climbers abseil simultaneously to the bottom. A third climber can always descend held by the first two from the far side.

Left top: Shannon Firmin on the Great Picket in Autumn 1972. Left middle: Ladram Bay stacks. Left bottom: Peter Biven sizes up the, then virgin, sandstone 'Chocolate Finger' stack in September 1971 (in due course, climbed using six-inch nails for protection). Facing page top: Climbers on the Great Picket in Nov 1971. Facing page bottom: Ladram Bay stacks, left to right, The Razor (100ft), Bonetti Tower (90ft) and the Tower of Babel (80ft). All were climbed in 1971 and '72 by Peter Biven and friends. Photos John Cleare.

The Parson by Steve 'Crusher' Bartlett

To the best of my knowledge, the most difficult stack away from the chalk areas is the Parson at Dawlish, and as this leans against the main cliff, its claim to be a true stack is rather dubious. However, the approach via a tunnel on the main London-Plymouth railway line and escape via very private grounds makes it a challenging and worthwhile day. Keith Darbyshire's first ascent in 1971 was well ahead of its time.

Mick Fowler

My dad retired to Teignmouth for the last couple of decades of his life (he lived around there when he was a kid), and I used to go visit him from time to time and wander the beach. I loved exploring beyond the popular beaches where you turn a corner and everything changes. When I was a kid, we would go visit my grandparents in Torquay. I loved scrambling around the rocks and cliffs of Babbacombe and Walls Hill and all over – boulder hopping and sea-level traversing long before I started actual rock climbing. I think these early explorations are my favourite childhood memories and have propelled me in my climbing and desert first ascents my whole adult life.

The Parson is one of a pair of sea stacks called the Parson and Clerk off the south coast of Devon. Most of the other tower, the Clerk, fell into the sea a few decades ago, so now it's just the Parson, preaching to the waves. This sea stack is near Teignmouth, a popular seaside resort bustling with tourists, children and ice creams. To get close to the spire involves a long walk along a sandy beach crowded with dogs, frisbees and sunbathers. Right behind the beach is a railway line, the main line from London to the southwest of England, and every few minutes, a train thunders by. At the far end of the beach, if the tide is low and the sea is in a friendly mood, you can scramble around wave-blasted black rocks crusted with limpets and find yourself in a different world.

The beach vanishes. Below, the fidgety sea burbles and sighs, slapping against the rocks. Above, unstable cliffs seep. The normal sounds – dogs, cell phones, cars, children, the background hum of civilisation – are gone. Instead, there is a cacophony of screams, trills and squawks: seagulls, cormorants and other birds at home. They don't like intruders. The Parson was first climbed back in 1971 by Keith Darbyshire, Peter Biven, Johnny Fowler and S. Nicholls. Darbyshire died a few years later, after slipping from the top of a sea cliff from wet grassy slopes. He was a thatcher, a person who made the straw roofs on those cute picture-postcard cottages.

The approach to actually get to the base of the Parson is through a sooty tunnel which carries those same express trains that have been rattling by every minute or so. Once inside the tunnel, you run to a hole. To get down to the sea itself, there are some shenanigans involving lassoing a spike, a rusty old chain or some such. The climb, if you actually get on it, has three pitches. The middle pitch utilises a feature referred to as the 'Brown Spider'.

Rumour has it that the last pitch, which ascends cobbles up the final spire, was protected on the first ascent by the judicious use of a kitchen knife stabbed into the rock. The belay under this pitch has no anchor. The 'descent' is a wild leap to the wet grassy slopes of the mainland.

Right top: The Parson by Steve 'Crusher' Bartlett.

Facing page: Johnny Fowler and Peter Biven gaining The Parson.

Right middle: Peter Biven leads the first pitch up to the 'First Mud Field' of The Parson.

Right bottom: Peter Biven and John Fowler making the Tyrolean traverse back to the foot of the main cliff from the base of The Parson. The Clerk stack is visible beyond.

Left: Keith Darbyshire.

Photos John Cleare.

Above: Emma Williams on The Watchtower Traverse, F5 S1, Torquay. Below: Bruce Woodley and Emma Williams on The Five Star Traverse, F6a S2, Torquay. Photos Mark Glaister.

Little Big Man by Arnis Strapcans

To the average informed climber, the name Pat Littlejohn means routes such as Eroica, Darkinbad the Brightdayler, Il Duce, America, Antiworlds and Savage God, to name but a handful. In other words, some of the best routes in the country, and virtually all the best hard routes in the South West. At the age of 25 years, he had already amassed a grander list of achievements than most leading climbers do in a lifetime. It would be difficult to imagine South-West climbing without him.

Through sheer energy and a deep passion for exploration, he has played a major role in prompting Devon and Cornwall from a minor holiday climbing playground to its present status as Britain's chief sea-cliff climbing area, and in doing so, has become the country's prolific first ascensionist of the last eight years. The actual number of first ascents stands pretty close to the 400 mark, but even Pat is unable to reach an accurate figure. In the eye of the climbing public, he is the big South West mystery man. In person, he seems even more enigmatic.

Nevertheless, he knows how to balance his enthusiasm with a sense of humour and consideration for the rest of the team. After a miserable winter's night in a derelict wartime lookout bunker on the Pembroke coast, Pat finally managed to get us out of our sleeping bags by making hot brews and bacon butties at the far end of the building. At other times his drive has been channelled into a more astute approach. One Sunday morning at the same bunker, he had everyone out on the crag by 6am. One of the victims later complained:

'The sly bugger had the only watch between us, and he'd wound it on by four and a half hours, so we all thought it was half past ten.' For full effect, read that in a Zummerzet accent.

Pat started climbing at the age of 14 on Exeter's local Chudleigh rocks, on an 80ft length of hemp rope knocked off from the local flower show (his mum wouldn't let him have her washing line). He quickly showed natural ability and soon became an able partner to big names like Pete Biven and Frank Cannings – in whose company he played an important role in the development of the then-virgin Berry Head, with new routes such as the classic Moonraker.

Torbay is an area for festerer and hardman alike. In high summer, one can doze in the sun at Long Quarry Point, one eye looking out for the Anstey's mermaid and the other watching Littlejohn committing suicide on overhanging scree to the accompaniment of a dirge from the squawking shitehawks.

Pat was not, however, climbing loose rock just for the sake of it. It's just that he was rapidly learning how to deal with it if he ever came across it, which proved a very important asset in some of his bolder leads later on.

Above: Pat Littlejohn and Anna Biven traversing into the Old Redoubt. Photo James Mann.
Below: Littlejohn and Keith Darbyshire on The Parson in December 1972. Photo John Cleare.

Ditching the Ropes by Pat Littlejohn

In the 1960s, climbers had everything to learn about sea cliffs – and being climbers, we learned the hard way. After a couple of parties were swept away at Swanage, we learned that belaying at the cliff base was usually a good idea. Another lesson was that if you fell into the sea with a hammer, pegs and other paraphernalia clipped to your waist-tie, you sank. Sea temperatures were another puzzle – you can check them in seconds now on the internet, but in those days, it seemed odd that the sea could feel warm on a bitter November day and freezing on a hot sunny day in April. Finally, we took a while to get our heads around the complexities of tides, currents and sea conditions, resulting in soakings, strandings, unplanned swims and even rescues by lifeboat – to my eternal embarrassment.

I admit I came into the game with a few advantages: having a dad with a lifelong association with the sea (Sea Scouts, Royal Navy, then amateur yachtsman). I learned to swim in the sea at the age of six, and just about all our family leisure time was spent on the coast, swimming and scrambling around the coves of South Devon and playing in the monster waves of North Cornwall, sometimes with our little 'belly-boards' made of plywood. So, it wasn't natural for me to see the sea as an enemy, which was part of the game in the early days of sea-level traversing. It was Rusty Baillie and John Cleare's Traverse of the Gods in 1963 which set the style destined to become the norm for the spate of 'sea traverses' in Torbay later in the 60s.

The rules of the game, if you could call them that, were that you used any climbing technique – free climbing, aid climbing, tension traverses, pendulums, etc. to traverse along the cliff base while avoiding contact with the sea. When a section was truly impassable, like a smooth zawn or sea cave, you would try to lasso something on the far side like a spike, or maybe a crack in which a nut might hopefully jam, and then cross by Tyrolean traverse. Only as a last resort would anyone take to the water, and then it would just be one person who would fix a Tyrolean to keep the rest of the party dry.

Now, it should be noted that many sea traverses were first done in winter when taking to the sea was decidedly unattractive, and for many of us, sea traversing became primarily a winter activity, to be undertaken when conditions were too poor for 'proper climbing'. Far and away, the leading exponent of Torbay sea traversing was Peter Biven, but interestingly, his first technical traverse was a deep water solo – the Pinnacle Traverse (VS) at Daddyhole linking Telegraph Hole Quarry to the Main Cliff. This became the standard way back to our rucksacks for Pete, Frank Cannings and me while developing Daddyhole Main Cliff in the summer of 1967. At low tide, it was gripping; at high tide, it was very tricky. At mid-tide, it was just perfect, with all the holds clear and dry and a reassuring depth of water underneath.

Later that summer, I soloed two other traverses: the Five Star Traverse, F6a S2, which led out from Torquay Harbour and Plimsoll Line, which I did to reach Daddyhole Main Cliff from Meadfoot. These were minor dabblings, however, compared to the epic Magical Mystery Tour, the sea-level traverse of Torbay's biggest cliff – the Old Redoubt – completed by Baillie and Cleare on the last day of 1967. Over the next two years, with various partners, Biven traversed nearly every bit of Torbay's coastline: at least 10 distinct sections in all. As a once-a-week climber by that stage, he was less interested in hard first ascents than in having a rip-roaring adventure with mates every weekend – and sea traversing certainly provided that. Eventually, the only remaining challenge was the traverse westwards from the Old Redoubt, where Pete had only managed a few pitches and dubbed his creation Bathos. That was the state of play at the end of the 1960s.

In 1971, I lived and worked in Cornwall for a spell, and having no regular climbing partner, I got deeply into soloing. It was a dangerous time, and I had some incidents that can give me nightmares to this day. I was following the logic that if I wouldn't fall off leading it, I could solo it – a flawed logic that completely ignored psychology and the unexpected. Luckily, a minor accident stopped the lethal trajectory I was on. Later that summer, back in Devon, I still had the urge to solo the classic hard climbs of my youth, like Machete Wall, Spider and Combined Ops, but fear and anxiety began to outweigh the pleasure, and it was time to look for a softer landing. That's when I got into the activity, which, a decade later, became known as deep water soloing.

At first, it was just the traverses around Torquay, from Five-Star to Pinnacle Traverse – backwards, forwards and with myriad variations. Throughout the early 70s, every time I visited my parents in Exeter, I would pop down to Torbay with just a pair of old rock shoes that I'd be happy to swim in. Normally, I love the shared experience of being with a climbing partner, but by being alone in the elemental world of rock and sea and climbing completely unfettered, I discovered that another form of climbing could be just as satisfying. Torbay is a paradise for this sort of climbing. I particularly loved doing The Watchtower as it was so accessible yet so remote-feeling and challenging. From the busy path of Rock End Walk in Torquay, you hop over the fence through some bushes, and suddenly, you are on a sea cliff totally hidden from the outside world. The problem for the soloist is Thunder Hole, a zawn which had only been crossed by swimming.

Frank Cannings Tyrolean traverse across the Great Cave at the Old Redoubt in 1967. Photo Cannings collection.

Climbers in action on the Magical Mystery Tour in November 1970. Photo John Cleare.

On my first couple of solos, I did swim this; at other times, the sea was colder and choppier, and I opted to climb out up Jekyll and Hyde and then down again on the far side of the zawn. In 1972, Keith Darbyshire and Hugh Clarke managed to climb across Thunder Hole using conventional climbing gear, and I repeated it in this style before incorporating it into my solo repertoire, just taking a short length of rope to lean across a cleft and reach the first holds.

Another favourite traverse was The Kraken. I had a bit of a history with this one as Ed Grindley and I had been rescued from it by lifeboat when I was just 15, two years before Pete Biven and party finally completed it with a rope move. I was going well in 1972 and managed to free solo it, adding another section to complete the link-up to Long Quarry Point. Again, it felt very remote whilst being, literally, a stone's throw from civilisation. The climbing on it is tough, the timing for tides crucial, and towards the end, there is a wall of superb limestone offering several different ways and a new challenge with each ascent.

By 1973, there was still one section of the Torbay coastline which hadn't been traversed – the very steep compact cliff leading onwards from Pete Biven's route Bathos. Two climbers from Plymouth: 'Mac' McFarlane and Deryck Ball took up the challenge and extended it by seven pitches to produce the superb Rainbow Bridge.

However, this still left one virgin section, the link-up to Biven's 1968 route Barnacle Traverse Continuation. Generally, I couldn't get Keith Darbyshire too fired up about objectives on the south coast – he was very much focused on the Culm Coast and North Cornwall. However, we hatched the idea of trying to free-climb Rainbow Bridge and complete the missing link-up.

In the spring of 1974, we did it, but over two visits and still leaving one short aid section. Such was the state of play on this route until the 1980s. During the late 70s and early 80s, my visits to the South West were more sporadic. I was living in South Wales, climbing two grades harder and had the sea cliffs of Ogmore and Pembroke to play on. These gave countless superb routes but didn't seem to offer traverses to match those of Torbay, so I had to get my fix when visiting Devon.

For me, going solo on sea traverses was not only incredibly enjoyable, but it was also the best possible training for the on-sight pioneering I was doing in Pembroke. I have no idea whether anyone else in the country was regularly soloing above the sea in this period.

In 1982, on a break from guiding in the Alps, I had the idea of soloing all the sections of Magical Mystery Tour in one push, about 5,000ft of climbing in all. I had previously soloed all four sections separately on several occasions, but there still remained the Tyrolean across the Blue Grotto, where all previous trips, both roped and solo, had involved a swim.

On the cave platform below the Old Redoubt, I looked at my watch and set off. Conditions were good, the rock flew past, and I seemed to reach the Blue Grotto in no time. The cave was dry, and while on previous trips I had taken it for granted that climbing through it would be impossible, this time I looked a bit harder – and saw a line. There was nothing to lose anyway, so I headed off into the gloom. It was strenuous and the outcome always in doubt, but the holds kept coming and, eventually, I pulled around into the brightness on the far side of the headland. After this you can scamper along easier ground until you reach Cradle Rock Buttress, and this is where you find out if you've timed the tides right. I was lucky and stayed dry, and soon there was just MMT IV to go.

This is quite a bit tougher generally and towards the end there is a sustained section across a steep wall – no place to run out of energy. On reaching the beach of St Mary's Bay, I checked my watch – two hours and 20 minutes for the whole thing. I made the mistake of recording this time and was soon lambasted in the climbing press:

'Littlejohn introduces speed climbing to Devon!'

A year or two after this, I had another bright idea – I would do all the Torbay sea traverses in a day (LOL...). This would also involve the first solo of Rainbow Bridge, and as the only gear I'd have would be rock shoes (no chalk for me in those days) it would also have to be the first free ascent.

Things began well. Reversing all of Magical Mystery Tour went fine; the Blue Grotto was dry again and seemed easier in that direction. Thence to Rainbow Bridge. The first five pitches were OK, just a bit awkward, but I knew it was all about pitch six. Keith and I had reduced aid on this pitch from six to four points, but that's quite a bit of remaining aid. I managed to pass the first aid point to where the other three crossed a steep, almost blank wall, and there I realised my chances of success were vanishing. I had spotted a possible lower line (where it was eventually free climbed) but couldn't get back to it; strength was ebbing until all I could do was throw myself at the moves. The fall, when it came, would have looked quite graceful had there been anybody to see it: a human torpedo perfectly upright falling 15m into a calm green sea.

I went very deep; the ring of bubbles where I entered the water looked miles away and labouring back to the surface took all the breath I had. There was no chance of regaining Rainbow Bridge, so I took a leisurely swim to the big platform at the end, hauled out, then after getting as dry as I could, pressed on along Barnacle Traverse Continuation to the Red Walls. I had blown it but wasn't quite ready to give up. After a hot drink and some food in Berry Head cafe, I jumped in the car to try to execute phase two of my plan – all the Torquay traverses. Five Star One and Two, Pinnacle Traverse Continuation, Pinnacle Traverse itself, then Plimsoll Line were dispatched with no problems; then it was over to Hope's Nose for Morning Town Ride. This was much harder than normal as the tide had risen by this time, but the Long Traverse to Long Quarry was just possible, and thus I reached my old adversary – The Kraken.

I crossed what is now called the Sea Slater wall and swung around Love Not War buttress onto the forbidding, shadowy cliff taken by The Wake. By now, most of the holds you normally use were underwater, and the climbing was hard. It got harder... and then too hard. I was off again, this time into the dull, choppy waters of Babbacombe. So the day ended with a swim and a foiled plan, but it remains the best day of traversing I've ever had and am ever likely to have.

I continued enjoying the Torbay traverses for at least another 30 years and watched the area transform into a mecca for deep water soloing. Had I been younger I might have joined the fray, but it's hard to break your old habits, and I prefer just to turn up with some old, comfy rock shoes and (these days) a chalk bag for one of the old favourite traverses. It hardly matters which one: the pleasure is to slip away from urbanised Torquay into the natural, soul-nourishing world of sparkling sea and sun-warmed limestone.

Facing page: Emma Williams on The Watchtower Traverse, F5 S1, Torquay. Photo Mark Glaister.

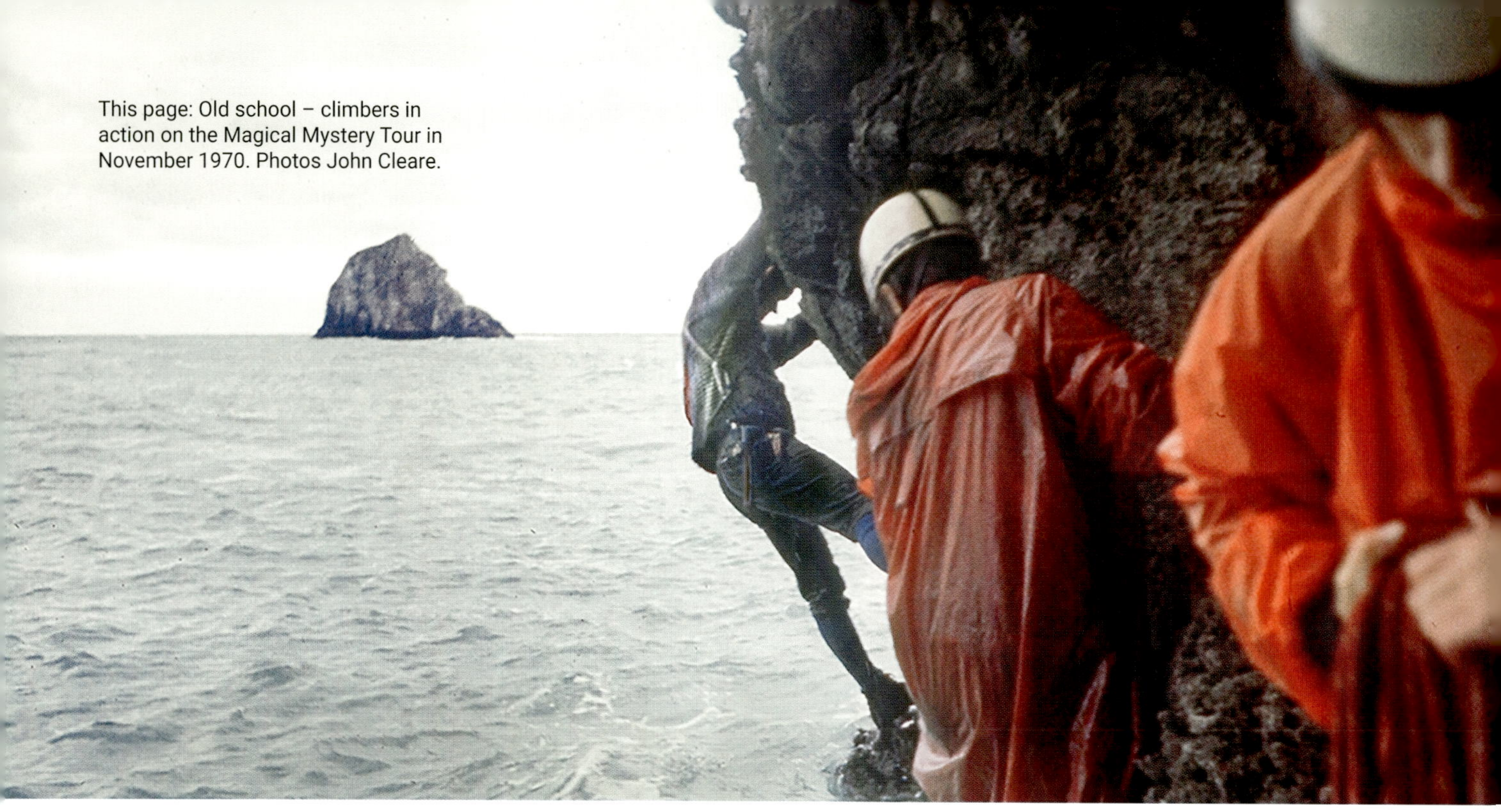

This page: Old school – climbers in action on the Magical Mystery Tour in November 1970. Photos John Cleare.

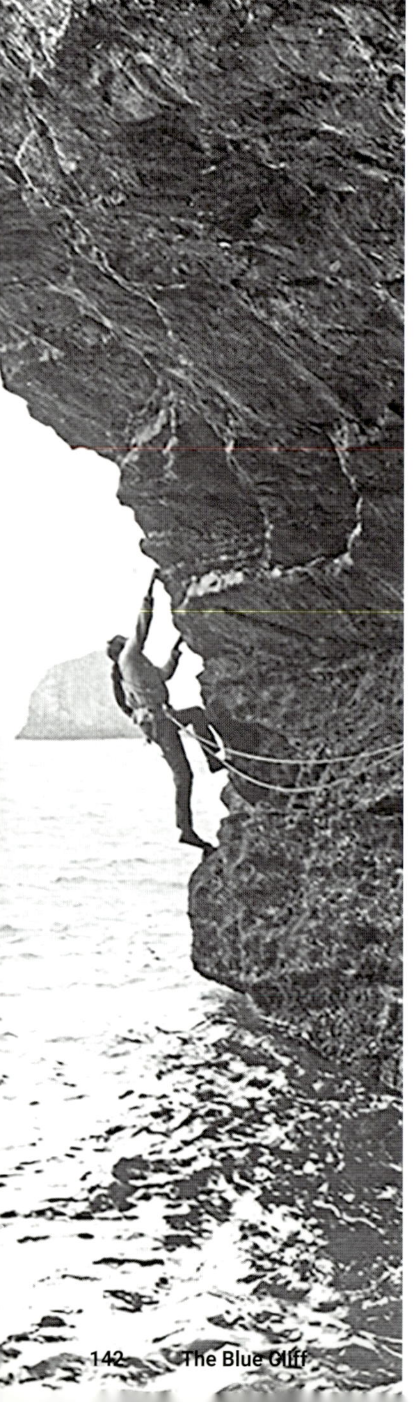

The Magical Mystery Tour

The Beatles' Magical Mystery Tour EP, the soundtrack to the 1967 film of the same name, was released in December of that year and peaked at number two in the UK's national singles chart. 1967 was also the year that John Cleare and Rusty Baillie started work on the complete sea-level traverse of the Old Redoubt. Cleare and Baillie were on the belay of Moonraker, waiting for Peter Biven and Al Alvarez to progress far enough up for them to climb: 'The first pitch is overhanging and quite intimidating, and we were freezing cold hanging around, so I shouted up to Pete asking what was round the corner. He didn't know – no one had looked. So, while waiting for Al to start climbing, we pushed around the corner to look. And with our Swanage traverse still in mind, we continued – and continued – and continued.

'You might say it was chickening out of hanging around getting cold and damp and seeing Alvarez fighting up those extremely strenuous-looking overhangs. And too much Christmas Pudding and booze the previous evening at Pete's place that it was the easy option. Naturally, we wore the rope. We usually roped for such routes, especially in winter, knowing only too well colleagues who'd been lucky to survive big waves – the winter sea is no place in which to go swimming off to France loaded with gear. Of course, we got wet – the footholds below the Overhanging Buttress are underwater, and we did use a couple of pegs on the Overhanging Bastion.'

They completed their first stage of this classic of the genre with ropes and aid pegs, finishing in the dark after Biven lowered a rope and Prusiks. Three days later, Frank Cannings and Biven did the first free and complete ascent of the Magical Mystery Tour from the Great Cave, finishing with a swim across the Blue Grotto. The remaining sections were polished off by Peter Biven over the next couple of years, resulting in four continuation traverses extending for almost one-and-a-half miles. In those days, the water was not seen as a safety net but as something to be avoided, and the routes were often done in deep winter. Sea-level traversing was, therefore, traditionally pitched with ropes, but The Magical Mystery Tour, F6a+ S0, has since become a classic DWS traverse. Pat Littlejohn:

'On the weekend of April 6th and 7th, 1973, Keith and his dad Brian Darbyshire hosted an Alpine Club meet on their farm in Devon. Pete Biven and John Cleare took charge on the Saturday, and a large contingent of fit and less fit people were taken along Magical Mystery Tour at Berry Head. It was a hilarious epic which none of us who took part will ever forget. People plummeted into the sea from various heights, were dragged out and had to continue to climb soaked to the skin – there is no upward escape. One of the local Devon team, Hugh Clarke, actually gave himself a hernia pulling one of the more portly AC members out of the sea.'

Transcending by Harry Sales

A truly Alpine meet. The Alpine Club at its traditional best. New transcents. Eight or nine on a rope. The mixture of enthusiasm and skill. The old barn to sleep in. Even the rope left on the rocks. It might all have been a hundred years ago in Switzerland. But it was this year in North Cornwall and South Devon, and it was an official AC meet. About 20 of us met in the upper floor of Keith Darbyshire's father's barn. The odd bales of straw which, late at night, we took to be provided for our comfort were really meant to mark the holes in the floor, but we survived for the climbing. This, on the Saturday, was at Lower Sharpnose Point, five miles north of Bude, North Cornwall; real wrecker country. But, the emphasis of the weekend was not on the up but on the along and the sideways. So on to Sunday and to Berry Head, Brixham. Some of us arriving there cast covetous eyes on some pleasant-looking slabby cliffs basking in the sun. But it was not to be: the master spirits drove us on, and we, about 12 in number, all made our way down to the start of the Magical Mystery Tour.

The route has everything. One descends from the car park on steep convex grass slopes, then steep but good rock onto an exposed traverse into The Great Cave. If the tide is right (it wasn't, but a Tyrolean helped), this can be traversed until it is possible to relax below The Goddess of Gloom. Then scrambling, albeit overhanging at times, but the holds are good, to the first crux, The Sump. This is a jutting-out overhang, just above water level, with one hard move in the narrow gap between the overhang and the clutching sea beneath. Communication is also a problem if help from a rope is needed, and it sometimes is. Next comes the real crux, The Overhanging Buttress, a full pitch of strenuous HVS. This is one of the rare uphill bits. Runners are superfluous because if one peels off, one's only escape is into the sea some 30ft below. The climbing goes over the prow by a descending hand traverse on well-spaced finger holds, completely overhanging. Peter Biven, having led this out of consideration for the size and variety of the party, fixed a pendulum. Keith Darbyshire tried it first but fouled the prow. John describes the results as hilarious, but Keith's expletives still ring in my ears as, with most of his body underwater, various helpers tried to give him momentum either backwards or forwards or both at the same time. The sight of him inspired the rest of us to dive in, in full gear, and swim for it.

And so on by more scrambling to The Green Grotto (a deep cave of what is thought to be freshwater of pot-hole origin) and more bathing. Then more scrambling to small stances by the mouth of The Blue Grotto. This is untraversable at any time. A spike had been fixed at the other side, but our lassoers were not on form, so Keith swam to fix a Tyrolean. Fixed means fixed because it jammed and is still there. Some of us, at least, were relieved to get across to the other side, which is fairly steep and leads to the arête of the promontory.

Above: New school – climbers in the Blue Grotto. Photo Greg Pittam. Below: Charlotte Obhrai and Max Dutson on the Magical Mystery Tour, F6a+ S0. Photos Grant Farquhar.

This and facing page: Ken Palmer on Rainbow Bridge, F7a+ S1. Photos Pete Saunders.

Rainbow Bridge

In October 1973, Andy McFarlane and Deryck Ball traversed doggedly rightwards from their bivvy in the Old Redoubt's Great Cave for 12 hours until dusk forced an upward escape. Their eight aid points on Rainbow Bridge (E3) were reduced the following year by Pat Littlejohn and Keith Darbyshire, who also added extra pitches all the way to the end of the rainbow. The pot of gold was eventually awarded to Crispin Waddy in 1989 for soloing the aided section at F7a+ S1, and Nick White soloed the entire line of Rainbow Bridge in September 1991.

Back to the Future by Dave Pickford

The best deep solo in Britain, and possibly one of the best in the world, is Rainbow Bridge at Berry Head in Devon. It's a traverse roughly 850ft long across a wall of the most perfect, compact, technicoloured limestone imaginable. There is barely a single loose hold along the entire thing. It features a tough crux section about a third of the way along, but most of the climbing is glorious jug-pulling above deep water. Even at dead low tide, it's still safe to climb it. I first climbed Rainbow Bridge in 1998, when I was 17. It was a perfect late summer morning, I remember, and the whole wall was bathed in blue and green light reflecting off the sea. After I left my friends on the clifftop, there was nobody else around; I was alone on the best piece of rock in southwest England. Gulls and a few oystercatchers floated on the water. The climbing was even better than I'd been told it was. Once I reached the big ledge after the Terminal Zawn, I tied my chalk bag around my head to keep it dry (I'd need it later) and swam back to the beginning. That afternoon, Mr Robertson and I did Caveman, the big spooky trad climb that takes on the challenge of the huge cave of Old Redoubt head-on. The combination of these two completely different routes made for one of the best days of sea cliff climbing I have ever had.

In the late summer of 2022, I went back to Berry Head on a cool September afternoon as cloud shadows moved across the water. I realised I hadn't actually climbed the whole of Rainbow Bridge, start to finish, since I first did it as a teenager. I'd been back to the cliff quite a bit since then, but for some reason, I hadn't repeated the entire traverse since I first did it more than half my lifetime ago. I clambered down the descent line and set off across that first gorgeous pocketed wall before the Pink Grotto, the strange sequence of upside-down solution pockets both familiar and at the same time far away. As I climbed, I thought about Philip Larkin's point that: 'Truly, though our element is time / We are not suited to the long perspectives / Open at each instant of our lives.' The route was still there, exactly as it had been when I first did it. I, of course, had changed. That's the thing about climbing; the routes stay as they are, but we don't. I thought about the blond-haired boy who'd barrelled along here that late summer morning at the end of the last millennium. Who was he? I still knew him, of course, but he was far away in time and space, adrift in the twilight realm between memory and forgetting. I quietly admired him, though. I smiled when I thought about his endless enthusiasm for climbing, his courage and his resilience. As the poet Geoffrey Hill said, 'The boy I was / shouts Go!'

The even younger boy who swam far out in Worbarrow Bay in an offshore wind that summer long ago prefigured the exploratory climber I'd later become. I made it back to shore that day because I didn't panic, understood the sea, and trusted myself at the same time. Reminiscence, of course, wasn't going to help me get across Rainbow Bridge without going for an involuntary swim. I knuckled down and made the slight descent to the crux pitch. It's all slopers and awkwardly-placed footholds; it was trickier than I remembered from the first time, 24 summers ago. Perhaps that's one of the things about age; we forget how much easier some things were with the lightness of youth and also how much harder life was back then. Philip Larkin, again, reminds us of 'the strength and pain / Of being young; that it can't come again / But is for others undiminished somewhere'. In climbing, as with so many things in life that involve risk and reward, Larkin's gnomic statement is profoundly accurate.

Then, all too soon, there was a brief tussle with the actual crux – a deceptive hand-match on a weird diagonal hold – and suddenly, the freedom of the groove that leads up and out to the next section of the climb. I'd crossed the mid-point of Rainbow Bridge. When I first passed this way, I was at the beginning of things; today, I was

Ruth Taylor on Rainbow Bridge. Facing page: Charlotte Obhrai on Rainbow Bridge. Photos Grant Farquhar.

somewhere in the middle of my life. What had gone before would not come again, and I couldn't know what might happen in the years to come.

If I climbed this route again in another 24 years, I suddenly realised, I'd be approaching old age. I'd be exactly the same age as my grandfather was when he died, quite unexpectedly, one late summer afternoon. An afternoon perhaps not so different to this one. So what? I just hope that if I make it that far, I'd still be able to find my way across this most beautiful of climbs one more time. Continuing on, I took the high variation before the Terminal Zawn, swinging footless and fancy-free along that glorious line of widely-spaced holds, my feet arcing out above the sea. The swell sucked and hissed in the unknown ventricles of the cliff's deepest heart. We cannot know what is deep within those places, those submerged clefts in the underwater stone, any more than we can know what lies at the centre of the Earth itself.

The way was clear to the Terminal Zawn: the finale. This conclusive section of the traverse is different to the rest of the climb; it's like a kind of coda. The movement is fine, but the rock quality isn't quite as good. And the very last section – a short down-climb at the apex of the zawn to the big ledge on which the route ends - is actually quite dangerous. It's as if Rainbow Bridge, in its final moments, wants to remind us of our own mortality after the surreal, serene, almost otherworldly climb we have just completed. I climbed that last section carefully – very carefully – and took the final step down to the ledge. I breathed in deeply and exhaled.

The Pied Piper of Brixham by Grant Farquhar

When I was first getting into DWS one habit I developed was always to jump off the top of a newly visited venue before climbing in order to extinguish any fear of falling from any height on the crag. After all, if you've already jumped off the top, then falling from any lower height should not be a problem, right? So went my logic. However, this plan backfired one day at Berry Head. I decided to jump off a suitable-looking ledge well above Rainbow Bridge which, I estimated, was about 60ft above the water. The thing is, not having been to the crag before, I was wrong. The crag, apparently, is a lot bigger than it looks and without any reference points, i.e. climbers on the crag or in the water, I seriously underestimated the height from where I was jumping which was more like 90ft.

So I jumped off... and seriously hurt my back, which would have put me off from any further landings in the sea that day if I'd had a chance to climb. But, as it happened, I did not. As I crawled, winded, out of the water, I looked up to see not only that it was very high but also Mike Weeks preparing to make the same jump, except he had his back to me and appeared to about to execute a backwards flip – he was in training to be a stuntman at the time. Stuntman training or not, I was just beginning to muster the breath to shout 'NO!' when he was off and twisting through the air to land... badly.

His shoes surfaced first, followed shortly thereafter by a distressed-looking Mike, who was clearly in pain and coughing up copious amounts of blood. By then, Mike Robertson was on the ledges at the base of the cliff and, having witnessed the landing, immediately jumped in, Baywatch style, to offer aid. I flagged down a nearby dive boat, and soon we were hauling Mike on board for a rapid trip to Brixham and a waiting ambulance. Mike's chest X-ray in Torquay Hospital showed a partially collapsed lung, and he was duly kept in overnight for observation.

A few years later, I was feeling confident after having done Rainbow Bridge many times before and was leading a procession of climbers who were less confident and wanted me, like a Pied Piper, to show them the way. I was slightly puzzled when the crux felt harder than I remembered it, with a massive slap into the groove, but I shrugged this off as me being hungover or something. One by one, the climbers danced their way along the traverse towards me as I confidently pointed out the (wrong) beta to them. One by one, they fell off and started the long swim back. As I waited, bridged in the groove, for the next victim, I noticed a finger-jug slightly higher than the line I had taken. Fuck is that in? I wondered. Steve Pack was next, and as he wavered where everyone else had fallen off, I pointed him at the hold, and he made it.

Above: Drone's eye view of the crux of Rainbow Bridge. Photo Richard Pollard. Right: Jonathan Woods on MC Navigator, F7a+ S1, in Berry Head Quarry. Photo Grant Farquhar.

The Wizard of Oz by Pete Saunders

The Wizard of Oz is a huge link-up of the finest deep water solo traverses at Berry Head in Devon. An early morning crossing is a magical experience, with the sun illuminating your private arena. Beneath your feet, small fish play, and the dark green sea playfully slaps the rock. Above you, peregrines dive bomb unwary pigeons. Stretching before you are 700m of limestone, stained red and pink by oxides of iron. This beauty is immortalised on the front cover of various guidebooks. Idyllic, yes, but you cannot afford to relax too much. The dark depths of five sea caves are yet to be negotiated, and you have to send four crux sections of F7a/7a+. Failure means game over and a soggy retreat.

Like its namesake movie, The Wizard of Oz is a technicolour journey of an epic nature. Its history involves a cast list, which reads like a Who's Who of British sea cliff climbing. Forty-two years in the making, this climb is a link up of 10 existing climbs. Magical Mystery Tour (F6a+ S0) is reversed into Rainbow Bridge (F7a+ S1), and Look Before You Leap (F6b S1) is down-climbed into The Cauldron (F7a+ S1) sea cave.

Deliverance to the second half of Rainbow Bridge is via a rope bridge where the wild jug rail of The Wave Variation (F6c S1) awaits you. To keep things safe, the route drops down to join the later stages of Gluteus Maximus (F7a S1) and deposits the climber on the Terminal Zawn ledges. After a good rest and a stroll, it's onwards with the Oz Wall Traverse (F6b+ S0). At its end, you drop down into Cavewoman (F6c S0) with its pumpy traverse and claustrophobic exit up a calcite chimney. After a little of Barnacle Traverse Continuation (F6b S0), the exhausted climber is faced with the monstrously overhanging White Rhino Tea (F7a S1) as a fitting finale. Local activist Ken Palmer had toyed with the concept for some time and eventually soloed it in one push on 18th August 2010 at an overall grade of F7b S1.

Caveman

I was there when Dave soloed Terracotta – I said he should have called it Redoubting Thomas.

Crispin Waddy

Dave Thomas's solo ascent of Lord of the Flies (E6 6a) on Dinas Cromlech stunned the climbing world, but he followed this up by soloing the first ascent of Terracotta, a multi-pitch E6 on Berry Head that would merit a DWS grade of SX, never mind S3. Dave already knew the crag well, having made the second ascent of Caveman in 1989 with Nick White. That same year, a back-of-the-car conversation with Crispin Waddy over which route would be harder to solo – Lord of the Flies or Caveman – planted a seed.

As well as soloing Lord, Dave also soloed Olympica on Lundy, and Call to Arms and Flaming Drambuie on Sanctuary Wall. At the time he was aware of DWS as a concept, but none of these solos were accomplished with a DWS-mindset. So his approach to Caveman was more as a pure solo rather than a DWS. He was interested in a new section of climbing straight up from the 'railway tunnel' of Caveman. So, with 'time pressure', soloing this direct new version replaced soloing Caveman as his project.

Crispin had opined that Lord would be the harder solo, but according to Dave it wasn't; Caveman/Terracotta felt much bigger. On the day that he did it, he was aware of the tide – it was a higher tide, but he was not particularly waiting for high tide:

'You just have to climb when you can. I ended up in the zone: that's for me.'

Dave would like to do it on a rope to see how hard it is, as no one has repeated Terracotta to date.

Call to Arms by Dave Thomas

He slumps with an arm through an in-situ sling under the roof, loudly proclaiming that he can climb no further and requires a rescue. I didn't know him well, though I now know the man to have a reputation for this type of shenanigans.

Glenn Robbins

I'd ambled along the bottom of the Sanctuary Wall to Call to Arms, which is a really steep, hard E4. I'd soloed that, reversed an HVS to Long Quarry Point and then carried on up a two-pitch E3 called Black Ice. It was a really beautiful outing, on my own, doing my own thing; I loved it. I went back a year later with Glenn Robbins, who wanted some photos. I hadn't been climbing much; I had a chest infection, and everything was a bit wet. But, stupidly, I said, 'I know Glenn, I'll solo Call to Arms it will make a great photo.'

When I set out, I was really having a hard time. I felt really unfit. I was climbing badly, and I didn't know what was going on. I thought, why have I agreed to solo this route? It was really stupid. I ended up two-thirds of the way up the route at a point where reversing it was a no-no. The route was much wetter than I thought it was going to be and, therefore, fragile. I was eight or nine metres from the belay, and I was completely freaked out. I wasn't in the right state of mind. I hadn't made the decision to solo the route with authentic desire at all.

There is a bit of a niche where I could almost get slightly bridged out, a bit of a bum-sprag sort of thing. My arm was through this sling on an old machine nut. It was really steep in this position, and I didn't want to weight the sling at all because it probably would have popped. The long and short of it is that I was getting really pumped, and I couldn't breathe properly because of the chest infection. The rock was in a terrible state. I thought I had really cocked it up – yeah, I'm going to die. Seriously, that's where I was. I was looking to the routes on either side, thinking, what E6 am I going to bail off onto as a route off? But I was too pumped for that option. I knew that Glenn was running around to the top. He was going to throw a rope down, but the route is too steep; you can't grab the rope. I was on my own.

There are situations that you find yourself in where either the decision you've made or the situation itself makes things so cut-and-dried that any amount of panicking and scary thinking just doesn't really get a look in because you know it's not going to help. That was one of those situations. I could be conscious of the fact that I was probably going to snuff it, but that idea didn't upset me too much.

I left it until the very last moment when I knew from redpointing exactly how much I had left in the tank. I looked at the next bit of climbing between myself and the belay, and I knew that – at that moment – I had to go for it. So, I went for it, and I just did not stop pulling. Every hold I touched, I pulled on to reach the next hold, all the way expecting – for sure – one of those holds was going to explode on me, and that would have been me off.

That was an experience that I don't want to repeat. I was completely out of control in terms of what I could physically do. It was one of those situations where the decision to solo something had gone very badly wrong.

Left: Pete Biven leads a reconnaissance on Sanctuary Wall above The Long Traverse between Anstey's Cove and Long Quarry Point in September 1969. Photo John Cleare.
Facing page: Dave Thomas soloing Call to Arms, E4 5c. Photo Glenn Robbins.

Journey Inwards by Dave Thomas

Once committed, however, there opens an experience of the simple joy and pain of breathing its spirit.

John Redhead's 'Authentic Desire' chapter in *...and one for the crow*

John Redhead's 'authentic desire' is a really interesting idea. I feel it when I go to Stanage on an autumn day, and there's this beautiful piece of golden gritstone in front of me. There's something so warm and cosy about it. Finding that sort of feeling, I think, is something that is wrapped up in all the questions we have about why we climb. What are the risks? What are we willing to accept?

I remember soloing Kinky Boots at Baggy Point, which is HVS 4c. It starts with a leap of faith. You just let yourself fall across the zone and catch the holds on the other side. It would be very difficult coming back from that position, or that's how it feels. There's this point of commitment; it feels like this totally decisive point of no return. And that route encapsulates how I see soloing completely.

I had to consider everything before setting off on that solo. My throat was probably a little bit dry, and it felt like a dangerous thing to do. That's the first thing I need to emphasise: it did feel like a dangerous thing to do. It wasn't something that I was getting on thinking, oh, I've done this a million times; there's no way I can fall off it. Thinking like that is stupid. So I set off, pulled across onto the other side of the zawn and just started climbing with this freedom that was like I'd stepped into my own world. It was me on the rock, nobody else and it was amazing. I'd dealt with the fact that it was a dangerous thing to do, or felt like a dangerous thing to do, before setting off. I'd made the commitment. So, therefore, I just had to get on with the act of climbing it.

Moving around the overlap, I got the moves perfectly. I'd never climbed it so well, but something about that took away: feeling close to and at peace with the rock, completely part of that environment, subtracted any hint of it being a test. There was no doubt. It wasn't, can I get up this route? I was also free from the accoutrements of rope, harness, and gear, and I associate all of those things with a test. Because if you're in the right environment – the place that's there for you to enjoy – you don't need all that safety equipment. I think that's what it is that I'm trying to express in soloing. If I was willing to make that first committed step, I could go to a place where I could experience myself. It was like going back to a place that I'd lost touch with.

In our earlier years, we put this story down. And it's all there, recorded, a story within us in real time. It gets laid down on a tape as we are experiencing it – unfiltered and undifferentiated. It's like a whole collection of different colours. Some are bright and happy, and some are dark and sinister. But it's there as a palette that we build, and everyone will have a different palette.

Often we seek out things that require that we see things differently, even if it involves putting ourselves in situations where we have less control. Listening to John Redhead talking about North Stack Wall, he talks about visiting yourself. There's something about doing something scary when you're climbing that requires that you let go of concerns about how things are going to make you feel. And I love that. The sense that you must be very deliberate in the moment, for instance, despite feeling scared. You don't feel scared all the time, but feeling scared is a really big part of soloing.

Fear, and feeling alone, was a huge part of my life as a child. If you're on the receiving end of some form of abuse, you experience yourself very much alone at the time when it happens. Whether you're four or whether you're 64 doesn't matter. In that moment of being abused, your experience is that you are alone. Which, of course, you are when you are soloing.

Lord of the Flies was a route that I wanted to solo. I'd lead it once, three years beforehand. So I knew, physically, I could get up it. I'd driven to Wales to meet up with my girlfriend at the time, Helen. I told her that I wanted to solo Lord, but when we got there, Lord had a wet streak running down it. The sun was full on that right

wall of the Cromlech, so it was a case of waiting and hoping that it would dry out in time to do the route. I left it as long as I could, and when I set out, I wasn't sure whether it was in sufficiently good condition.

The big question is: why did I want to solo that route so much? Where did I feel that I needed to go in myself? Perhaps for me, it's a question of being able to maintain my integrity despite a very difficult situation. And I think that's a theme in my life. I feel that if I put myself in a very challenging situation, I'm able to perhaps heal wounds from when I've lost integrity in the past, say as a young kid. I really went with the uncertainty that I didn't know how it was going to go. I reasoned that it was no longer a hard route – it was roughly 10 years after the first ascent, so no longer cutting edge, and it's only about F7a+.

I was firmly embracing the idea of negative outcomes, not just positive outcomes. I wasn't thinking, yeah, for sure I'm going to get up this route. One of the ideas I definitely had in my head was that it was all going to go wrong. I wasn't actually thinking, oh, it is going to go wrong. But the idea that it could go wrong was a possibility that I had to accept in embracing the challenge of setting out on that route; making the commitment to go for the solo.

But when I set out, everything just sort of opened up in front of me. It just felt like this long line of really nice crimps and pockets. It was fantastic, absolutely wonderful. But, when I was around about the crux where there used to be a peg, and it's kind of a bit more thin, there was this guy on Cenotaph Corner, and he was having a complete nightmare. He rattled down the corner, pulling out his gear and landed on the ledge at the bottom. And he did the same from a little bit higher up. That really distracted me, and I had to remind myself that I was on Lord: come on, concentrate. I had to get myself back in the zone.

Above the girdle ledge, there's a bit where you share on a slightly sloping finger-rail. I loved the experience of holding and touching those holds and having this real sense of the beauty of where I was at the time. I thought, what a wonderful place to be: a long way up this fantastic wall high up above the valley floor, and I absolutely loved it.

Where I was mentally before I set out on the route, compared with my experience of being on it, couldn't have been more different. Soloing Lord was one of the single most enjoyable moments of climbing in my career. The sheer enjoyment of being on that route is something that I've been able to relive and keep and treasure. So little of that was about how other people might have seen it.

I got a bit of flak about that solo, which was completely unjustified. It was the Easter bank holiday of 1990. Yes,

Left: Charlie Woodburn on Caveman, E6 6b or F7a+ S3 as a DWS. Facing page: Climbers on Caveman. The thought of deep water soloing at that height is definitely stretching the concept. Photos Grant Farquhar.

the crag had been busy earlier in the day; we had to wait for the route to dry. But it wasn't as busy as people have since made out. For me, it felt no different than setting out on a solo anywhere else at any other time of the year.

A lot of us find ourselves in climbing. We're looking for those wonderful moments, those sweet spots. Everyone at different points in childhood has experienced those moments where everything was about them. Of course, the real world, as an adult (except for narcissists), isn't like that. But there are moments in climbing where you find everything seeming to go in your favour, and you feel at peace; you're experiencing flow. The problem is that only tends to happen when you are willing to commit. Commitment is a very big part of the equation if you want to find those flow states.

A lot of my soloing has involved a great deal of introspection and revisiting trauma that a much better experience has taken away. The risk I'm taking is that I fear that I might find myself back in that very dark place where I was as a child. I've definitely found as I've got older that it's the darker stuff, the more painful stuff, that people experience that is the really important stuff. The range of colours that you want to work with will tend to change as you get older. Picasso said that all children are artists. The problem is remaining an artist when you grow up.

I had a rope access business. I didn't seem to have a problem getting jobs; from a business point of view, you could say that I was quite successful. There was the positive feeling of doing a good job and getting paid for it and having more money in the bank than I needed. So those are positive colours. But, ultimately I found that business unsatisfying. My dad always emphasised money – he was always about money. That's what he saw as his role. He would be home for dinner in the evening. He would often be in bed when I left for school. His work didn't require a particularly early start, but he, my dad, was there to perform the function of work and bringing money home. But it just didn't feel as though he was part of the family. He certainly performed a function – he provided for us. But he wasn't present emotionally.

If you have a fear of being hit as a child, forever wondering what is going to be coming your way, that leaves a lasting impression. That was a very strong theme in my life as a child. Fear of what was going to be coming my way – what am I in for? And I express that when I'm climbing for sure. If you had that experience as a child you also experience the fear of abandonment. You no longer have that sense of being cared for by somebody; you are very much on your own; you are being abandoned to your fate.

I was a bit timid in the sea, at least in my dad's eyes. He'd lose patience and push my head under the water. Unfortunately, it's still part of the relationship that I have with my father – he was not supportive at an early age. There was never that sense of nurturing and someone genuinely caring. The best parts of our relationships often get forgotten because we tend to remember the more painful things that occur. As we get older we can, hopefully, rediscover the good parts.

There have been times when I've set out on a bigger solo that I've relived some very dark things. So when I think back to Terracotta, I can access some of the feelings that I was aware of, that I was carrying around with me at the time, and that I just took for granted.

I've never really been conscious of it, but those feelings were a huge part of my motivation for setting out on the route. Soloing was about dealing with the conclusions I'd come to about myself from experiences in relation to people like my parents. There's no intended judgment. So hopefully nothing of that sort will be drawn from it.

I was sat at the top of Moonraker putting my boots on. I had my harness on; I can vividly remember doing my laces up. I was about to clip into the ab rope to abseil down Moonraker to the start of what I was setting out to do, which was to solo a new route at the Old Redoubt, a sort of direct on Caveman. In my mind, I was going back to a time when I was a child. I would have been about four years old. The treatment was harsh, and it was a really dark experience. My father was much more powerful than me, and I was getting beaten.

Psychologically, I was going back to that place. For me, the consequences of what I was setting out to do were mixed up with the abuse that I'd experienced as a child. I clipped onto the ab rope – that was the point of commitment.

What set the wheels in motion was that I'd been climbing at the Old Redoubt an awful lot over the summer, and I looked at this line that was a way of straightening out Caveman. I didn't really want to practice it first, but I could see that there was a block on the lip of the overhang, and this looked like a typical Old Redoubt stuck-on loose block. I was thinking in terms of soloing it. I always had fantasies about soloing stuff – that was what I loved to do full-stop. That's what my climbing was about, but I thought that I couldn't justify soloing it if that block was loose.

So, I abseiled Terracotta to check it out. And, of course, the block *was* loose. I cleaned what was left and thought I'd better check to see whether or not the move goes. I had my weight on the ab rope when I tried the moves, and I thought, yeah, that goes.

When I committed to start up Caveman everything just flowed. I had this sling that I was planning to clip into the belay to give myself some respite. But, when I got there, it wouldn't reach the belay because it was too short. It didn't really matter too much. I was bridged out in the niche at the belay. There was a moment when I had to drop down onto the rail that runs through the, iconic, what I call 'railway tunnel' part of the route. It was shifting like a pack of cards balanced on top of the rail in lots of little bits. But it was the hold I had to use, and I just had to, literally, hold it together and hope it didn't disintegrate.

Thankfully it didn't, and I basically just aped along the rail. I stuck my foot in the break at the end and hung out backwards, laughing. This is the thing: making the commitment and taking the chance that you might discover something really horrible; you find yourself going to places that are absolutely amazing. There I was, in the middle of this amazing sea cliff, hanging upside down, dangling from a leg jammed in a break high above the sea. Absolutely awesome.

What concerned me was that for a significant part of the time, I was going to be way above anything that might be considered a safety zone. So, I'd set out on this new route, absolutely not thinking in terms of being able to fall off. I didn't expect to fall off the first part of Caveman in any case. And what I find is that if you really commit to the climbing, that's less likely to happen, but you've got to be firmly committed. As soon as you start worrying, am I going to fall off here? You start climbing really badly, and you are much more likely to fall off.

So, I prefer climbing without the sense of contingency just to be completely committed. The other thing I was really worried about was a boulder that was submerged at that point, but only just, and I thought about it as this huge coffin-shaped boulder. It had all the drama of signifying my demise: if I was to fall off, I'd hit it and drown.

Fear is one of the prices we have to pay for feeling more authentic. If you want to be yourself – if you want to feel that you can experience you for who you are – you also have to accept a lot of negative thoughts and feelings. I've always had the feeling that those painful experiences that I had when I was younger took something away from me. Soloing, especially on something big – with the commitment required – is part of trying to get that something back. When I solo, I'm setting out to touch base, psychologically, with the past trauma but to arrive at a different outcome. That was the test: it was about accepting what life would mean for me in that situation, and not running away because I felt scared.

My friend Frank Ramsey had agreed to take some photos, and I think he didn't know what to make of this at all. He was getting a bit freaked out, actually. As for me, I was really enjoying that moment in a really playful way, not dissimilar to the way that a kid might feel happy if they were being spun around in the air, for example, by their dad.

There's a sandwiched wall between the roof of Caveman and the next roof that Lip Trip traverses. I stepped up and left onto that wall to look at the next bit of climbing which was the bit of climbing I'd not done before – the new bit. Frank was spinning around out in space behind me. I'm bridged, just making sure I'm recovering, getting a shake out. The only hold available to recover on was a block with a little capping stone on the top of it, but the top of the stone was loose.

There was a big pinch at the back of the roof, that was really hollow when tapped, which I would hold with one hand. Then I would take the top of the other hold off and sort of squiggle something in the mud underneath it, just to show how ridiculous it was. I'd put the top back on and pinch it together, matching on it, to carry on shaking out. The weird thing is, I don't quite know why I was doing that. I think I was just trying to distract myself before the next bit of climbing, but it was all a bit much for Frank who was, sort of, freaking out.

Then I went for it, out across the next section of roof. There was quite a nice, easy flake out to the lip. The climbing wasn't very hard, but it was very airy at that point. And, of course, I was that bit higher and more committed. I felt like I was very much on the lip of everything at that point. I had my friend Frank with me, but I was on my own. I got to the lip and thought I knew exactly what I had to do when I got there: sling my foot on a heel hook and pull into this undercut. It was really small but I was planning on using it to slap up to the next break.

But, when I got there, it felt like I had zero chance of executing the move this way. My ascent was later written about in a way that made it sound more dramatic than it probably was for me in that moment: I just made a snap decision. I didn't even think about it and just laid one on. I pulled in using that little undercut and and sort of did this big swinging crossover with my right hand to some large flat holds above my left which I hit, thankfully, and carried on to the Dreadnought belay.

Dave Thomas on Terracotta.
Photo Francis Ramsey.

Being Seen by Grant Farquhar

A long time before learning about the work of Jaques Lacan, or about Object Relations theory, I used to speculate that encounters with routes like Breakaway at Henna were a way of returning and processing the fractious, frightening and unpredictable experience of being breastfed. That's actually quite a serious contemplation. Loss of attachment for the infant is far more traumatic than the prospect of hitting the deck – or the salty brine.

Dave Thomas

It was 1986, and I was sitting at the base of Dinas Cromlech. I spent that summer camped in the Llanberis Pass, so I spent quite a bit of time up on that crag, and the others, doing the routes and enjoying the atmosphere.

So I was sitting there, below Right Wall, when a very young Milky-Bar-Kid lookalike turned up and started feeling the holds at the bottom of Lord of the Flies. WTF? I thought.

'Too wet!' he pronounced.

This was my first introduction to Dave Thomas. Not fancying Lord on that particular day, he went off to climb Hall of Warriors which in the guidebook at the time was given the intimidating grade of E6 6c.

I forget what route I climbed next, but while on it, I heard a high-pitched shriek from around the corner – Dave had fallen off Hall of Warriors. I then bumped into him on a few occasions over that fantastic North Wales bluesky summer when he was usually climbing something hard. This was well before his soloing exploits, but his precocious strength and talent were obvious.

Dave appears to be someone for whom climbing meets a deep psychological need. He finds redemption and solace in his climbing. He exorcises demons. Soloing is the ultimate fulfilment of that need.

In his seminal book *The Body Keeps the Score*, Bessel Van Der Kolk writes about the concept that traumatic memories are stored in a different place from normal memories. Telling the story is important, and the basis of psychotherapy, but talking about it does not guarantee that the trauma will be processed simply by doing so:

'When people are forced to submit to overwhelming power, as is true for most abused children... they often survive with resigned compliance. The best way to overcome ingrained patterns of submission is to restore a physical capacity to engage and defend.'

American psychologist Martin Seligman researched the concept of learned helplessness in 1967 at the University of Pennsylvania. This concept describes the behaviour exhibited by a subject (dogs) after enduring repeated aversive stimuli (electric shocks) beyond their control. It was initially thought to be caused by the subject's resignation to their powerlessness by way of discontinuing attempts to escape even when an escape route was presented.

In humans, learned helplessness theory is the view that clinical depression and related mental disorders may result from a real or perceived absence of control over the outcome of a situation.

In Seligman's hypothesis, the dogs do not try to escape because they expect that nothing they do will stop the shock. To change this expectation, the scientists physically picked up the dogs and moved their legs, replicating the actions the dogs would need to take in order to escape from the electric shocks. This had to be done at least twice before the dogs would start willfully escaping on their own. In contrast, threats, rewards, and observed demonstrations had no effect on the 'helpless' dogs.

It may be that for some people, action is required to process trauma and this may be what Dave was seeking out – to exorcise his demons by exercising his capacity through solo climbing; to place himself in life-threatening situations which activate those parts of the brain where the trauma has been sequestered.

Trauma-focused therapies such as Eye-Movement Desensitisation Reprocessing, that go beyond simply talking, are theorised to work by accessing and reprocessing the traumatic memories described by Van Der Volk. It may be that experiential therapies, such as climbing therapy, work in a similar way and help the subject overcome their learned helplessness through action, not just talking.

In climbing, there are the mutual positive benefits derived from a trusting relationship with your climbing partner – your life is literally in their hands and vice versa – and the acceptance gained from the larger climbing community. Of course, the harder and bolder – and more dangerous – your climbing then the more you will be recognised and accepted into the fold.

Influential psychoanalyst Melanie Klein (1882–1960) observed that infants recognise that their achievements, such as crawling and walking, give their parents joy and results in the giving of attention – being seen.

So, when we strive to perform difficult and serious challenges we might be fulfilling inner needs through processing trauma, and we might be meeting the need to be seen – and accepted – not only by our parents but also by our tribe.

Andy Meyers on the third pitch
of Caveman. Photo Mick Fowler.

Caveman by Ben Silvestre

It is a great and darkened hole,
And jagged teeth run through it.
Jutting as icons of fear,
In constellations of steepness,
They are scars in the rhythm of hope.

Hang up your doubt, if you dare; this
Is the Great Redoubt. Redouble your will
And enter this hall of mirrors,
Where the phlegm in your soul
Must be greater than pristine stone.

Here, the placid eyes of seals
Steal from you the most secret of feelings.
The steady dripping of time ferments
With dour jokes, sick laughter,
Cramping forearms.

You may find that all which you loved is gone,
That you remain abandoned here;
Suspended wanderers of this prehistoric church,
The final members of a broken cult,
Hanging on to the smallest of truths.

If brute strength wanes
You will be swallowed by the waves,
But with patience you may find
All the space of a million years.
And I shall remind you, lest you forget –

A caveman lives in each of us.

White Rhino Teabag by Julian Lines

My personal opinion is that Magical Mystery Tour and Rainbow Bridge are the two best traverses in the world, and that's up against some pretty massive competition. Caveman is one of the definitive S3s. High enough to scare, and upside-down enough for maximum entertainment. It's just brilliant. And it has a whole pile of great history attached to it.

Mike Robertson

Tim had climbed Caveman and was psyched to go back and solo it. He was looking for some 'spotters', which in high deep water solos means swimmers who would dive in and fish you out of the water if you are unconscious or get a punctured lung. Mikey and I were in the vicinity and went down to Berry Head to watch the spectacle. Tim chose the morning high tide and duly swung his way up through the cave with a smooth competence. Now, Caveman is over 200ft, maybe 300ft, and too high for deep water soloing, so the ethical tick for soloists is set at 65ft at the end of the lip trip where Caveman joins Dreadnought, and where the climbing above is only 5b. So when Tim reached Dreadnought, he climbed down a few feet and launched off into the water. I was in shock at his nonchalance at a 60ft plunge as if it was child's play. He popped up and swam out, laughing and buzzing.

I don't exactly remember how our commitment shaped itself, but Mikey and I were formulating a plan to come back on the evening high tide to try it ourselves. The trouble with the evening high tide is that it was on the brink of darkness, so timing was quite critical. Mikey had climbed Caveman previously, and so it was decided that he would go first. We traversed around the back of the cave and up onto a gearing-up ledge from where I could see the whole spectacle unravel before me. The briny bowl was filling up nicely and the light fading as Mikey set off. He traversed in and diminished in size as the reality of the horizontality of the roof became apparent. Some shouts and footless manoeuvres saw him on to the lip trip, where he slowed down a little and made it into Dreadnought.

A few minutes later, there was a splash as he jumped through the grey from a great height. After ascertaining he was OK, that was my cue to set off before it got too dark. I traversed in on big holds and into the roof. It didn't feel too bad on chunky horizontal strata, and I felt quite confident until I got to a section where the holds got smaller and the steepness made itself felt. I scrabbled into the niche and took a breather to gather myself. The huge horizontality of the dark roof flew out and fizzled into the greyness of the sky. The water didn't look too distant, and the darkness surprisingly helped rather than hindered. I didn't really know where I was going, but all I had to do was follow Mikey's chalk rightwards along a massive jug rail, which led back into the vertical at the start of the lip trip. This tight, sandwiched wall was sustained and exposed; I climbed quickly and felt relatively calm because I didn't stop to receive any lactic acid burn. When I reached Dreadnought, unlike Mikey and Tim, I wasn't too keen on jumping from great heights, so I agonisingly climbed all the way down to the belay, stuffed my gear in my dry bag and jumped into the darkness and swam out.

The following morning, I went down to the cave again with Mikey to ponder Caveman in the light of day. I think Mikey traversed off to do a speed ascent of Magical Mystery Tour, and I decided to go for a swim in the cave as it was warm and sunny. On swimming around, I saw this incredibly steep line close to the water on the left-hand side of the cave. I thought I would give it a try, so I set off naked as there was nobody around. Nearing the end of the climb it got difficult; I was fighting, and then, to my astonishment, someone was traversing below me. I felt vulnerable and self-conscious as my balls were dangling out above his head. One last burst of energy for the finishing hold... I just couldn't hold it and fell off. I certainly felt less self-conscious and relieved, clothed in the water. It was Dave Henderson, the Devonian guru. I knew the nearby White Rhino Tea was his route, but today he almost experienced a white rhino teabag. Dave duly dotted my chalk and completed the climb – Killa Gorilla. You win some; you lose some.

Facing page: Tim Exley off White Rhino Tea, F7a S0. Photo Brian Hannon.
Right: Charlie Woodburn on White Rhino Tea. Photo Dave Pickford.

The Art of Climbing Without Climbing by Neil Gresham

Do I bother you?

Don't waste yourself.

What's your style?

The art of fighting without fighting.

Show me some.

Dialogue from Enter the Dragon (film)

Cutlass is one of those DWS test pieces which is hard to ignore, being easily accessible, bouldery in style and the perfect height for a siege campaign. A compact orange wall looms above the popular Rainbow Bridge sea-level traverse, perfectly sculpted, salmon-streaked and minimally splattered with edges and pockets. It's a chunk of your favourite Euro-sport crag, leaning over the sea and with your name written on it.

Local hero Ken Palmer had originally top-roped the line, grading it F8a and laying down the gauntlet. After a tip-off from the DWS Godfather, Mike Robertson, in 2007, I started attempting it. The route was short and steep, with five hard-ish moves into a really hard move and then a move that I just couldn't figure out. My idea was to try it ground up because that's the purest way.

Word soon got out, and a couple of strong climbers also threw their hats into the ring. The pressure was on, especially as the other suitors had worked the moves on a rope. After two all-out sessions and half a dozen splashdowns, the top crux still felt like an insurmountable barrier.

The ground-up ethic can be infuriating, however, leaving you feeling a slave to your own ethical standards and questioning whether they are justified. It's always on your mind that if you were to work the route on a rope, you would probably go home that day with dry shoes and chalk bags, but where's the fun in that?

Ultimately, if you can hold your nerve and stay true to the cause, it can make an ascent feel all the more rewarding – if it comes. But therein lies the catch. I'd been trying the top crux with two different methods on rotation: it was either a slap to a sloper or a roll-over from a pinch to a flat rail. But each method felt equally implausible, and time was ticking on.

I returned to the drawing board at my local gym and built two different replica boulder problems to simulate the two possible sequences. Of course, I spent a lot of time thinking about the route and visualising the moves – the art of climbing without climbing. With my strength and confidence topped up, I returned for a third session on the route but was completely shut down once again.

Moving into my fourth session and double figures for attempts, it was time for desperate measures. After some deliberation, I bent the rules and asked my pal, Rich Heap, who was filming from an ab rope to feel the holds and offer an opinion on the beta: 'It's the cross-over for sure!'

He may as well have given me a rocket pack. I fired through the crux on my next go and battled onward to the top. It was the 13th attempt, but no jinx for me. Instead, relief, joy and disbelief that I was on the top and, this time, not in the sea.

Above Matt Cox on Cutlass. Below: Matt Cox off Cutlass. Photos Brian Hannon.

Facing page: Neil Gresham on Cutlass, F8a S0. Photo Mike Robertson.

Unroped Riviera by Ken Palmer

May I ask what you were expecting to see out of a Torquay hotel bedroom window? Sydney Opera House, perhaps? The Hanging Gardens of Babylon? Herds of wildebeest sweeping majestically?

Basil Fawlty

When I think of deep water soloing in Devon, I am reminded not only of the warmth of the sunbaked English Riviera limestone but also the warmth of friends and friendships that have formed a backdrop to my days playing above the sea. And playing, I think, is the correct word to describe the multifarious sandbagging, piss-taking, whooping (if you must) and general hilarity that has accompanied my trips into this particular verticality. I am lucky enough to have grown up by the sea, both in Plymouth and the moderately more exotic realm of Mauritius, where my sister and I would swim every day after school until the dinner bell rang, and we would wander drippingly back to the house.

Nowadays, I've swapped the lollies we used to suck all afternoon for an artisan ice cream on the Torquay seafront, but that thrill of the sea has never really left me. It is always a place where I feel at ease. It was in the late 1980s that I first began climbing ropeless above the sea. It wasn't really a deliberate thing; it just came from a general desire to explore the cliffs around Berry Head, and it seemed the most sensible way to explore a lot of the new route potential that existed at the time. One of my first forays was on the overhanging wall that hosts Rainbow Bridge; a cut-out in the base of the cliff gave a freefall into the sea that seemed an alluring way to make safe the ascent of the steep wall above. Nick Hancock and I had been looking at the route for a while and decided to give it a go on top rope before risking the plunge. Nick went first and quickly cruised the 6a crux at 30ft before cutting loose and nailing the lunge for big holds on the upper wall.

At the top, he lent back in the expectation of a gentle descent, but unbeknown to him, I'd decided to tie a knot in the rope in order to give him a bit of rush on the way down. My intention was that the knot would pull him up short, just above the water. Detail wasn't necessarily my strong point as a 25 year old, so as Nick leant back and I let go of the rope, I was horrified to see him plunge straight past me at high speed. I'd tied the knot in entirely the wrong place. He hit the water full force before submerging in a puff of chalk and expletives whilst I stood and watched a string of air bubbles thread their way upwards from his hair, which stood momentarily erect in the crystal clear water. His surfacing was a nervous laugh, 'Are you OK?' Relief at his indignation being all that needed tending to. 'What was that?' Nick quite reasonably asked. 'That was a hands-off whizzy', I replied, unsure of how he would take it, but also quite pleased with the outcome as I now knew a fall from the top of the route would be an entirely safe, if slightly exhilarating, event. My chance to solo the route would come later, but not before my own hands-off whizzy, which I think is understandable given the circumstances; it was moderately terrifying.

Soon after this, I was part of a South African climbing exchange where they showed us the sun-drenched majesty of the high veld, and we were keen to show our African counterparts some of what good old Great Britain had to offer. Of course, what it did not have to offer was any sun-drenched majesty. It was a wet and windy summer's day when I thought my favourite stomping ground of Berry Head might offer the best of any available shelter. In a Force 8 gale, it was not going to be a day for any hard repeats, and by lunchtime, most of the roped ascents had been abandoned for the fuggy comforts of a nearby cafe.

A few diehards had heard of the potential for deep water soloing in the area and were keen to try what was then a burgeoning interest in a new and apparently fun form of climbing. It is sometimes hard to appear cautious when you're young, especially when there is a heady mix of rock climbing and visitors who definitely need to be impressed. So I ignored the gale and the powerful rise and fall of a steadily increasing sea and suggested a sea-level traverse as a way to pass an increasingly stormy afternoon.

Ken Palmer on Hands Off Whizzy, F6c S2. Photo Rich Mayfield. Facing page: Ken Palmer on the first ascent of The Barrel Traverse, F7c S0. Photo Nick Hancock.

The next two hours were some of the most unpleasant I have ever spent rock climbing. As we set off, there were squeals and shrieks of enthusiasm at the thrill of high seas and the potential for an unexpected soaking for anyone unfortunate enough to stray too close to the waves. As the route progressed, and we passed areas we might not be able to reverse, it became increasingly clear that enthusiasm was morphing into a deeper-seated sense of unease. The wind was whipping up a greasy spray that soaked the rock whilst a grinding swell began to creep in on the persistent southerly that seemed ignorant of the fact that this was supposed to be a summer holiday destination.

The South Africans clenched their buttocks as swells rose waist-deep before draining with a deep sucking sound from the caves and buttresses that barred our progress. I tried to reassure them that a chalk bag full of brine was quite normal and that shivering and fear of imminent drowning were all part of the day's planned activities, but it was a day when my ease and comfort in the presence of the sea were subject to some misgivings. Later, I reassured our guests that I would have dived in to effect a rescue had any of them looked likely to succumb to the waves. They shrugged, heads down; there was no way they would have gone in after me, and they didn't much appreciate my quip that we were like Tom Cruise in a foot spa – out of our depth.

Since then, I've spent many a day traversing the multi-coloured limestone that rises above the inviting waters of Torbay, though with an increasingly practised eye on the weather conditions. Most of this area faces east and sits out of the prevailing wind, nicely protected by the rainshadow effect of Dartmoor. On a summer's morning when the sun is still low and the sea is silky smooth, there are moments of clarity, flow and detachment that can be hard to find in a world that is always clamouring for space in your head. Of course, there are sometimes crowds. For example, a stand-up paddleboarder parked beneath your trembling attempts at a high crux can give rise to thoughts of deep water slaughtering, and then there are atmospheric events which occasionally bloom the sea with uninviting swarms of jellyfish, but in recompense, the rock can give everything you need in the moments when it all comes together.

In the rusts and creams, browns and even greens of the South Devon limestone, you can lose yourself in the dance between buckets, pockets, scoops and blow holes that allow a climber passage through this magic place. If I had to highlight some of the best South Devon has to offer, it would be routes pictured to the right.

Right top: Ken on The Wizard of Oz, F7b S1: 'This feast of climbing would make it into my top five desert island climbs. 2,500ft. Of superb ocean-sculpted limestone that is strawberries and Devon cream smeared on the face of a smiling emerald sea. This yellow brick road of caves and pockets, slopers and tip-toe traverses requires technical cunning rather than muscle, although somewhere, maybe over the rainbow, there is a surprisingly bumpy jug fest to finish. A rockover finish has also got the better of many a Dorothy or Tin Man.' Photo Pete Saunders. Right bottom: Ken on Killa Gorilla, F7b S0: 'One of my favourite Dave Henderson routes. Forget the grade; do it at half tide. It's a juggy romp close to the water with a killa finish.' Photo Andy Thomston.

Above: Ken on the first ascent of Blue Planet, F7b+ S2: 'A steep pockety wall close to the water gets you onto the upper section of Christine, F8a S2, which is a superb technical roof with a top wall that is easy to slip off unexpectedly. If you jump out when falling, the water gets shallower. The first ascent was made at 7am in drizzle, wearing a buoyancy aid. An ecstasy-induced paranoia convinced me that any later a start and I'd be beaten to the prize). Photo Mark Glaister. Bottom: Ken on the first ascent of Soft Cell, F7c+ S0: 'If you love a good dyno up an extraordinary piece of orange limestone then this one's for you. Sent after spending 10 minutes psyching up one early morning. I was sat on a small ledge up in the roof joined throughout by a pigeon a foot away, magical.' Photo Pete Saunders.

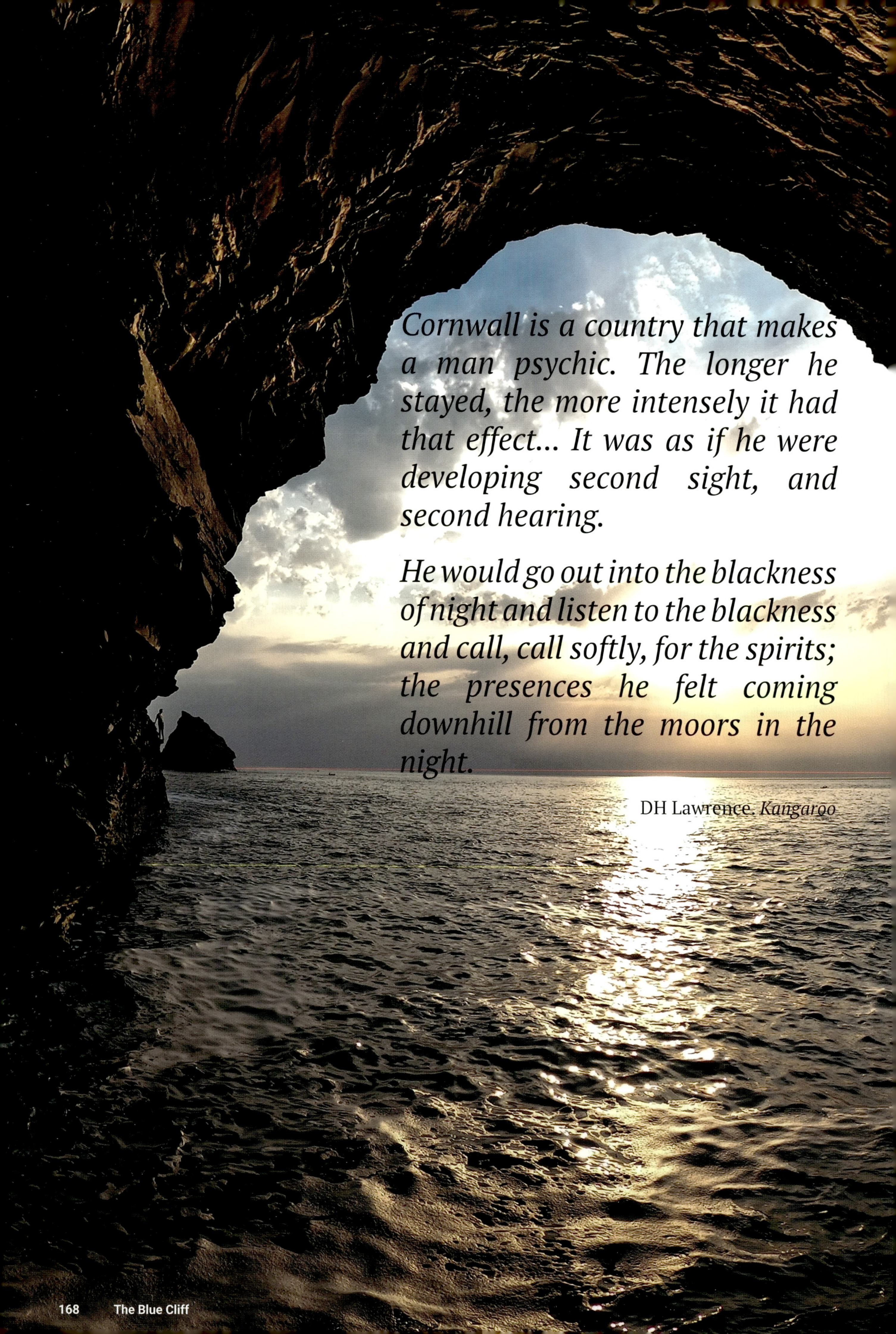
Cornwall is a country that makes a man psychic. The longer he stayed, the more intensely it had that effect... It was as if he were developing second sight, and second hearing.

He would go out into the blackness of night and listen to the blackness and call, call softly, for the spirits; the presences he felt coming downhill from the moors in the night.

DH Lawrence, Kangaroo

Cornwall

A sense of wilderness and isolation persists; the sea is moody, the moorlands mysteriously conceal a wealth of prehistoric relics, the farms are remote, and the place names – Mousehole, Chair Ladder, Woon Gumpus, Brandy's Zawn and Ding Dong – are those of an enchanted, make-believe world. The moors are covered by gorse, the headlands by thrift, honeysuckle and blackberry; the beaches are sandy and pleasant, and over all hangs the aura of the ending of the land.

Frank Cannings in *Mountain* magazine #15, May 1971

Walter Parry Haskett Smith (1859–1946) noted the climbing potential of Cornwall in his 1894 book: 'To the true-souled climber, who can enjoy a tough bit of rock, even if it is only 50, aye, or 20ft high, the coast of Cornwall with its worn granite cliffs and bays has much to offer.' He was right.

The Road Between Tides by Des Hannigan

The Elizabethan traveller Celia Fiennes visited the barren rocks of Land's End in 1698. Fiennes established a healthy precedent for access when she wrote: 'I clambr'd over them as farre as safety permitted me.'

What makes her particularly appealing to me is her documenting a right of access from historical times. Celia Fiennes was a potential rock climber too, or at least was possessed of the insatiable curiosity and adventurousness that drives people to explore below the cliff edge.

I am a cliff native of the north coast of the Land's End Peninsula, but the linked sequence of south coast granite cliffs that runs from Pordenack Point to Gwennap Head and beyond is every bit as compelling as the black-browed zawns of the peninsula's north coast; with added sunshine, too. Such lofty headlands as Pordenack Point, Carn Boel, Carn Les Boel, Gwennap Head, Hella Point, and Treryn Dinas (Logan Rock) are superb viewpoints, easily accessible to walkers and coast path wanderers. What lies below them is even more compelling.

Traditionally the ground at the base of these cliffs, the intertidal zone between mean high water mark and mean low water mark, belongs to the Crown, although the Queen is unlikely to make a Royal Progress along such a rocky road. To wander about the base of great cliffs such as Chair Ladder or over and around the fantastical cluster of pinnacles and spires of Treryn Dinas is to experience a world as fascinating as that of the mountain passes and corries of the Cairngorms or the sinuous river gorges of Crete.

Dead low water of a spring tide in quiet weather is the time to wander here. Traversing the base of sea cliffs takes you into an extreme world, to that sometimes uneasy but always intriguing frontier between land and sea, to places that each lunar

Facing page: Looking out of the back of the cave at Penally with Lee Bartrop on the sea-level traverse into the cave. Photo Richard Pollard.

Above: Bosistow Island, Carn Lês Boel, Land's End. Peter Biven returns from an ascent of the island/stack by Tyrolean traverse, originally fixed by lasso.

Below: Belayed by Alison Chadwick-Onyzkiewicz, Peter Biven ascends the Diamond Horse tidal pinnacle at Nanjizal, Land's End. Carn Lês Boel rises in the distance.
Photos John Cleare, 17 July 1971.

day are made inaccessible by flooding tides and by the unimaginable mayhem when storm seas and immovable granite meet at any state of the tide. At calm, low water the air is salt-laden and fluid; early in the day, the first 50ft of cliff oozes greasy salt; even the strongest morning light seems subdued at first.

Boulders within the tidal zone are sea-polished or are covered with acorn barnacles and algae. Those higher into the splash zone are coated with the black lichen of *Verrucaria maura* that seems painted on and is often mistaken for polluting tar oil. Overall, there is an astringent odour of the undersea as if the ebbing tide has left its ghostly shroud across the rocks in expectation of its return.

There is no straight road between cliff and sea; the way leads maze-like over and between boulders, along ragged edges, above scooped-out basins of trapped sea water, in and out of narrow chanks and dripping corners, often inescapable except by technical climbing. Always there is the vast openness of sea and sky on one side and the towering cathedral walls of granite on the other. To explore these extreme environs is a privilege as long as no damage is done. The compulsion to explore is irresistible for as long as strength and spirit allow, and for as long as you are confident and capable of 'clambr'ing', like Celia Fiennes as far as safety permits, and then, always a little further.

Leslie Garnet Shadbolt (1883–1973) was a mountaineer who is particularly known for his first ascents on Skye. He is credited with inventing the 'limpet technique' – whereby after a sharp tap a limpet could be relied on as a foothold – during his explorations on Sark in the 1920s. He was an enthusiastic exponent of sea-level traversing despite damage to his arm after he was severely wounded at Gallipoli during the First World War.

The Limpet Technique
by Leslie Garnet Shadbolt

Sea-cliff climbing has its limitations, but under the best conditions can afford really fine sport. I never understood the lack of enthusiasm for it until I tried it in Cornwall not long ago at one of the few parts of the coast which seem to be at all well-known to climbers. It is true that there are some good climbs there, but there is a fearful lack of continuity about them, and one never has time to get really steeped in the spirit of the thing before it is over. The adhesive limpet takes a prominent place.

The more robust specimens may be used as footholds if due warning is given of the coming strain. Lest anyone should be tempted to take out a patent in pocket footholds, let him be warned that it has already been tried. In more youthful and optimistic days, I carried out an exhaustive series of experiments with quantities of the largest size in limpets, but, in spite of careful nourishment, they refused to stick to unknown rock faces at the word of command, basely dropping into the sea the moment they were released.

After all, it is hardly playing the game to use such adventitious aids, and the true mountaineer should scorn to stoop so low. The rise and fall of the tide is so great that the character of the coast is quite altered. Some places, which at high tide afford splendid climbing, at low water are nothing but rocky foreshore. Low water, however, has to be utilised to overcome obstacles which are impassable at other states of the tide, and the general idea is to climb each section at the highest tide possible. The rope is not of much use for this kind of work, and generally on the steepest walls there is an unobstructed drop to the water – to fall in is, however, considered a breach of good taste.

It is not, of course, always possible to start climbing on a falling tide, and as there are very long stretches of coast where there is no possible way of getting to the top of the cliffs, the risk of having one's retreat cut off should be borne in mind. A hot day, with a calm sea, might well be chosen for such an expedition. This horrid thought suggests the necessity for wearing clothes which are comfortable to swim in – to wit, shorts and a shirt with sleeves ending above the elbow. Rubber gym shoes instead of boots are advisable.

And as happens sometimes when the weather is very fine, the cliffs looked as if they were conscious of the ships, and the ships looked as if they were conscious of the cliffs, as if they signalled to each other some message of their own. For sometimes quite close to the shore, the lighthouse looked this morning in the haze an enormous distance away.

Virginia Woof, *To the Lighthouse*

Sir Leslie Stephen (1832–1904) was a Victorian English 'notable man of letters' and a prominent figure in the golden age of alpinism. He was educated at Eton and Trinity Hall, Cambridge, where he had also been ordained as a priest. Later, in 1862, he had a personal rejection of being a clergyman; he resigned from his position at Cambridge and published a book on this change in belief, *Agnosticism and Atheism*. He joined the Alpine Club in the year of its formation, 1857, served as

president and edited their journal. He published, along with many philosophical works, *The Playground of Europe* in 1871, which has become a mountaineering classic. He bought the lease of Talland House in St Ives at the end of 1881, and this house was used as a family retreat during the summer over the next 13 years. Stephen was known for his prodigious walking on the moors and occasional climbing; for example, nearly every climbing history has him 'gangling and prehensile' up a chimney on Gurnard's Head. Sir Leslie Stephen's daughter, Virginia Woolf (1882–1941), created a posthumous portrait of him in the character of the atheist philosophy professor patriarch of the Ramsay family in her famous 1927 novel, *To the Lighthouse*, in which she caricatures him as the archetypal Victorian Dad:

'What he said was true. It was always true. He was incapable of untruth; never tampered with a fact; never altered a disagreeable word to suit the pleasure or convenience of any mortal being, least of all of his own children, who, sprung from his loins, should be aware from childhood that life is difficult; facts uncompromising; and the passage to that fabled land where our brightest hopes are extinguished, our frail barks founder in darkness (here Mr Ramsay would straighten his back and narrow his little blue eyes upon the horizon), one that needs, above all, courage, truth, and the power to endure.'

The location of the story in *To the Lighthouse*, the house on the Isle of Skye, was based on Talland House, with many features from the house in St Ives Bay carried into the story, including the gardens leading down to the sea and, of course, Godrevy lighthouse. Later on, at the end of the book, during the actual visit to the lighthouse, Woolf hints at her father's earlier climbing prowess:

'He rose and stood in the bow of the boat, very straight and tall, for all the world, James thought, as if he were saying, "There is no God", and Cam thought, as if he were leaping into space, and they both rose to follow him as he sprang, lightly like a young man, holding his parcel, on to the rock.'

After the First World War, Virginia Woolf revisited Cornwall, including Talland House under its new ownership, and describes Cornwall's early climbers in her diary entry of 30th March 1921:

'I'm not in the mood for writing but feel superstitiously that I should like to read something actually written in Cornwall. By looking over my left shoulder, I see gorse yellow against the Atlantic blue, running up, a little ruffled, to the sky, today hazy blue. And we've been lying on the Gurnard's Head, on beds of samphire among grey rocks with buttons of yellow lichen on them. How can I pick out the scene? You look down onto the semi-transparent water – the waves all scrambled into white round the rocks – gulls swaying on bits of seaweed – rocks now dry, now drenched with white waterfalls pouring down crevices. No one near us but a coastguard

Left: Virginia Woolf with her father Sir Leslie Stephen.
Photo: Bloggingwoolf.org. Creative Commons Public
Domain Mark 1.0.

sitting outside the house. We took a rabbit path round the cliff, and I find myself a little shakier than I used to be. Still, however, maintaining without force to my conscience that this is the loveliest place in the world. It is so lonely.

'Occasionally, a very small field is being ploughed, the men steering the plough round the grey granite rocks. But the hills and the cliffs have been given over as a bad job. There they lie graceful even in spite of all their stones and roughness, long-limbed, stretching out to sea and so subtly tinted; greys, all various with gleams in them; getting transparent at dusk; and soft grass greens; and then one night they burnt the heather at Tregerthen, the smoke rolling up over the crest, and flame shining. This we saw from Ka's. The Eagle's Nest stands up too much of a castle boarding house to be a pleasant object, but considering the winds, firm roots are needed. Endless varieties of nice elderly men to be seen there; come for climbing.'

Tregethern and Eagle's Nest lie close together in the St Ives hinterland, a few miles along the West Penwith coast road towards Zennor. Eagle's Nest was bought by Professor John Westlake In 1873 and, after his death, was sold to the painter, politician and writer Will Arnold-Forster. His wife Ka, formerly Katherine Laird Cox, was once one of the poet Rupert Brooke's 'Neo-Pagans', a term coined by Virginia Woolf. Westlake and his wife Alice held regular house parties at Eagle's Nest, bringing together their extended families including his nephew Arthur who first went there as a young boy of six. Arthur Westlake Andrews later came to be – with good reason – regarded as 'the father of Cornish climbing'. By the 1920s, after serving in France in the First World War, he was living at Tregerthen, in the large, less ornate, granite house immediately to the southwest of Eagle's Nest known as the Old Poor House, and this 'nice elderly' man remained there until he died in 1959.

During the First World War, DH Lawrence lived down the hill at Lower Tregerthen in one of the cottages, known as the Tower House although he named it 'Rananim' which apparently means 'Rejoice' in Hebrew. Alison Symons described the scene in her book *Tremedda Days*:

'Mr Arthur Westlake Andrews, a nephew of the judge, came to live in the Old Poor House. He was an enthusiastic climber and started the climbing club at Bosigran on the high cliffs beyond Zennor. He was also a photographer, being the first person to bring colour slides to Britain from America, and a topographer. He wrote poetry, made marmalade, and played tennis well into his 80s, practising with an enormous racket in a small high-walled court built near his house. Towards the end of his life, his sister, Miss Andrews, joined him. She was a sculptor and drew beautifully. After Mr Andrews's death, she collected an entourage of elderly ladies with ear trumpets and walking sticks and looked after them.'

Born in 1868, Andrews spent his childhood in Wiltshire, where his father was an Anglican curate and a keen amateur geologist. He was educated at Charterhouse near Guildford and Magdalen College, Oxford, where he made his first contacts with the world of mountaineering, having rooms below those of American historian and mountaineer WAB Coolidge (1850–1926). He became a teacher and lecturer in a variety of subjects, especially geography. He was also a noted athlete, representing Oxford in cross-country running and the high jump. Tennis and badminton were also interests of his; between 1889 and 1913, he competed at the Wimbledon championship on seven occasions. But the other great interest of his life was climbing.

His first of many visits to the Alps was in 1890, and the following year he was making first ascents in Norway. In Britain, he made his first visit to North Wales in 1890, and in 1892 he visited Skye. He reconnoitred the climbing possibilities of St David's Head in Pembroke around 1900. Andrews came to know Lliwedd particularly well, and his collaboration with JM Archer-Thomson produced the climbers' guidebook on Lliwedd in 1909, the first Climbers' Club guide. Andrews was elected a member of the Alpine Club in 1899 and joined the Climbers' Club in 1901, serving the latter enthusiastically as editor, president and guidebook writer. Due to him, the Climbers' Club acquired the Count House at Bosigran in 1938 – Andrews's nephew had purchased Bosigran Farm in the mid-30s – and he became the first custodian.

But of all the places he climbed, it was West Penwith that he cherished the most. After climbing elsewhere, he was able to appreciate the magnificence of the wild, rock-bound coastline and would have been aware that references to climbing in Cornwall had already appeared in print. So, in 1902, when he embarked in earnest his exploration of the Cornish coast it was with knowledge of the established attitudes and rules adopted in mountaineering at that time.

He was often solitary, but his sister Elsie, who was about 12 years younger, often accompanied him on the early climbs near Bosigran. Almost immediately, he saw the possibility of horizontal climbing: the sea-level traverse. This was slowly to dictate the course and trends that modern sea cliff climbing took and was to become one of the defining differences, and involving new techniques, between coastal and inland climbing. In those early days when the Alps were the focus, the British hills were merely regarded as a convenient form of Alpine training. Outcrops were dismissed as 'scrambles'.

It is interesting that both Andrews and Tom Longstaff independently opted for long coastal traverses. Possibly the vertical height of the cliffs appeared ridiculously short, whereas a long coastal traverse might offer undertakings approaching Alpine proportions in terms of length and commitment.

In those days of heavily nailed boots Andrews was one of the first to climb in plimsolls or 'rubbers'. Vibram soles appeared in the 1930s and replaced nails during the 1950s before the first rock boots (PAs) arrived in 1956.

Andrews's great-niece, Anne, recalled that: 'The reason he climbed in plimsolls rather than boots was that he planned to traverse all the way round the cliffs from St Ives to Land's End between high- and low-tide marks. Plimsolls dried out quickly when they got wet with seawater, which boots didn't… also they gripped reasonably well on the granite even when wet.'

Andrews actually harboured the ambition, never to be realised, of traversing the whole coastline of Britain in the intertidal zone – the basis of the coasteering traverse as we know it today. If he fell short of this achievement, he nonetheless covered considerable sections of West Penwith. By 1905 he could write: 'For more than 25 years I have examined the district which lies on the north coast of the Land's End Peninsula and I have not nearly exhausted the possibilities.' His interest continued for another 50 years – truly a lifetime devoted to climbing and West Penwith.

One of the cruxes of Andrews's traverse was to find a way across the mouth of the Great Zawn at Bosigran via the Green Cormorant Ledge. Andrews solved this with rope tactics, solo, in 1923. On the first visit he had to build a pyramid of stones on the far side in order to escape. The route was not repeated until the clean ascent in 1956 by the Biven brothers and Trevor Peck at Hard Very Severe.

Andrews was instrumental in introducing George Mallory, Geoffrey Winthrop Young and others including WP Haskett Smith who wrestled with an early coasteering ethical conundrum:

'For most of us to cling to a perpendicular cliff while the raging surf boils 150ft beneath us is quite sufficiently exciting, but Andrews has had to relieve the tameness of it by inventing a novel gamble of his own… His hobby is to pass round the cliffs below high water mark, which means that you are always racing against time and tide. It is a thrilling game, and the worst bunkers in it are the zawns. These are great recesses with sheer, parallel walls and often a considerable depth of water. The accepted practice is to jump in and swim across, but can it be justified to a tender conscience? Can a man honestly claim to have climbed round the cliff if he had done part of the distance by water? If that is legitimate, why not go the whole way by boat? It must be left to the casuists to decide.'

During the Second World War, Andrews served as a special constable. It was during this period that he began to write poetry, eventually publishing *Selected Poems on West Penith and Reflections* in 1957. Andrews wrote the first account of Cornish climbing in the *CCJ*, 1905, and this was followed by more articles. He fully described his traverses in his 1950 CC guidebook to Cornwall. These were glossed over by the authors of subsequent editions with the following explanation:

'These are no longer in fashion so have been omitted.'

Above: US climber Henry Barber crosses to Porthmoina Island by Tyrolean traverse from the foot of the Bosigran Face in June 1976.

Below: Composite image showing the ascent of Bosistow island stack at Carn lês Boel, Land's End, by Pete Biven and Alison Chadwick-Onyszkiewicz on 17th July 1971.
Photos John Cleare.

The Cornish Cliffs by AW Andrews

God gave all men all Earth to love.
But since our hearts are small.
Ordained for each, one place should prove
Beloved over all.

Sussex by Rudyard Kipling

For more than 25 years, I have examined the district which lies on the north coast of the Land's End Peninsula, between St Ives and Cape Cornwall, and have not nearly exhausted its possibilities. The climbing is, in many cases, as hard as anything in the British Isles, and though none of the routes exceed 200ft of actual work, they are so full of variety that actual height seems of small importance. The best climbing is on the granite, which is most prominent at Wicca and Bosigran.

At Bosigran, the headland breaks off in steep cliffs with some pitches of 200ft towering over the Bosigran Pinnacle that rises 150ft or so from the sea below. The rock is absolutely safe if care be taken to test the holding properties of the feldspar crystals, which are possibly the cause of the lines of weakness in the granite and which are especially prominent on all practicable routes. The following statement as to the size that these accommodating crystals attain is taken from a geological work of high reputation; it affords a pleasing confirmation of the climbers' belief that all rocks were created for his use and enjoyment and suggests the real interests of man as scientist who veils his private ambitions and interests beneath the formal structure of an official report: 'Some of the granites contain porphyritic feldspars in great abundance and of large size; near Lamorna Cove they are occasionally six inches in diameter.'

Above: AW Andrews. Photo EC Pyatt, courtesy David Medcalf.

I fear that there are no such giants on the north coast. Crystals built on such a scale as this would almost provide individual climbs and would at least take the place of the iron pegs occasionally driven into Alpine peaks; here, however, though less pretentious, they often supply a long and anxiously felt want. We shall certainly be in complete agreement with the continuation of the same writer's remarks:

'The first impression that the Cornish granites produce is apt to be that of a general monotony, but they present so many problems of their own that that is only transient and soon gives place to an absorbing interest.'

Surely we may conclude that the writer had ventured to leave the somewhat limited foothold of the scientific student for that of the climber, with his many-sided capacity for attaching himself to rocks. As Besant tells us in his charming description of granite scenery in Armorell of Lyonesse, nature has raised no Alpine peaks in Cornwall, but when she addressed herself to the construction of these cliffs, 'She brought to the task a light touch: at the moment she happened to be feeling the great and artistic effects which may be produced by small elevations, especially in those places where the material is granite.'

No one could call the scenery monotonous who has wandered along the coast, climbed down to the margin of the sea, and turned the heads of the caves that lie far below, unseen and unsuspected from the cliff edge. There is a whole world there that the tourist can never reach, far from the excursions, which are all focussed upon two points, the Land's End and the Gurnard's Head, both inferior in beauty to Bosigran and many parts of the coast; a world tenanted by seagulls, who, with a wisdom born of experience, build on the more inaccessible ledges.

Rabbits abound on the less steep slopes, and a fox may often be seen climbing the granite rocks with the skill of an old mountaineer. From the sea below, a seal occasionally inspects the scene in the intervals of fishing and visiting his friends in neighbouring caves. Colour is by no means absent; when the gorse and heather and thrift are in full bloom, nothing can equal the gold and purple brilliance of the hillside. Below all lies the sea, ever moving even on the stillest day, for behind it is the force of the open ocean, and fretting with its line of foam the half-hidden reefs that edge the shore, and in a storm driving its spray halfway up the cliff. Fishing is almost non-existent on the coast, as it is only possible to launch a boat at a few

Facing page: Alan Rouse soloing at Bosigran. Photo John Cleare.

places, such as Porth Zennor and near the Gurnard's Head, and even there, there is, I think, now no boat. Even the coastguards omit their nightly walks over most of the coast as smuggling is obviously impossible.

Even in summer the north coast is never relaxing, unlike Penzance, as it's exposed to westerly and northwesterly winds from the open Atlantic. The ocean winds bring rain, and the hills are sometimes shrouded in mist, but there is also a large amount of sunshine. I have often found that the sun was shining brightly on the cliffs when a thick curtain of mist was hanging over the edge.

There are no thousand-feet precipices in West Penwith, or even hundred-feet pitches, and in most cases, it is almost as easy to escape from one climb to another as it is on the side of Tryfan. But there are compensations. The sea is always with us, and though not so permanently blue as on the Riviera placards, is often still more intense in colour. Besides its artistic value it has a psychological influence on climbing, and the difference between a traverse above an unruffled mill pond and a foaming surf is not easily forgotten.

To the author, certain climbs under the less favourable of these conditions call back memories of struggles and discomforts quite as vivid as those of occasional Alpine peaks in bad weather or of Lliwedd in a cold and wet Easter. These adventures were not sought but came unheralded, like the seventh wave that rolls out of a relatively calm sea and sweeps the cliff for 20ft or more above the normal level, just as it does on a larger scale at Ascension Island. On one occasion, the leader received a shock on finding a past president of the Climbers' Club on a ledge well above him when he expected to be hauling him in from the depths. The rest of the party finally emerged from the abyss. For this reason alone, a rope is desirable.

The fact that certain parts of the coast can only be traversed between the high- and low-tide mark, at low tide, or even a low spring tide, and that the margin for reaching a point where the cliff can be reasonably ascended is often extremely small, imposes a time limit on the operations. The actual steps are not, except in a few cases, of great difficulty, but they have to be taken with a certain freedom and speed, 'touch and pass', over rocks which are liable to be wet and slippery. Here you really do find 'a wetter water, slimier slime'. But the appeal of climbing on the Cornish Coast is not, on the whole, one of severity. There is good scrambling in delightful surroundings and a few pitches which reach the Severe standard of modern rock climbing.

Variety, however, does exist. Granite has its faces, ridges and chimneys. The faces are seamed with ledges and cracks, broad and deep or minute, but the general roughness, due to the disintegration of the feldspar crystals, allows great variation of route as finger holds can often effect a hand traverse at seemingly impossible places. The ridges are broken into pinnacles, and the elevation is sometimes almost vertical. The chimneys can be deep or shallow and afford practice in different styles of workmanship.

Serpentine by AW Andrews

There is a pinnacle that few could scale.
Though he had twenty hands and twenty feet
That stands upon the shore in Kynance Vale.
On a long neck of sand where the waves meet:

And you can reach the bottom at low tide
And look at it, and start upon a climb,
And then come off with haste, because the side
Is smeared with sliminess beyond all slime,

On which a limpet well might make a slip.
And this upon a face of serpentine.
Upon whose polished wall no pointed tip
Of claw can hold, and certainly not mine;

So I have called it in my category,
(A Mild Severe) and left it, to my shame.
And hope some ace will come, and climb, and try
The first ascent, and win immortal fame.

But please remember if you make a fuss.
That we are not the men our Fathers were.
And those that follow will be less than us.
And though you may not call this climb severe.

Homer had put it in a higher class.
As in his story Scylla waited there.
To snatch a hapless climber who might pass
And take him off his balance from the rear.

And though you find no cave in which to rest,
I dare say Scylla's absence will be best.

Sunset on Zennor Hill by AW Andrews

Below a buzzard hovers in the air.
With wide-spread wings, so still they seem to be
Cut out of stone, and waits and watches where
His prey hides from him, but I watch the sea.

Lacking the hawk's keen glance, with eyesight poor.
And strive to pierce the impenetrable veil
That dims my view beyond the darkening moor.
And as the light fails, so I also fail.

While sinking 'neath its load the winter sun,
A tired traveller, with shortening stage
Relaxing strain, at last, his respite won.
Opens the prison door of night's dark cage;

When, lo! beneath the ball of dying fire
Shows a faint speck before my wondering eyes,
Not there, that's Scilly! On the right, look higher!
The Islands of the blest, where Lyonesse lies.

Facing page: Stacks off Gunver Head, Padstow. Johnny Fowler returns to the mainland by Tyrolean traverse from Middle Merope Island on 4th Sep 1971. Photo John Cleare.

It is a magical place, but the magic is black... the north Atlantic coast... with its desolate moors strewn with Druidic monuments and fallen cromlechs and ancient abandoned tin mines going back to the times of the Phoenicians, seems to belong to an entirely different world. In fact, I am sure that it does, and that it represents a corner of the lost continent of Atlantis.

Altogether it is like entering the kind of country described by Algernon Blackwood or Arthur Machen – a land in which the boundary between the subjective and the objective becomes vague and indecisive. You begin to distrust the evidence of your senses, and to realise uneasily that things are not always what they seem to be, and this feeling becomes steadily intensified as you leave St Ives and – passing Tregerthen Farm and the cottage in which DH Lawrence lived during the years I knew him – approach the village of Zennor, where the innermost periphery of this spiritual Black Country begins.

From there to Gurnard's Head and Bosigran Castle (where I lived) the crescendo continues and reaches its climax beyond the village of Morvah: a name of which the dark, sinister sound still strikes a chill into my very marrow.

Musical Chairs by Cecil Gray

Black Magic

Cornwall is very primaeval, great black jutting cliffs and rocks, like the original darkness, and a pale sea breaking in, like dawn. It is like the beginning of the world, wonderful.

DH Lawrence

Cornwall is a fertile landscape for folklore and mythology. Enduring stone remnants of ancient rites and customs are a subliminal part of the landscape and the psyche. Ancient attempts to fashion portals to other worlds and communicate with the gods abound in the form of stone circles and standing stones. There are many Cornish tales of a bizarre, occult nature about Aleister Crowley including that he and his, mainly female, followers took powerful narcotics, danced naked in the stone circle at Tregaseal, held orgies on the moor, conducted pagan rites at ancient sites including reviving the druidic practice of human sacrifice, performed a black mass in Zennor's church and that he summoned up the Devil himself in Carn Cottage.

On the hillside above Eagle's Nest near Zennor Carn lies Carn Cottage which – to this day – has a sinister reputation amongst the locals. The notoriety of the 'Aleister Crowley house' most likely stems from the evening of 21st May 1938 as the *West Briton* reported at the time:

'Mrs Katherine Laird Arnold-Forster, aged 51, died early on Monday morning at Zennor, in a cottage near her residence, the Eagle's Nest. She was taken ill there on Saturday when she was visiting a sick friend. She was suddenly taken with a seizure, and she never regained consciousness. Her death came as a terrible blow to her many friends because, except for a slight cold during the last few days, she seemed to be quite well.'

Katherine Arnold-Forster, usually known simply as Ka, lived at Eagle's Nest below Carn Cottage. At the time, according to the Penwith Local History Group, Carn Cottage was being rented out by Gerald and Ellaline Vaughan. On that evening, Ka was asked to go up to the cottage because, depending on what account you read, either Mrs Vaughan was very ill, the Vaughans were in a state of panic because the cottage was haunted, or the Vaughans were concerned about a satanic ritual taking place in the cottage involving Aleister Crowley. Whatever happened that evening, by the end of it Ka was dead. According to the information from various sources, it appears that Ellaline Vaughan was mentally ill, and I would speculate that her psychotic symptoms incorporated Aleister Crowley who at the time was infamous as 'the wickedest man alive'.

The satanic rumours from that night at Carn Cottage were later fictionalised in Frank Baker's *Talk of the Devil* and AL Rowse's *Night at the Carn* before being written as fact in Denys Val Baker's *View from Land's End* and then thoroughly investigated in Paul Newman's *The Tregerthen Horror*. However, such publications seem to have only fuelled wilder sensationalist rumours about the cottage including that Crowley was using powerful dimensional gateways around Zennor to summon and communicate with an ancient reptilian extra-terrestrial race from the constellation of Draconis.

At some point, Gerald Vaughan had become fascinated with DH Lawrence, and knowing that Lawrence had lived for a while near Zennor, he moved to the area in late 1936 or early 1937. He rented Carn Cottage, and it was while he was living there that he married Ellaline Agnes Norwood. Gerald and Ellaline had become friendly with Katharine Arnold-Forster, and Gerald invited Ka to dine with them on that fateful evening in order to help ascertain the extent of Ellaline's hallucinations – which may have incorporated Crowley.

Ka fell asleep after the meal, and Gerald let her sleep. At some point, he realised that she was not sleeping but had died. The local doctor was called, and he concluded that she had died of a cerebral haemorrhage. Another source suggests that she has also had a history of heart problems and died of heart failure.

The Vaughans moved out of Carn Cottage shortly after, and Ellaline's mental health continued to deteriorate until she was eventually committed to St Lawrence's Hospital in Bodmin. Carn Cottage was left abandoned until after the Second World War, until it was bought and renovated by the artist Bryan Wynter who was among the influential St Ives group of British painters, along with Patrick Heron who moved into Ka Cox's former home, Eagle's Nest, in 1956.

In reality, the only verified visit of Aleister Crowley to Cornwall occurred from the fourth to 14th August 1938. He was in Cornwall, apparently, to visit his young son at the invitation of his son's mother. In 1934, in the course of the celebrated 'Laughing Torso' libel case, in which Crowley sued the artist Nina Hamnett for calling him a black magician in her 1932 book of that name, the 59-year-old Crowley was introduced to a 19-year-old from Newlyn, Cornwall, named Deirdre Patricia Doherty. Legend has it that when the libel suit ended and Crowley exited the courthouse, she ran up to him and said, weeping, 'This verdict is the wickedest thing since the crucifixion. Is there anything I can do to help?'

Three years later, on 2 May 1937 in Newcastle, she gave birth to the boy Crowley considered his son and heir, Randall Gair – also known as Aleister Ataturk. Newlyn was his family home, and he lived there for several years.

Aleister MacAlpine: Ataturk Crowley: Randall Gair: Count Charles Edward D'Arquires (1937–2002) by Des Hannigan

When I first met Aleister, he was living with his mother, Deirdre, at the house called Wheal Betsy at the top of Chywoone Hill. Aleister lived in a caravan on the grounds. Wheal Betsy was an Arts & Craft house that was built in 1910 by Deirdre's grandfather, the Pre-Raphaelite painter Thomas Cooper Gotch. Next to Wheal Betsy was the building site for a number of bungalows that were the antithesis of Gotch's house. The site was known as the 'Pink Estate' because of the hideous colours of the rendered walls. Throughout one hungry winter in the late 1960s, I worked as a labourer there along with the artist Russ Hedges. Aleister Ataturk turned up as a fellow labourer one day. The developers decided to suspend work over the winter and laid off most of the workforce. Since Aleister lived next door and was considered the most reliable (he referred to Russ and me as 'The Ruffians'), they appointed him as 'winter watchman' and gang boss of our team of three.

We were left to look after the site, and we worked hard in often lousy weather. A contract team of asphalters had breezed in to lay out the estate roads. They were long gone before we found that they had 'accidentally' buried the unconnected ends of the service pipes to the houses. Our job was to dig cross ditches until we found the buried ends.

Aleister was very conscientious. He was a kind and likeable man beneath his rather formidable exterior – tall and powerfully built and with a great Crowley head and cropped black hair; he usually wore dark glasses when out socialising and dressed eccentrically; jodhpurs and riding boots often featured. Aleister had great physical presence, but if you recognised and understood his vulnerability he was unthreatening. His size and his demeanour, plus a touch of showmanship, belied this of course.

I took him into the old Wimpy Bar in Penzance's Market Jew Street one Saturday morning, and I swear the place emptied in seconds. There was definitely something of the natural performance artist about Aleister. He was a touch prim and righteous, although he had a sense of humour and, surprisingly, a sense of irony. He nagged Russ and me, alleging that all we were interested in was 'drink and women'. (Outrageous suggestion, Aleister!) He also despaired of our alleged 'left-wing' views. We teased him relentlessly about the coming Revolution. He had a well-developed sense of privilege. Ultimately, Aleister made it very clear that he considered us to be his intellectual inferiors. We didn't mind a bit. He was certainly a fairly harmless fantasist.

Stories trickled out concerning his achievements. He told us that he was an accomplished pianist and had been the star of several performances in Russian concert halls. For some unexplained reason, he claimed that he was not 'permitted' to perform in England; a fairly watertight insurance against our good-natured scepticism. We were invited round for tea to Wheal Betsy once or twice. Inside, the house seemed pleasingly chaotic, and it contained a lot of valuable paintings and heirlooms. Aleister himself had an impressive collection of swords and rapiers. He served tea very formally – dainty china cups and teapot. Once, he played the piano for us – rather badly, I'm afraid.

I knocked Aleister headfirst into a muddy ditch, by accident, while driving our brakeless dumper truck. He rose with the wrath of Zeus upon his great shaven skull, and for a brief moment I braced myself for a paternal thunderbolt at the very least. But, actually, he was something of a peacekeeper – particularly at Wheal Betsy.

Every now and then, all hell broke loose as Deidre's enormous and varied flock – adopted children and several remarkable daughters, all with long black hair with dyed white streaks – broke into eldritch shrieks and squabbles. Aleister would look increasingly pained as the racket increased and would then trudge off wearily to the house. A few seconds passed, and then the screeching would stop abruptly. Aleister had arrived!

Above: Ataturk and the elder Crowley on a beach in Cornwall.

Below: Clearly a chip off the old block. Ataturk in his 'Supreme Council' regalia in 1980.

Photos: Cornwalllive.com. Creative Commons Public Domain Mark 1.0.

He would come back shaking his head and would apologise to us. Russ and I thought it was great fun. I lost touch with Aleister soon after and heard only vague rumours and reports. A year or so later, I was at a party at Doug Cook's house on the moors above the village of Madron. It was packed with a crowd of art students and poseurs, all pretension and face paint, a fad that had just come in. They were dancing around like dingbats when, suddenly, the crowd parted, the dingbats all fell back in awe and adoration, and there – like a vision from central casting's villains' wardrobe – was Aleister, in riding boots, jodhpurs, white polo neck, Fair Isle sweater, dark glasses and an SS officer's cap, plus riding crop in hand; referencing Indiana Jones years before its time.

The crowd worship was potent. Aleister preened outrageously. Then he caught sight of me just as I said, with good-natured scorn: 'Aleister! What the **** are you playing at!' The crowd shrank back in horror. Aleister stared for a few seconds and then said, 'Oh bugger! It's you, Hannigan!' He then turned tail and disappeared rapidly back into the crowd with his acolytes wailing in his wake and giving me furious, hate-filled looks.

It was several years before I saw Aleister again. He was living in the old Madron Workhouse, a remarkable granite building with some grand features, the nearest that Aleister could get to full Scottish Baronial, I expect. Much of the building's finest stonework had been pillaged from the Iron Age site of Chun Castle above Morvah. Its provenance was just right for Aleister – misty Iron Age primitif from Heath Stubbs archetypal 'hideous and wicked country' of Penwith's high moorland. (When the Madron workhouse was eventually demolished, I managed to get hold of a finely cut lintel to incorporate into an extension on my house back in Morvah. I have always hoped it was part of Chun Castle returned home).

When he lived at Madron, Aleister had a family (including two young children) but had declined, sadly, into very delusional ways. He styled himself Count Charles Edward D'Arquires, a title bestowed on him by his father at an early age. He had set up a 'Supreme Council of Great Britain' with himself as the 'Adjudicator'. His Acting Private Secretary, Peter Bishop, believed that we should all sit at the bottom of mine shafts and be transformed into super beings when a shaft of sunlight struck us. This was nothing new for Cornwall. During his stay at Zennor, DH Lawrence had friends who were said to lower themselves down a nearby mine shaft where they sat naked in an underground stream, unquestionably glowing in irradiated ecstasy.

I was walking down Madron Hill one day when Aleister stopped in a fairly smart car. I had been fishing out of Newlyn for a number of years by then. I was delighted to see him although he had become very pompous and even more otherworldly. He gave me a lift into town and right there and then offered me the job of Fisheries Minister in his 'Government': 'You would be ideal, Hannigan,' he said, 'not just because of your fishing background but because you have the right appearance. Blond hair and blue eyes...'

I declined the exalted position – with huge reluctance, of course. Aleister was very serious about taking over the governance of the UK – by persuasion. Eventually, in 1976, Aleister hired a posh limousine, complete with Supreme Council pennants and, with Bishop, was chauffeur-driven to London. They went in their finery: Aleister in his dress uniform jacket with gold trimmings, epaulettes and velvet cape. They tried to get into Downing Street for an audience with Prime Minister Harold Wilson in order to persuade him to join the Supreme Council. The message was delivered to Wilson who, unsportingly, declined the offer.

I never saw Aleister again. Sadly, life treated him badly in his later years. He had mental health problems for most of his life, and there are harrowing reports of his later decline. He died in a car crash in 2002.

———————————————————

There is a black-and-white photograph of Crowley with his young son on a beach near Newlyn. The boy looks slightly bemused, perhaps at the identity of his overweight, elderly companion. A black dog stands behind Crowley looking away from the camera suggesting an animal familiar into which he had deposited his evil side for that day. Crowley wrote about this visit to Cornwall in his diary: 'Sun 5th August 1938. A perfectly glorious day... JBJ joined us out to Morvah, after photographing at Paul. Rock climbing again. Wooed Greta on cliffs. She is a comedian; will one day come and snatch. JBJ to dinner, talked Qabalah and Pythagoras. PM at The Dolphin, Newlyn. Early to bed.'

'JBJ' is John Bland Jameson who was an actor and pupil of Crowley's who was in rehearsals for *A Midsummer Night's Dream* at the time. 'Greta' was the beautiful 31-year-old socialite Greta Mary Sequeira Valentine who was familiar with Crowley from the noisy London Café Royal set of bohemians where he was known as 'the Magus'. One source has it that Crowley followed her to Cornwall, but she had already formed an attachment to Lamorna Birch, the landscape painter. Crowley and Greta's 30-year age gap apparently created quite a stir at the Lobster Pot restaurant in Mousehole where they sat sharing a bottle of Chablis at a table by the window. Greta with her 'mesmerising' looks and Crowley heavy set and balding; the pair were, literally, the beauty and the beast.

Crowley's extensive and meticulous diaries extend from the time of his initiation into The Hermetic Order of the Golden Dawn in 1898 through to his death in 1947. His diaries record in detail his acts of 'magick', sexual or otherwise, and supposed supernatural occurrences, but he documented none such during his time in Cornwall. However, he did record that he went rock climbing. We don't know exactly where or what he climbed although, presumably, this was without supernatural assistance. And, looking at his portly figure in that photograph, it was unlikely to have been anything particularly difficult. Then again, that didn't seem to stop John Dunne.

West Penwith Coastal Traverse by AW Andrews

On traverse, in the evening, on the coast
Of Cornwall, with a cliff upon your right
You cannot scale, and on your left
Nothing but sea, America and night.

AW Andrews

British mountain rock climbing expanded rapidly from a practice for Alpine mountaineering into a separate sport having its own traditions and literature. When climbers came to explore the possibilities of our sea cliffs, therefore, they found that the granite of Cornwall possessed all the desirable attributes of a good climbing rock – good holds, rough surface, and the requisite shattering which provides cracks, chimneys and ledges and therefore climbing routes up natural lines of weakness separated laterally by areas of unclimbable rock.

The old chalk climbers in their literature did not fail to draw attention to the attractive situation of their climbing, and this, perhaps, has been one of the major assets of sea cliff climbing. The sea provides a more interesting background than the expanse of moor, stony hillside or lake, which confronts most mountain crags. It has often been written that small-scale outcrop or practice climbing should never become an end in itself but should always be regarded as part of a larger whole. But, just as British rock climbing became finally independent from Alpine mountaineering, so could any aspect of outcrop climbing grow into a sport independent from British rock climbing and from mountaineering if it had sufficient potential for development on separate lines.

In most cases it is difficult to appreciate that suitable conditions exist, for most outcrops are situated in hilly country with surroundings very similar to those of mountain crags, but a notable exception is provided by the climbing described here. The climbing is entirely dissociated from mountains so that it can in no sense be called mountaineering. The sea forms unique climbing surroundings, and the weather is good. There are no long walks to crags, and there is no necessity to be miserable in order to feel that the sport is being suitably indulged. We are surely justified in the speculation that cliff climbing may one day come to be looked upon as a separate sport. This is important because it is wrong to come to Cornwall solely to practice for climbing on bigger crags elsewhere. This you can do, of course, but in addition, Cornwall has features of its own which can make it enjoyable in a far wider sense to a mountaineer.

The granite is a particularly good climbing rock – one of the best in the world. In other respects, an entirely new outlook has to be developed. The sea, sometimes vertically beneath and within jumping distance of very hard moves, may be a comfort to the climber who can also swim, even though quite a considerable swim might be needed to reach a suitable landing place. For the non-swimmer, or where swimming is impracticable, it forms an additional emphatic reason for not falling off even in quite low places.

The question of boots versus rubbers has not yet been settled. The roughness of the rock makes either equally suitable, and under the circumstances, it is probable that mountaineers will choose boots at first, particularly as they are also well suited to the general terrain of the cliffs – steep grass and large areas of prickly undergrowth at the top, seaweed at the bottom. However, the climber who feels that this is more than a mere practice ground will in time prefer rubbers which make rock climbing more pleasant and are more in keeping with the local climbing garb, which varies from nothing to as little as possible. The rubbered climber must treat with respect the slimy rocks at sea level and is perhaps at a disadvantage in the prickly undergrowth. Elsewhere he can go as he pleases. The possibilities of the bare foot, too, should not be overlooked in some of the super-slimy places.

Facing page: Rich Pollard on Astral Stroll, E1 5b, Carn Gloose, West Penwith. Photo Mike Hutton.

Climbing between high- and low-water marks offers its own problems. At some parts of the coast, long traverses between these limits have been worked out, including places only passable very near low tide or even at special periods in the year when the tides are particularly low. The zawns are a serious complication. The problem arises of the passage of seaweed and slime-covered rocks for which no precise technique can be recommended. This is a subject not treated in the standard technical works and will have to receive special attention when a book on technique here comes to be written. The traversing of some passages submerged in inches of water or of others rapidly between successive waves has also to be negotiated. Climbers should remember that periodically there is a wave larger than the rest, usually just at the time selected for one of the moves in question.

By local rules, the climber is considered to be climbing as long as his nose is above water, but hereabouts ability to swim becomes more important than ability to climb. On such traverses, slowness or special difficulty may mean the loss of the race-against-time between successive easy ways down the cliff, and it then becomes necessary either to swim off or to climb straight up. The latter course can lead sometimes to exceedingly awkward terrain, where the good granite at the cliff base is crowned by almost vertical grass slopes.

Traverses, though some of the earliest routes made here, form only a small part of the climbing as a whole. The most interesting climbing between Bosigran and Pendeen consists of traverses. On all the big well-weathered granite crags on which normal bottom-to-top routes have been made, the rock is above reproach. And the sea is always there to be enjoyed in the myriad ways it can be enjoyed – the permanent background to climbing in Cornwall.

NORTH COAST

Between St Ives Head and Cape Cornwall lie some 12 miles of coastline facing approximately northwest. This stretch is usually referred to as the 'North Coast'. The cliffs are principally granite though St Ives Head and Gurnard's Head are greenstone, and some of killas also occur. Immediately above the cliffs is a platform which rises again to a range of hills – Rosewall Hill, Trendrine Hill, Zennor Hill, Carn Galver, Watchcroft, Chûn Castle, Carn Kenidjack, etc – in places 800ft high and many of them capped by tors. There are numerous pleasant coves, often with small sandy beaches, of which Treveal, Ver, Gurnard's Head East and Portheras are the best known. The zawns – steep-sided clefts, vertical or overhanging at the landward end and formed by the removal of relatively soft rock in the joints by sea action – are an outstanding feature and, as will be seen, create special problems in the climbing.

As a long continuous traverse is often impossible, very good climbing can be found by descending at the most interesting places and traversing as far as possible in either direction. The description begins at Treveal, three miles west of St Ives. The rock is mostly killas until Wicca Pool is reached where the granite first appears at the popular Wicca Pillar. The traverse can be continued by small ledges to Pendour Cove, and then easy shore leads to Carnelloe Headland. Deep rifts of black rock lead to the cliff opposite Carnelloe Island where there is a remarkable blowhole.

Below Long Carn is a cauldron with a traverse on a slab just above the water. An amusing place with only friction holds. This is reached by a face on the south or a short buttress on the north. Good scrambling leads on to Porthmeor Cove. The granite comes in here and continues to Pendeen. A series of ledges, some of which are exposed only at a dead low tide, afford a straight and narrow way. There is often no alternate route or escape up the cliffs if the sea does not allow the passage. Consequently, a party trained in 'touch and pass' has a distinct advantage.

From Pendeen to Cape Cornwall is metamorphosed rock, some of the cliffs exceeding 200ft. They are not very suitable for climbing as deep-cut zawns make traversing impossible near sea level, while higher up, the debris of old mining operations makes it unpleasant. There are isolated patches of excellent rock but not continuous enough to make it worthwhile to describe climbs.

WEST COAST

Between Cape Cornwall and Land's End, the coast runs southwards for five miles; this is called the 'West Coast'. The cliffs are not so prominent, and at Whitesand Bay there is a considerable stretch of sand. For some two miles south of Cape Cornwall – almost to Aire Point – the cliffs are high and similar in quality and climbing opportunity to those between Cape Cornwall and Pendeen. From Aire Point to Sennen Cove is a wide sweep of sandy bay with low cliffs sloping back at an easy angle above. The village of Sennen Cove faces north across this bay, sheltered from the west by the headland of Pedn Mên Du, where high granite cliffs start again.

SOUTH COAST

From Land's End, the coast swings round almost in a semi-circle for 14 miles to Penzance. In the first stretch to Treryn Dinas are the finest and highest cliffs, practically entirely granite with a succession of small coves and zawns similar to the north coast. After this, the cliffs are lower and slope at an easier angle, though still granite as far as Mousehole. This stretch is called the 'South Coast'. The country is cut up by shallow river valleys and rises gradually to the north. The majority is cultivated, but coast hills, the cliff tops and the hills form wild moors in which antiquarian remains are plentiful.

Menlove

Waves, they are woven in a burn
To fit a rock obstruction.
They may be planted in cornfields
Or fields of man's destruction:
And while they thirst or roll or burst
Still grindingly they turn.

Waves by Menlove Edwards

Above: John Menlove Edwards by CWF Noyce. Photo courtesy of the Climbers' Club.

Facing page: Still images from 'Rough Weather Landing', the classified-at-the-time film recording of Operation Brandyball. Images courtesy 131 Commando Squadron Royal Engineers.

John Menlove Edwards (1910–1958) was known especially for his first ascents of many now-classic routes in Snowdonia and as a guidebook author. He started climbing in the late 1920s with his brother, Hewlett, in the Lake District, but prior to that many childhood family holidays had focused on adventures at the coast such as to the Isle of Man where Menlove recalled:

'Many exciting bathes in tempestuous seas, when we would come out all covered in cuts from being thrown up and down on the rocks... There were many exciting caves in the cliffs, through which you could take a boat. We calculated that we could swim through the caves, leaving our boat to float on the tide, and catch it when we emerged on the other side – and we did! I still remember the swell in the caves, being right up against the walls one moment and right down in green, lighted water the next.'

The brothers also climbed on the sea cliffs near Tenby during a family holiday around this time. These childhood experiences set the scene for his later climbing and aquatic exploits. Menlove first climbed in North Wales in 1930 and rapidly became one of the leading figures in Snowdonia. He graduated in medicine from Liverpool University in 1933 where he had founded the Liverpool University Rock Climbing Club with his brother.

As well as climbing, Menlove excelled at aquatic activities, and his remarkable feats of rowing, sailing, and swimming are legendary. In the Easter of 1931 he swam through the narrow gorge of the Linn of the Dee in full spate. In January 1936, he rowed across the Minch to the Isle of Harris. The outward journey took him 28 hours before he reached land at a deserted cove. Three days later he rowed back in 24 hours. He explored the Cornish coast during the war, and this brings us to one of the most famous, or infamous, tales about Menlove that goes to the heart of the subject matter of this book. This story has often been repeated, and the earliest source I could find for it was in *Samson*, the collection of his posthumous writings published in 1960:

'He did little climbing during these two years but visited Skye and traversed the main ridge of the Cuillin on his own. He also spent a holiday in Cornwall, where he made several climbs, some of them started from the sea. He developed a technique for taking advantage of big waves to wash him high on the rock without hurting himself. However, three soldiers who watched him at it and later tried to emulate him were drowned, and the technique must really have been a matter of applying his great strength and his swimming and climbing skill. But the old art of relaxing and enjoying himself simply on regular climbing weekends had by now left him, and it is significant that his madness really starts from the time he lost this means of sublimation of his ambitions and vital energies.'

I could not find an earlier source for this dramatic story which must have occurred in 1942 or 1943, so I turned to the records of the deaths of servicemen. One such, the online commando archive, lists all those who died during service. The Second World War Roll of Honour makes for harrowing reading. Not many died of 'natural causes', with most 'killed in action' or 'died from wounds' in their 20s in various campaigns ranging from Norway to Burma including, notably, Operation Chariot at St Nazaire, Operation Jubilee at Dieppe and D-Day. Many were missing and presumed dead after HMS Fidelity was torpedoed and sunk in 1943. Some died in prisoner-of-war camps such as the seven men of Number 2 Commando who were captured after Operation Musketoon in German-occupied Norway, and later executed, under Hitler's infamous Commando Execution Order, at Sachsenhausen Concentration Camp in 1942.

Most died overseas and are buried there, but many died in accidents in the UK, either during training or otherwise. Some of these were drownings, but they usually occurred in rivers and lochs in Scotland during training exercises. There were some with no cause or location of death given other than 'died in the UK'. None of these cross-referenced with each other to suggest multiple fatalities. One candidate for the Menlove story is Private David Henry Hill who died during training after 'a fall from cliffs near Camborne' on 4th December 1942. However, the closest-sounding match comes from 1943 during the most hazardous and challenging training exercise that was ever attempted by Number 4 Commando in Cornwall and codenamed 'Brandyball'.

Operation Brandyball was a mock seaborne raid on an inland objective that was protected by the 300ft cliffs known to the commandos as Brandys after the nearby island of that name. The landing on the cliffs was to be followed by a climb with full kit up the cliffs to seize the target. It was the brainchild of Lieutenant-Colonel Robert Dawson and was filmed and witnessed by senior army top brass including General Bernard Montgomery.

The demonstration was planned for 7th June 1943 with a final rehearsal the day before. Unfortunately, the weather on the sixth was not good with rough seas. C Troop's boat capsized at the base of the cliffs depositing the commandos into the sea. Some managed to scramble onto the cliff, but others were not so lucky. Lance Sergeant John Albert Chitty (21 years) and Lance Corporal Donald Samuel Hoodless (22 years) both drowned.

This incident was witnessed by the mountaineer and author Frank Smythe.

Mountain Warfare Training by Frank Smythe

I would like to describe an extraordinary incident and a very brave action. It began with the philosophy expounded by Mr AW Bridge. He is not only a skilful climber but an adept in the art of falling. I do not mean that he ever falls off accidentally, though I have seen him do so deliberately from a mere 25ft or so. Like all of us, he abhors falling off, but he maintains that if a fall does occur then the climber should use his best endeavours to save himself the worst consequences. If possible, he should jump for a ledge, even though it be 100 and more feet lower, rather than allow himself to strike some projection and be turned upside down. Mr Bridge was assiduous at teaching this to his pupils, and one of them undoubtedly owes his life to it.

The first cadre trained in North Wales came from Number 4 Commando. Some time after the course ended, it was detailed to give a demonstration in assault landing on the cliffs some 20 miles west of St Ives to various officers from the War Office and Admiralty. The landings were to be made direct from the sea onto the base of the cliffs in light folding assault craft, a most hazardous business.

I travelled to Cornwall to witness the exercise. There was time for only one rehearsal. The boats put off and approached the cliffs. The tide was at flood, and suddenly, a biggish swell began to manifest itself. I, who was near the water, soon realised that a fragile boat could not survive in the surf at the base of the cliffs, and I hastily scrambled up to warn the OC of Number 4 Commando. He at once signalled to the boats to stand off. All did so except one boat under the command of Lieut Haig-Thomas, which did not observe the signal. This endeavoured to land its crew on a ledge at the base of a promontory, but directly it touched the rocks it was overturned and its occupants, all of them wearing assault equipment, were thrown into the surf.

One man was dragged down by the anchor and drowned immediately, and another presently succumbed to the weight of his equipment. A few were hauled to safety on ropes by some men who had been detailed for just such an emergency, but the remainder with the upturned boat were washed into a narrow rift in which the surf raged furiously. Meanwhile, I raced along the top of the cliffs, followed by Corporal Meddings and the OC, since the best chance of succour lay in a descent of the promontory to the far side of the rift.

In one place we had to cross some steep grass, and here Corporal Meddings slipped. He was unable to stop himself and slid down and out over the edge of the cliff, which at this point was fully 100ft high, straight into the narrow jaws of the rift. Yet he still had sufficient control over himself when he reached the edge not only to remember Mr Bridge's advice but to act on it: he jumped. He jumped clear of all projections and disappeared from sight into the rift. I saw him for an instant in mid-air, then above the beat of the surf there was a horrible sound far beneath.

Ropes were procured, and I was lowered down to the ledge where I found not only Haig-Thomas and the remainder of his crew but, to my astonishment, Corporal Meddings who had no worse injury than a severely dislocated shoulder. As he was in great pain, a Neil Robertson stretcher was sent for, and he was hauled up the cliff. The others followed one by one, but it was not until long after dark that the rescue was completed.

It was then that I heard the story of Corporal Meddings's extraordinary escape and his bravery following it.[1] He fell into the rift at a moment when it was deep in water and missed the upturned boat and the struggling crew. Finding himself still alive, although with a dislocated shoulder, he at once seized hold of a drowning man and, after a desperate struggle, managed to fight his way to the ledge. Because of his outstanding bravery, he was recommended for a decoration but, I understand, did not get one since the gentlemen who decide these matters held that his action arose out of fortuitous circumstances.

Above: Frank Smythe from *The Modern Review,* 1937. Public Domain.

Francis Sydney Smythe (1900–1949) was an English mountaineer, author, photographer and botanist. He is known for mountaineering in the Alps and the Himalayas. He wrote a total of 27 books, mostly about mountains and climbing, and gave many public lectures, including to the Royal Geographical Society.

In 1930 Smythe was a member of an international expedition to attempt Kangchenjunga. In 1931 he was the leader of the first successful expedition to climb Kamet (7,756m) the highest peak climbed at the time. On this expedition he identified a region that he named the 'Valley of Flowers', now a protected park in the state of Uttarakhand, India. He was also a member of three Everest expeditions during the 1930s.

During the Second World War he served in the Canadian Rockies as a mountaineer training officer for the Lovat Scouts and led the Commando Mountain and Snow Warfare Centre in Braemar, Scotland.

Mount Smythe (10,650 ft) in Jasper National Park, Alberta, Canada was named in his honour.

---

1   Corporal Sidney Meddings was wounded during operations in Western Europe (Flushing Holland) but survived the war.

The Art of Falling by Alf Bridge

It does not seem a long time since I eagerly devoured the contents of borrowed journals. I wanted to understand the technique of walking, the technique of climbing and the doctrine of descent. The men who wrote in the journals of these subjects satisfied me. But I was very puzzled and yet had not the pluck at that time to ask why there was not an article on what seemed, to me, to be the most important technique of all: the technique of falling. I have waited for a long time for a more worthy man to take up his pen and write a thesis on this subject, but it is evidently thought a rather dangerous subject for a climber to discourse on or put down his views in cold print before the outside world.

However, I am confident that if climbers will only give the art of falling their serious consideration it may be of great help to them at some time, and so I am prepared to risk the criticisms which are sure to come my way and write the article in the hope that it will stimulate interest. No one, least of all myself, makes a habit of getting into such positions that they can neither get up nor down, and therefore they eventually have either to jump off or fall off, but everyone who climbs runs the risk of at some time or other finding themself in that predicament. I have had quite a number of falls. It does not really matter where I had them, or how. Sufficient it is to know that if I had not trained myself in the technique, I would have been a most unfortunate man. As it is, I have managed to get away fairly lightly.

A friend of mine, who is also a member of my club, told me that as a lad, he had a great fear of fire, so much so that he set about preparing an escape for himself and chose as an exit the bedroom window. He practised the art of falling at an early age. For before he could jump from the bedroom window, he had practised falls from the roof of the coal shed and then a little higher from the garden wall. Eventually he could jump from the window and make a happy landing. And so it must be for all of us who wish to be proficient in the art of falling. At first I was very frightened of my rehearsals, but I persevered. A height of only six feet was good enough for a start. Gradually, as I gained confidence, I took bigger falls, and at present I can drop myself 25ft, and when in form, I can manage about a 30ft drop.

Most of my practice falls were off gritstone climbs in Derbyshire, and more often still, I had one or two practice nights a week in a quarry not very far from my home. It depends on the type of pitch as to which falling methods I use. For instance, if I am on a slab, I deliberately flatten myself against the rock and friction my way down to the stance. In another instance, I may be on a broken face of rock, and in that case, I try and decide very quickly which is my best landing place and jump. But in both these instances, I must land with my feet together with relaxed muscles and quite prepared to roll myself into a ball and not fall backwards onto my head. It is most important that the leader should let the second know if he is in extremis so that his second can be prepared for the fall. I have seen about five climbers fall in about the same number of years, and of these five, only one fell in a respectable way.

The majority will keep the rest of the party in ignorance of their dilemma, and then the first warning seems to be that horrid, convulsive scratching of edge nails against the rock and then a very unhappy landing. One of our leading climbers pointed out to me that the energy used in jumping a bad pitch might have been used much better in getting up it. That is partly true, I agree. But no one is going to make an unorthodox descent of a climb unless he is compelled to, and it is for such occasions that one uses the technique to make a happy landing. Personally, if I feel that I am going to come off, I fight the feeling as strongly as I can, and it is only when I am beaten that I jump or friction.

I think that guts and coolness are the most important parts of the technique to develop. I remarked earlier in the article how frightened I was of my first practice falls. These days I tackle them with much more confidence, and I think I have almost got over the fear of falling and suffering very severely from the results. There have been cases where climbers have had a fall and never had a chance of saving themselves, and I do not suggest in this article that they had the remotest chances of doing so. At the time of writing, I read with the greatest regret about a climbing accident. It brings home to me the fact that we are all liable to have a fall at some

Part of the training of the troops was to learn the art of falling from the climber and steeplejack Alf Bridge (1902–1971), who became a part-time civilian mountaineering instructor for the army during the Second World War. A native of Manchester, Alf Bridge was an outstanding rock climber and a member of the famous Bogtrotters Club. He performed many incredibly arduous walks including his celebrated 30+ hour circuit of the Peak District, in 1927, from Greenfield to Chinley covering 70 miles and 13 routes up to HVS on Laddow, Stanage, Cratcliffe Tor, Robin Hoods Stride, and Castle Naze.

time, and at such a time a technique of falling may mean all the difference between a slight accident and a very serious one. A sprained ankle is much to be preferred to a broken leg.

Of course, I cannot say the technique is an art unless I can teach it. I am afraid I am failing miserably at teaching (in print) the art of falling. It would be much easier for me if I could demonstrate a few practice falls and prove that I can make a happy landing – and on one occasion, a fairly happy landing after I had dropped myself 60ft. But that was no practice effort. If, however, I have aroused any interest in this technique, it will be infinitely more than worth the while, even though it may lead to people having a wrong impression of my climbing technique.

I do not think I would have tried to write this paper if it had not been for the fact that the rock experts in this country told me that they are confident that the art should be developed by all climbers. These men have given me the confidence to write my views about the art, and I hope my efforts have not been in vain.

Brandy's Zawn is just west of Bosigran where Menlove is recorded to have made the first ascent of West Ridge Direct, coincidentally or not, also in 1943. One speculates that Menlove was observed climbing on these cliffs by some of the commandos, perhaps Lieutenant-Colonel Dawson himself, and his climbing provided the inspiration for the location of this exercise with the ensuing unfortunate tragic results.

Such cliff assaults were critically important for the later success of the D-Day landings. Robert Dawson spent the entire war with Number 4 Commando. In August 1942, he commanded C Troop during the ill-fated raid on Dieppe. In April 1943, he was given command of Number 4 Commando, and they began training in earnest for D-Day. On June 6, 1944, Dawson was wounded twice during Operation Overlord but, despite his injuries, continued to lead his men in combat. The citation for his subsequent decoration states: 'It was due to his leadership and direction that the attack was successfully pressed home.' With the support of HMS Warspite, Roberts, and Erebus as well as rocket-firing landing craft and a squadron of Typhoon fighter-bombers the commandos were ultimately successful.

In return for my guardianship of their integrity [the mountains] offered me a sanctuary for all the higher impulses, all the less sordid hopes and imaginings which visited me anywhere through the years.

Geoffrey Winthrop Young, *On High Hills,* 1947

While well-regarded as a clinically excellent psychiatrist Menlove's theoretical ideas, which he regarded as the most important work of his life,[2] failed to gain traction. Further major stressors including being a conscientious

objector during the Second World War and having same-sex orientation (during a time when this was illegal) precipitated a severe depressive illness which resulted in his hospitalisation and eventual suicide, which was a tragedy that in many ways parallels the life story of his contemporary, Alan Turing.

In his seminal essay, 'Scenery for a Murder', published in the *CCJ* in 1939, Menlove writes a tale of an unnamed narrator and his climbing companion, Toni:

'The fact is, there has been murder done, done under their noses, and the fools can't spot the murderer. So I will set it all out: but I will have to be a little careful, for the murderer is a gentleman and very well connected; one is supposed to talk of him carefully, so to speak, as if he wasn't there and didn't do anything... But murder, you say. There was no one else present, you say, no murderer. So? Nobody else? Have you forgotten the singing, have you forgotten the scenery, the wild scenery? And how are you here to tell the tale, you ask? How! Do you not understand? But the boy himself, Toni, my friend? Did he die? Was he killed? Or I, or was I alone? Well, those are quite different matters, and really I must confess I don't know. But take it which way you like, that does not alter the facts; and make no mistake about it: that does not justify that very wild scenery; nor does it justify murder.'

Scenery for a Murder is a crucial piece in understanding Menlove's psyche and the stressors that resulted in his depression and suicide. In his superb biography of Menlove, Jim Perrin discusses this essay as an allegory of Menlove's intimate relationship with his climbing partner Wilfrid Noyce (Toni) and the impact of this and the homophobic culture in which he lived on his emotional health:

'He finally rounds on the reader and asks who was killed, himself or the boy? The answers lie back in the allegory. The boy's potentiality to feeling was killed off by his education and background, against which he could not rebel, so his natural inspiration was sublimated into a fixed barren obsession acceptable to his social code. And his siren singing having drawn a response from the narrator, his inability to come across has killed off the narrator's capacity for feeling, frozen him by a lack of reciprocity in response. Both the scenery and the murder are the society in which Noyce and Menlove must live... It remains one of the most grimly sardonic jokes in climbing literature that what has been regarded for over 40 years as one of its central texts was written only peripherally about the sport, and is a personal and social allegory which offers up a scathing, angry, yet still humane critique of the effect on the individual of the society in which Menlove lived and in which we, to a lesser extent, still live.'

In *Samson*, Noyce, probably due to his proximity to the source, is more circumspect: 'Perhaps the most perfect of JME's works... the editors will not spoil enjoyment and interpretation with their own comments.'

---

2   Unfortunately, these were all destroyed after his death.

Most people, especially Freud, misunderstand the Freudian position.

Aleister Crowley

'Take it which way you like,' said Menlove. My take on it is that he is prophesising the 'scenery' for his eventual suicide or 'self-murder'.

***

The inter-war years were very quiet times on the sea cliffs. Economic factors were a severe restriction on the small numbers climbing, and when they could go most people climbed on the more well-known mountain crags and the gritstone edges and the limestone of the Peak. But the Second World War changed that: the military came to Cornwall in order to train raiding parties as quickly as possible in the art of vertical assault. At the beginning of the war Professor Noel Odell toured the coasts of Devon and Cornwall seeking suitable areas, finding that the superb granite cliffs in Cornwall were readily accessible and the mild weather permitted climbing during all seasons. By December 1943, the commandos had moved their base of operations to St Ives in order to train in cliff assault techniques in preparation for the invasion of Europe.

Not only do sea cliff climbers claim to have introduced climbing in rubbers to mountaineers, as AW Andrews did in 1912 on Lliwedd, but the requirements of the military, in training for surprise attack, resulted in the introduction of the moulded rubber 'Commando' sole. Many National Servicemen were posted to camps near the coast, particularly those who joined the Royal Marines, the Navy and radar establishments in the Air Force.

One example was Tom Patey. From 1957 to 1961 Patey served in the Royal Navy attached to 42 Royal Marine Commando. He continued to climb at every opportunity. While stationed in Devon he climbed on the outcrops and sea cliffs of the South West and in 1958 he made the first ascent of Rakaposhi (25,550ft) on a military expedition to Karakoram with Mike Banks. A training course in Norway gave him the opportunity of the winter ascent of the Romsdalhorn along with Arne Randers Heen. Compulsory ski-training visits to Aviemore with the Marines allowed him to pursue his grand passion of Scottish winter climbing.

Cliff assault involved more than pure climbing, with much training done in small boats leading to the ability to land through surf on a rocky shore at night in poor weather. The commandos had their own climbing methods such as fireman's ladders, rockets and mortars with grappling irons and used their own names for features, for example renaming Bosigran Ridge to Commando Ridge. Alongside their military tactics servicemen climbing at a high standard, such as Joe Barry, Mike Banks and John 'Zeke' Deacon, laced the cliffs with outstanding and often challenging conventional rock climbing routes. The most popular of the Cornish commando cliffs still bear the marks of their heavy traffic.

Above: Peter Biven in action in the awkward chimney of the Direct finish to Ochre Slab on the Bosigran Face.

Below: Trevor Peck, notable Bosigran pioneer in the 50s and 60s – usual rope-mate of the Biven Brothers – and inventor of much now-standard technical climbing gear on the Ochre Slab, Bosigran. Photos John Cleare, March 1967.

Commando Climbers by Mike Banks

Given the necessary equipment, no cliffs are insurmountable.

Combined Operations Pamphlet number 24: Cliff Assaults, 1945

The war changed everything. In 1943 the situation was that Nazi Germany controlled the whole European coastline down to the Spanish border and had erected massive defensive works on every likely landing place. They called it their 'Atlantic Wall'. The commandos were therefore charged with finding chinks in this armour. Accordingly, in St Ives, they formed the Commando Cliff Assault Centre, whose mission was to land on a rocky coastline at night, scale the cliffs and carry out a raid. Special boats and equipment were devised, and cliff climbers trained. These techniques were used very successfully by the US Rangers on D-Day when they surmounted a cliff and, after a bitter fight, immobilised a gun battery at the top.[3]

St Ives, in Cornwall, had long been the stronghold of the climbing Commandos, but in 1949, for administrative reasons, the unit was incorporated in the main Commando training unit on the fringe of Dartmoor. The Cliff Assault Wing, as the sub-unit was called, settled into a steady annual routine. Every Royal Marine Commando (and this amounts to hundreds of men yearly) was taught the elements of amphibious raiding.

This was carried out in Plymouth Sound and included jumping out of small boats in the darkness and shinning up and down a 100ft cliff. Descent was made by the Geneva method of rappel. The main effort of the Wing, however, was concentrated on training rock climbers, and to this end, we had a small hutted camp near Land's End. Tricounis were used in preference to vibrams for two reasons: they were far superior on the slippery, seaweedy foreshore, and they were better for teaching a neat initial technique. Vibrams were worn occasionally during the course.

In the spring, or more accurately from January to March, the Wing gave itself over to running the Corps Cold Weather Warfare training in the Cairngorms and Norway. The next major item on the programme was the annual Exercise Run Aground when we stormed the 300ft Culver Cliff in the Isle of Wight before a large audience of senior NATO officers. Finally, about one year in three, we put on a display at the Royal Tournament, when a steep 60ft artificial cliff (cunningly furnished with just the right jug-handles) was stormed from the arena, and the subsequent cliff top battle won with ruthless military certainty. And then we returned to Cornwall for eight months or so of dull, old climbing.

The role of the Wing was very wide, and one cannot fail to be impressed by its competence in fulfilling that role. It did not aspire to be a mountaineering club; it was a military organisation and, as such, had inevitable limitations. The climbing was strictly governed by the needs of training, and few of the instructors (with certain brilliant exceptions) fully developed their own climbing potential.

The Commandos were to stay on in Cornwall for decades after the war and made a major contribution to the development of cliff climbing. They climbed in tricouni-nailed boots – which pretty well limited them to the grade of Severe, but they put up a considerable number of routes, particularly at Sennen. Their lasting memorial is, of course, Commando Ridge at Bosigran. Three Commando instructors were pre-eminent: Zeke Deacon, Mac McDermott and Vivian Stevenson. They climbed in the friendliest of rivalry with Peter and Barry Biven and Trevor Peck; the guidebooks are witness to their rich harvest of routes.

Mike Banks (1922–2013), above, was an adventurous climbing soldier and author. He transferred to the 3rd Commando Brigade during the Second World War prior to the brigade's amphibious assaults in Burma. After the war he was appointed to the Commando Cliff Assault Wing at St Ives initially as a climbing instructor, later becoming the commander. Photo courtesy David Medcalf.

---

3   The operation carried out by the US Rangers to take 'the high ground' at Pointe du Hoc was the only actual cliff assault that took place on D-Day, but if the 225 men from the 2nd Ranger Battalion had failed, it was believed that all of D-Day would have failed. The rangers scaled 100ft cliffs on ropes secured by rifle and small rocket-propelled grappling hooks and ladders (on loan from the London Fire Brigade) under downward heavy fire and grenades. Once they scaled to the top of the cliffs, the guns that were supposed to be on-site had been moved inland approximately one mile, a measure implemented by the Germans to protect the guns from sabotage, so the Rangers had to push through bunkers and endless lines of enemy troops to get to the gun emplacements and set the charges in order to complete their mission. Out of 255 from the battalion, only 90 survived, and that doesn't count the wounded.

When they started climbing, few routes had been done at either Bosigran or Chair Ladder, and Carn Barra was virgin territory. Cornish climbing really blossomed in the 50s: I look back on this as a golden era. Nylon ropes had become standard; rock boots had been invented, and climbers were beginning to use car wheel nuts as rudimentary protection devices. But the flavour of the era – which opened up new vistas was entirely artificial climbing, using pitons and etriers. Wherever there was a crack, there lay the possibility of a route. The commandos tended to fade out from the 70s onwards when their role changed from cliff raiding to mountain and Arctic warfare, and they were less in evidence. Following the development of the nut, the climbing wall and modern training methods, the predictable tidal wave of formidable new routes washed over the honey-coloured granite of the cliffs.

***

The figure silhouetted against the skyline on the Bosigran cliff top was clearly that of an elderly man, his white hair blowing in the wind. The year was 1947, and I was being trained to become a commando cliff-climbing instructor. The Cornish coastline was a very empty place in those days; strangers were noticed. I was told that he was Arthur Westlake Andrews, who lived a few miles away. He was, I gathered, the founding father of Cornish climbing.

Not long afterwards, I joined the Climbers' Club and realised that Andrews was the custodian of the Count House, the club hut at Bosigran. He was a friendly and hospitable man, always interested in what was developing in Cornish climbing. Despite the disparity of our years – I was 25, he was 80 – we got to know one another. I would be invited to his house at Tregerthen, next door to Eagle's Nest above Zennor, where he would show me faded photographs taken in the Alps in the previous century. Together with his sister, Elsie, we would hold grave discussions on the etiquette of the Cornish cream tea. He held strongly to the view that the jam should be put on first, then the cream on top.

Moving on to 1957, I was in charge of the Commando climbers and often teamed up with the redoubtable Zeke Deacon on the lookout for new routes. On one occasion, we were virtually provoked into a new route by Andrews. In 1923, he had traversed into the Great Zawn of Bosigran and had written that if the ledges on which he had seen cormorants perching could be connected, then a notable route might be created. He urged us to try it. The Green Cormorant Face route resulted. It should be remembered that we were in the pre-nut era but, thankfully, rock boots had been invented. Artificial climbing was the flavour of the month. Zeke, who of course led, used several pitons with etriers on the first pitch. Even so, I found it at the limit of my ability. In any case, the second's job was no doddle. His task is to get the pitons out and doing this swaying on a wobbly ladder, walloping a peg at ankle level, was both acrobatic and exhausting.

We duly reported back to Andrews and received his congratulations with an accompanying cream tea. Now Andrews was also a poet; a few days later, a poem arrived in the post in which he parodied Fletcher's famous lines on the pilgrims setting off on the Golden Road to Samarkand. He was making a wistful, but not judgmental, comment on 'progress' in climbing technique; a lament at the loss of innocence perhaps. Looking back to his early Cornish years, he asks:

Who are these beggars in their ragged clothes
And old felt hats and roughly mended hose
Who wear old pairs of worn-out rubber shoes?

He gives the answer:

We are the pilgrims of a bygone age
Who traverse on the shore from stage to stage
In lust of seeking what no man has seen
Between Bosigran Head and far Pendeen.

But what of these commando climbers?

Laden like Tartarin with all the gear
For storming citadels that climbers wear
Pitons and karabiners, ropes and slings,
And everything except pair of wings.

They reply:

We are professionals and this crusade
Is but the usual practice of our trade.

Andrews died, ripe in years, in 1960 to be succeeded as custodian of Bosigran by another quite splendid old gentleman, Rear Admiral Keith Lawder. Keith had retired from the Navy at the age of 55 and taken to climbing with infectious enthusiasm. He was the most amiable of men, always ready for a joke or a prank. In fact, the Admiral had never ceased being a Midshipman.

When he was well into his 60s, Keith set about exploring Lundy with Ted Pyatt. His memorial as a climber must surely be his first ascent, at the age of 68, of the incomparable Devil's Slide. It is arguably the finest slab climb in Britain. Whatever their talent, it is a route that any visitor to Lundy, who has an eye and a heart, must feel impelled to climb. One of the most joyous climbs I have ever done was to repeat The Devil's Slide with Keith Lawder, then in his 79th year. The third on the rope was his grandson, Iain Peters. Since his early youth, instead of playing children's games, Iain had been taken out on climbing adventures by his adored grandfather.

So, from my youthful acquaintance with Arthur Westlake Andrews to my avuncular climbs with Iain Peters, I seem to have spanned West Country climbing from its inception to the present day. It is a kaleidoscope vision.

Cornwall remains a very special place where, on a languid summer's day, your dreams and fancies can take wing. Leave the guidebook behind and wander where you will. Take some of the questing spirit of AW Andrews with you.

My Shag Song by Nea Morin

Bosigran will forever retain that elusive magical attraction which it has exercised over me since my first visit there nearly 30 years ago, but the thrill of that first visit can never again be recaptured. Few people came to the Count House then, often shrouded in mist swirling back and forth, breaking and reforming. It had a secret withdrawal, an air of mystery, almost of menace, enhanced by the great gaunt chimney of the old tin mine pointing a sinister finger to the sky and by the plaintive background moaning of the Pendeen lighthouse borne on the rise and fall of the wind. At Rosemergy Farm, they told my daughter Denise it was 'tingly' at the Count House and weren't we afraid to sleep there?

There was no electricity, and water was from the well, but at that time there was a magnificent great slow-combustion stove in the hall which kept the whole place warm and where clothes could be dried. After supper, gathered round a roaring fire in the sitting room with the Tilly lamps lit and the shutters closed, any sense of eeriness was soon dispelled. I had never before climbed on granite sea cliffs; the compact rose-coloured rock, the brilliant azure of the sea with its snow-white edging of surf extending far into the distance made a picture of breathtaking beauty. That first impact left an indelible impression that has not faded with the years. Climbing was exciting, for there was as yet no guide book, and few routes had been pioneered.

The first sight of Porthmoina Isle is dramatic indeed. It lies like a great ship at anchor in the tiny bay of Rosemergy Cove, encircled by the cliffs of Bosigran Face to the east and Bosigran Ridge to the west. The first indication of the nature of the coast is the jagged outline of the pinnacles of Bosigran Ridge which is visible from the Count House, gives some 700ft of climbing, and is one of the first routes we did. At that time, there was a very elegant alternative to the ordinary first pitch.

Instead of traversing round the nose of the ridge at sea level and up the chimneys of the east wall above the narrow little zawn, the alternative route consisted of a delicate traverse across a slab on the west wall of the nose. The advantage of this start was that it could usually be done when the tide was too high to traverse around the nose. Unhappily, during a severe winter storm, a great slab of rock split away, removing the first 20ft or so of the alternative start. The final pitch of the ridge (which most people used to leave out) is an intriguing problem, the answer to which I found was to climb facing outwards.

Crossing to and from Porthmoina Isle can be hazardous at high tide or in rough weather at the seventh wave; many a party has been marooned there for several hours, and it remains the haunt of gulls and kittiwakes. On our first exploration of the island we kept to the ridge climb over the top and back the same way. Then we followed the Guillemot Traverse from south to north returning by the middle East Face traverse, thus making a circuit of the island well above the tide-mark.

Sea cliff climbing takes getting used to. At first the constant movement beneath can be most disconcerting, especially when the wind and the noise of the waves breaking just below make communication between members of the party well nigh impossible. One quickly realises, too, the fallacy of any idea one may have had that a fall into the sea would be less lethal than onto dry land. Mr Andrews, the discoverer of the cliff-climbing of West Penwith, told us about his traverse of Green Cormorant Ledge from Cave Zawn to Great Zawn. We went to explore, but at the first attempt did not even succeed in finding the approach.

On a later occasion, Denise and I made a more determined attempt and spent hours trying to get across the crux move into the Great Zawn. This is where Mr Andrews, climbing solo, had lassoed a 'rock hitch' above, and so swung down into the Great Zawn. We could see his bollard, but there was a gull nesting on it. We managed to insert a nest of pitons, hopefully for a belay, but we weren't very expert, and the pitons – all three of them – were most decidedly rickety.

After repeated attempts we retreated sadly knowing that the Biven brothers Peter and Barry were coming shortly and would surely succeed as indeed they did with Trevor Peck. Nevertheless Barry coming last, and without the security of a rope from above, preferred to take off his clothes and jump into the sea.

The early days of scrambling and exploring were tremendous fun and full of excitement. At Carn Les Boel south of Land's End near the beautiful bay of Nanjizal we made the first lead of the Grooves, a splendid one-pitch climb of 80ft on a thin curtain of rock jutting out several hundred feet into the sea. On rough days the waves break against the sides sending sheets of spray shooting 60ft and more into the air, completely submerging the climb.

Further south still near the southernmost tip of West Penwith, we spent idyllic days on Chair Ladder unaware whether or not we were on known routes. I climbed a chimney crack (just right of the magnificent South Face Direct), but lacking the courage of my conviction that it would go, I was tempted by a top rope and the climb was subsequently christened 'Nearly'.

One elegant climb in the Chair Ladder group is the Terrier's Tooth which gives most delightful climbing with a hard start and a fine exposed finish. Many of the best climbs are on Bosigran Face Doorway, Doorpost, Autumn Flakes and Nameless, all of which are splendid natural lines in not too high a category. In 1953, Dennis Kemp and I made the first ascent of Nameless. At the time, we were in a party with Menlove Edwards, and he had something in mind on Window Ridge, but after setting out in this direction we found the wind at gale force and Dennis and I, who both felt we would prefer somewhere more sheltered, proposed Bosigran Face.

Menlove probably thought we were just chickening out, and he continued on his own and made a direct finish to Western Ridge. As we had foreseen, the Bosigran Face climbs were sheltered, but the weather was deteriorating and without Menlove our enthusiasm had waned somewhat. However we prospected along looking for a good line and hit upon the easy beginning ramp of what is now Nameless.

Once having made a start we continued, with Dennis in the lead to the top of the ramp. After an awkward move to the right, Dennis disappeared for what seemed ages. When I eventually came to follow, I was shattered to find he had made a very delicate move in an ultra-exposed situation with only a wee chockstone for a runner (a la Menlove) in one of the twin cracks immediately below the final crux pitch. By now it was drizzling, the rock had become very slippery, and our position wasn't a happy one.

The crux move above was extremely awkward with no alternative and our only belay a couple of very wobbly pitons high on the wall out to the right. It was a fine lead by Dennis, and we were only sorry that Menlove hadn't been with us, for this was one of the earliest VS routes 'through the heart of the most interesting region of Bosigran Face'. Now, of course, the face is criss-crossed with far harder and far more spectacular climbs. But it was fun to have been in on 'a good line' before the proliferation of VS and HVS routes.

In 1967 whilst standing on a ledge at the foot of Terrier's Tooth, I was hit on the head by a large lump of granite falling clear for 30ft or so. Luckily it was a glancing blow, for I was not wearing a helmet. I was only knocked out for a few seconds and with the help of my son was able to climb up to a grassy ledge. I was then whisked off by ambulance to Penzance and after having several stitches was told to go home to bed.

A couple of days later Trevor Peck turned up at the Count House and suggested that what I needed to set me up was to do a small climb. I agreed, so off we went to Alison Rib which was fine. But then Trevor suggested we go along to the main Bosigran Face where he said he knew of a little climb which would be just the thing for me. I was quite sure there was nothing on the face suitable for me in my still rather shaky state, but along we went, and I found myself embarked upon what turned out to be Autumn Flakes.

The first pitch was a strenuous layback. I had no strength and found it really hard going. Our belay before the second pitch was one of the then-new Peck crackers, a minute chock about the size of a child's thimble with a wire loop attached. Off went Trevor apparently with no qualms as to whether I would be able to follow. The next belay was at a good piton at the top of the first pitch of Nameless with which Autumn Flakes shares this stance.

Above this came the crux of our climb – a delicate traverse. I could feel my strength ebbing alarmingly and was really scared. However, Trevor seemed to have no doubts at all, and it was his confidence that enabled me to make the grade. But Autumn Flakes in 1967 seemed to me 10 times harder than Nameless in 1953. Well, I was one hip operation and 1.5 years older, and I guess Trevor's and my combined ages must have been at least 130.

As soon as we had finished on Autumn Flakes, Trevor and Peter Biven made a new climb, Paradise, HVS, which took a line up the face close to Nameless. It was fascinating to watch these two splendid climbers working the route out. Just two years later, Trevor died. In 1970 Dennis Kemp took me up both Terrier's Tooth and Doorpost, the latter a superb climb and at 220ft the longest on Bosigran Face, and that I fear may well have been what in these parts might perhaps be called my shag song.

Facing page: Peter Biven and partner crossing onto Porthmoina Island at low tide, March 1967. Photo John Cleare.

Coasteering by Edward Pyatt

Edward Charles Pyatt (1916–1985), above, worked at the National Physical Laboratory in Teddington. He started climbing on Southern Sandstone in 1936 before venturing elsewhere in Britain and the Alps.

He was editor of the *Alpine Journal* for 12 years and produced a number of guidebooks and books on climbing, including the Climbers' Club Guide to *Cornwall* (1950) with AW Andrews, and his seminal *A Climber in the West Country* (1969), below. Pyatt also wrote extensively about walking, including a guide to the Cornish Coast Path.

In the 1950s, Pyatt and Admiral Keith MacLeod Lawder (1893–1987) explored the Cornish coastline, walking hundreds of miles of coast, mapping features, investigating crags and recording possibilities as well as making some new climbs of their own on the Culm Coast. EC Pyatt photo courtesy UKC.

Facing page: Andy Pollitt and Chris Hamper traversing at Bosigran. Photo Glenn Robbins.

Coasteering, which may be defined as the applications of the principles of mountaineering and rock climbing to the scenic features of the coastline, offers walking on varied terrain, climbing on a variety of rock types, the ascents of virgin summits, novel techniques including complex problems of access and all in novel surroundings.

Edward Pyatt, *A Climber in the West Country.*

The 1970s have seen a rise in interest in rock climbing on Britain's sea cliffs and the standards of the routes on Anglesey and at certain locations in the West Country are as high now as anywhere inland. However, it is of considerable interest to find that there is a historical thread running back for three-quarters of a century and more, during which time the work of a few outstanding individuals has passed almost unnoticed. Sea-cliff climbing is not, therefore, just an extension of mountain rock climbing into non-mountain areas, but rather an independent entity having had its own innovators and having developed its own traditions and methods.

Features of a shoreline are defined by the positions thereon of the high- and low-water marks. These are not fixed absolutely rigidly but depend at any particular time on the relative effects of the principal tide makers – the moon and the sun – and on the ellipticity of the orbit of the earth. Above high-water mark is known as backshore, between the marks is foreshore, and below low-water mark is offshore. The cliff line marks the boundary between land and sea, the point currently reached in the erosion process; it can vary from a low bank in soft rocks to a steep crag, perhaps as much as 300m high, inland. The craggier sea cliffs obviously offer conventional climbing sport to mountaineers either on the rock faces or on residual features left behind by erosion in the form of stacks, pinnacles and so on. There are problems of approach. If a feature rises above the backshore, it can be reached by walking; on the foreshore its foot is only accessible at some lower states of the tide; offshore the approach has to be by boat, by swimming or by a traverse from an adjacent foreshore.

There is, however, one other climbing activity available on sea cliffs which is unique to this particular terrain. This is to traverse the cliff foot below steep and unclimbable cliff sections from one easy way down to the next; the quality of the rocks matters little since that portion within reach of the sea will always be sound. The cliff has, of course, to include substantial sections of the foreshore type; too much backshore cliff and the expedition is merely an arduous walk; too much offshore cliff and a traverse can only be carried out by conventional rock climbing. Most lengthy routes will involve a mixture of all three, but it is the foreshore sections which provide the widest variety of problems.

The bigger the tidal range, the vertical distance between high- and low-water marks, the more sporting the traverse becomes. Sometimes the climber faces a problem which can only be passed at a particularly low tide, so that he is forced to time the whole expedition to arrive precisely at that moment. The nature of the foreshore may vary from tide to tide – sometimes, for example, small beaches of sand are scoured away completely to be replaced when the conditions revert to normal. Route finding may be intricate through boulder fields, over and round ribs and buttresses, and broken up by pools and arms of the sea. Often there is a treacherous slime or seaweed coating on many boulders, calling for a technique which one of the pioneers dubbed 'touch and pass'. Above all, the climber must know exactly the position of the 'easy way' for which he is aiming and be sure that he can reach it in the time at his disposal.

Some years ago, I suggested the term 'coasteering' to cover the climber's activities on the coastline. Coasteering has much in common with mountaineering. It uses similar techniques and equipment and leads into similar situations. It has similar objectives – the ascent of summits, the climbing of rocks for their own sake; the exploration of relatively inaccessible and relatively unattainable terrain. It is carried out on the border of the familiar world and an alien world so that the more violent mood of nature can be observed and admired without the actual need to be involved with them; the mountaineer, on the other hand, has to actually enter an alien world in order to practise his sport. The cliff foot traverse and the problem of access in this specialised terrain are peculiar to sea cliffs.

Project Penwith by Matt George

In the mid-80s, about the time I was enjoying my first forays into my own coasteering playground near Land's End, it began to be offered as a commercially guided activity for the first time in Pembrokeshire in Wales. The man whose name has gone down in coasteering history as the forefather of the coasteering industry is Andy Middleton. He introduced guided coasteering at his newly founded business at the Twr-y Felin Hotel, near St David's. Within the space of just a few years, a handful of providers were offering coasteering in this region of southwest Wales.

From there, commercial coasteering's spread across the UK was rapid but albeit quite haphazard. Unexpectedly, it seems some of the more remote regions of the UK were the first to adopt coasteering outside of South Wales. Whilst it took the best part of 10 years to reach North Wales, The Hebrides of Scotland were offering the activity within a few years of Pembrokeshire, with Hebridean Pursuits running coasteering sessions on the Isle of Mull by 1990. The Outer Hebrides were soon to follow, with North Uist Outdoor Centre taking up the torch in 1992. In these pre-internet days, networking amongst outdoor instructors would have played a huge part in helping to spread the word. Indeed, across the UK, and as we look at coasteering's growth in other parts of the world, the majority of cases began when they had seen or heard about the idea elsewhere and realised they could apply it to their local coastline.

There are exceptions, however. As Hebridean centres were developing coasteering in the northernmost fringes of the British Isles, something was cooking in the UK's southernmost islands. John Fox founded Jersey Adventures. in 1993, unaware that coasteering was being offered as a guided activity elsewhere. Locals had been sea-level traversing the coastline of Jersey for decades to reach its secluded coves. With one of the highest tidal ranges in the world, either impeccable local knowledge, a willingness to get wet, or both, would have been necessary to reach the coveted Venus pools. These pools, only exposed at low tide, were said to be frequented by local women who, according to folklore, believed that bathing in these pools would increase their fertility. Amazingly the southwest of England was seemingly asleep as all this coasteering was springing up in relatively obscure parts of the UK. Sooner or later, coasteering in Cornwall was inevitable, with surf schools in the UK surf capital of Newquay realising there was a new activity they could offer their clients.

Once again, it seems the idea was born independently rather than emulating what was going on in other parts of the country. Mine exploring legend, Pat Moret, says that coasteering in Cornwall started around the turn of the millennium: 'We came up with the idea by mistake. Myself and a pal went looking for tunnels on a Newquay headland with wetsuits, lamps, and helmets. We swam around a bit, climbed up and down, etc, and jumped off some rocks. We then saw some kids tomb-stoning, so we got a bit braver. I was working at Lusty Glaze (an outdoor activity centre) at the time. The next day I told them about it, and we went out and recreated the trip. Over a few weeks, we came up with some routes and started taking groups out. Early the following year, Lusty Glaze found another place in Wales that was doing similar. They were calling it 'coasteering'.

It didn't take long for the Cornish surf schools to begin to diversify and include coasteering in their offerings, and the activity spread across the county. Newquay remains the coasteering epicentre of Cornwall, with the highest concentration of Cornish providers offering coasteering on the Newquay coastline. Devon, also late to the game, had established coasteering by the mid-noughties. Fast forward to today, and Devon has the highest number of National Coasteering Charter providers of any region. Significantly more than the entire nation of Wales, where coasteering began.

Coasteering gradually spread to all corners of the UK mainland, as well as most of its groups of islands. From the Isles of Scilly to the Orkney Islands and from Jersey and Guernsey to the Isle of Lewis, coasteering is now available. It's probably safe to say that no area of suitable coastline in the UK remains where a provider hasn't moved in to capitalise on what has been described as one of the UK's most popular adventurous activities.

In the spring of 2020, in between the first pandemic lockdown and the mayhem of one of the busiest summers Cornwall has ever known, 'Project Penwith' quietly came to its conclusion. As I ascended the crumbling cliff below the iconic engine houses of the Crowns Mine at Botallack, I completed the final section of coasteering the entire coast of the Penwith peninsula. That is coasteering every 'coasteerable' section of coastline from St Ives, heading west towards Land's End, and continuing along the south coast of West Cornwall, beyond Penzance. Penwith is the name of the district occupying the far west of Cornwall. It is Kernow Coasteering's homeland and is where we carry out many of our activities.

Some people are extremely impressed upon hearing this, mistakenly thinking that I did it all in a single session. I must stress that this is not the case. Project Penwith was many years in the making, chipping away at each individual section when time and weather conditions allowed. Those not familiar with coasteering, and just what level of detail you cover the coastline in when doing it, would not be able to grasp the amount of coastline covered.

The southwest coast path covers some 43 miles around the Land's End peninsula, as measured from St Ives to Penzance. The path straightens out the impossibly convoluted coastline below and chops off promontories and headlands at will. On the other hand, a significant portion of this distance can be deducted from the terrain covered in the project: sandy beaches and endless stretches of boulder beach are not coasteering terrain. Examples of these include the three miles of boulder beach stretching from Cape Cornwall to Gwynver Beach, near Sennen Cove. Similar terrain excluded shorter sections such as immediately west of Porthmeor Beach at St Ives until the real coasteering starts at Clodgy Point. On the south of the peninsula, the long stretch of boulder beach at St Loy also offers nothing in the way of coasteering. That leaves an awful lot of coasteering on the remaining cliffs – years worth, in fact.

Like many things in life, Project Penwith started by accident. The first sections were discovered and explored back in 2012 whilst I was searching for routes for what was to become Kernow Coasteering. Those initial explorations revealed some early gems, including routes we still use with our customers to this day, such as at Pendeen Lighthouse. I do remember on a winter's day, assembling with some mates in the car park at Porthgwarra, as light snow was falling, to go and explore the coastline around Gwennap Head.

Descending the cliffs at Carn Guthensbras led to a committing start, with a sizeable jump to get things started. As we headed along the coast towards the tiny beach of Porthgwarra, given its recent moment of fame as Ross Poldark's bathing spot of choice, a good chunk of coastline had been explored. The scenery was undeniably stunning, but the quality of coasteering wasn't the best, so the route wasn't developed. West Cornwall is particularly exposed to swell coming in from the Atlantic, relative to the rest of Cornwall. This has been instrumental in shaping the stunning coastline here but does create problems for coasteering.

A memorable early exploration was seeking an answer to this problem. Facing east, and therefore potentially offering some shelter from winds and swell, a friend and I set out one stormy day to coasteer the stretch of coast from Kemyel Point, near Lamorna, back towards Mousehole. The hope was to find a route that offered shelter from those westerlies. Alas, we bit off rather more than we could chew, and it turns out that that location offers no noticeable shelter from the Atlantic swells whatsoever. A very challenging but fun coasteering session ensued as we battled the swells all the way back to Mousehole. It wasn't a bad route, but its lack of jumps, caves and convenient exit points, not to mention its complete lack of shelter from the swell, meant that no further research was undertaken at the site.

The headland of Land's End was explored back at this time too. It remains one of my favourite places to coasteer in West Cornwall. It has a few tasty jumps, so long as you hit them at the right stage of the tide, but what makes this route so special is its very committing nature, the quality of the cliff scenery, and the caves... so many caves, from cathedral-esque caverns, down to the claustrophobic grotto we lovingly refer to as the 'Ghost Train'. As time went by, a desire to find more coasteering routes to share with our guests gave way to a more personal interest in indulging the simple love of coasteering and exploring the coast in coasteering the whole peninsula. So that was the primary motivation. I simply love the feeling of getting out there and exploring the coastline.

It's suitably remote to give one a feeling of genuine exploration without having to go to the Himalayas. It's also an amazing way to connect with your local area. Living on a small peninsula in West Cornwall and being interested in the outdoors, it is a no-brainer, at least for me, that I should get out there and explore the coastline. What genuinely surprises me is that no one else is doing it. You have to make the most of what's around you, so that's what I try and do. It always feels like exploration to me. There is always something new to see, or a new way in which to see it, as weather and sea conditions are never the same twice.

However, the feeling of exploring a new section of coastline is another level. Having no idea of what's around the corner, whether it's a jaw-dropping cave system or a challenging swell feature that needs to be negotiated, is a very rewarding experience. As mentioned above, no one is doing this. Who knows how many caves I have visited and gullies and cracks I've squeezed through that possibly no other human being has ever been in before? As the idea to coasteer the entire peninsula took shape, it was necessary to apply some logic to how it could be approached and to cut down the coastline into chunks that could be realistically achieved in a single push. This would often be to link up to obvious landmarks, such as two headlands, typified by the very remote bay and headland coastline between St Ives and Zennor.

It was inevitable that the south coast would be completed first, being significantly more sheltered from swell and prevailing winds than the north coast. On any given day, the chances of getting out on the south coast are far greater than being able to access the north. The south coast was all but completed when most of the north coast remained unexplored. Its propensity for rough conditions means that for much of the time from autumn to spring it remains totally inaccessible. That is, of course, when the owner of a a seasonal coasteering business has the most spare time. During the summer, when I'm coasteering all the time anyway, the motivation to coasteer additionally during free time is inevitably lower. It was, therefore, in the downtime between the first lockdown of 2020 and the start of the summer season that huge chunks of the north coast were explored.

The weather was unusually settled for a prolonged period which, combined with small swells, meant that it was the optimal time for north coast exploration. It was this period, coasteering in the strange times of 2020, that allowed Project Penwith to be concluded, possibly years sooner than it otherwise might have been. The other factor that had marked much of the north coast as 'low priority' was its remoteness.

As we head east, towards St Ives, we enter very far-flung terrain. The coast becomes further and further from the B3306, the West Cornwall coast road. This is where the walk-ins start to become brutal and weed out all but the keenest coastal explorers. Furthermore, ways to easily access the coastline, or maybe more importantly, to escape, become ever more scant. Only with a detailed local knowledge of this coastline should one try and coasteer in this area of Cornwall. With the coastline being very far from the road, especially between St Ives and Zennor, the huge walk-ins and walk-outs were not very appealing. These walk-ins were up to an hour in some cases, which may not sound like much, but in full coasteering kit and on either end of a tiring three-hour coasteering session they certainly felt like a lot at the time.

All in all, the peninsula was explored in 25 individual sections of coasteering. Most of these could be tackled with about three hours of coasteering, not including walking. That became the optimal length, and it's possible to cover a fair distance during that time. Any longer than that, fatigue tends to get the better of you. Conveniently, that was often the length of time needed to coasteer between the major landmarks that seemed the obvious places to divide the individual sections of coast. This particularly holds true for the north coast of Penwith, with the major headlands of Clodgy Point, Pen Enys Point, and Carn Naun Point, all being about three hours of focused coasteering apart from each other, for example.

Some sections were certainly shorter, but some were definitely longer too. After the long walk-in to Wicca Pool to tackle the section back to Zennor Head, I barely had the energy to make the long climb back up the cliff to reach the coast path once more. Certainly, the most demanding section was undertaking a very long stretch of coastline from Portheras beach, at Pendeen, heading east to reach the nearest known safe exit at Porthmoina Cove beneath Carn Galver. This took a remarkable five hours in the water. I vividly remember the staggering scenery of Porthmoina Cove, with the high granite cliffs of Bosigran main cliff and Bosigran Ridge on either side. However, I also remember being so fatigued by that point that I could barely swim, and my interest in the outstanding scenery was minimal. Needless to say, I have returned to what is some of the most impressive coastline in west Cornwall since then to enjoy it in a more manageable chunk.

Many of these coasteering sessions were undertaken with my fellow guides at Kernow Coasteering, as well as with friends. However, a number of the legs were done solo. I actually found some of these sessions the most rewarding; it certainly is much more adventurous going out into unknown territory by yourself. This began in instances when the conditions were too good to miss, but no one was available. But, as time wore on, I started to acquire a taste for solo coasteering. Whether or not coasteering alone is a good idea is a tricky question.

It's fair to say that every section of the coastline is unique. Probably each section of the coast has some feature that makes it stand out. Almost all of the coastline between Gwennap Head and Land's End is awesome. I'm particularly fond of the granite coast of West Cornwall, with its golden, castellated crags forming particularly beautiful cliffs and features. The granite, in general, seems to lead to the formation of the larger sea caves in the area, which is always a huge tick in the 'what-makes-an-awesome-coasteering-route' box.

The Tin Coast from Cape Cornwall to Pendeen has some amazing stuff. If coasteering here at low tide, the crazy pink seaweeds make a psychedelic contrast to the wave-worn rock, which can cover the full gambit of colours from black to bright green. This coastline is characterised by steep, narrow zawns. Some of these are claustrophobic squeezes, whilst some are towering behemoths that leave one totally gob-smacked. For much of the way between Portheras Cove and Porthmeor Cove, we're back in granite. With its northerly aspect, it differs subtly from the granite coastline of the west and south of the peninsula. Often shady and often rough, it's definitely a step further in terms of its isolation. It does include some of the most stunning cliff architecture, dare I say it, in the whole of Cornwall.

So, whether it's the giant caves of the granite coastline south of Land's End, the spitting backwash caves of the Tin Coast, or the precipitous zawns of the secluded north coast, each section offered up something memorable. And, of course, the experience made all the difference, whether it was having a great trip out with friends or taking an isolated trip, alone, in perfect conditions.

Facing page: Matt George exploring the Porthcurno section of Project Penwith. Photo Kernow Coasteering.

From Sea to Shining Sea by Henry Barber

Bosigran and Great Zawn are among my favourite cliffs, mainly because they're by the ocean, and I love that, and also because the climbing is similar to Cathedral Ledge: little knobbly square holds, rounded holds, square-cut holds, polished rock, rough rock and every crack and face combination imaginable, with some pretty good overhangs too. In fact, there are more good overhangs at Bosigran than there are at Cathedral.

I first went to Cornwall in 1974 with Al Rouse, Al Burgess and Al Harris: a good team. We had a really good time at Chair Ladder and Bosigran. I went into Great Zawn with Frank Cannings and did Green Cormorant Face, and then I soloed Variety Show. I saw that overhanging corner line to the right of Liberator that Livesey did, Fool's Lode, and wanted to do that badly. Omen as well, that looked good tucked away in the back there. I never did them, but they looked like my type of climbing – really weird climbing in sustained situations.

I thought Déjà Vu was pretty scary because, for me, first ascents have always been frightening. Frank Cannings said he had a really good line to show me and just launched out. I was a little disappointed with the route, as it's jammed in between Green Cormorant Face and Dream – if it was by itself, it would be fantastic. When I did it, I didn't have any pitons, but since then, pitons have come and gone on the route. I think that's bad because what I was into, in fact what I've always been into, is using nuts. England is famous for introducing nuts to world rock climbing, and when I first visited, I was a little disappointed that the concept hadn't been carried to the extreme i.e. developing the nut past the point of just being a type of protection to being the ultimate type of protection. When I started bypassing pitons on routes, using nuts right near the pitons, it was looked on as an extremist attitude, but it was just to prove that if the piton was missing, it shouldn't be an excuse not to do the route.

On Déjà Vu, I put two Number 2 Stoppers in sideways and launched out up the face – it was a pretty bold lead. That's the type of route that I'm well known for in the United States. I don't have that many first ascents there, but the ones that I do have get very few repeats not because they're grotty but because for the most part they're necky. That's the type of climbing I like, and I like to be in the shape where I can do that well. I don't like to do a climb, having fallen off it once, and then justify having done it better than the next guy because it's, say, the second hardest route in the country. I like to be in the shape where I can do, say, the 20th hardest route in the country which is so necky that it's only had two ascents – that to me is much more like climbing.

I liked Cornwall. I love the water, the situations and the history of that jump across the gap, unbelievable! That's what I came here for, the history and the stories. I got into climbing from hiking and just being in the mountains. I ended up in a situation where I was excelling because of the people around me and started looking towards the Whillans and Brown stories to excite me to go somewhere else. I wanted to come to Britain and climb some of the famous Welsh and Lakes routes. When I first visited, in 1973, I was an oddity to English climbing because I was too keen. The problem was an inability on both the British climbers' part and my part to adjust to the fact that, whatever it might have been, there was a difference between our climbing scenes.

To the British, I was too interested in climbing and seemingly not interested in the people – that was the way I came off. What I was into, though, was climbing to be there and meet the people. I picked Britain firstly because of the stories and the history and secondly because it is an English-speaking country. That's also the reason I went to Australia, and from there, I got more confidence in myself. It takes a lot out of you to go to an area if you're not in tune with it. If you don't have a girlfriend or friends to have a good time in the pub with, then you're not really primed when you head out on a hard route. You just sort of have to have that feeling; if you feel good about yourself, then you're going to do good things – or at the very least you have to have friends so you have some competition going, spurring each other on.

Left and facing page: Belayed by Pete Livesey, 'Hot Henry' Barber makes the intimidating leap across to the Green Cormorant Ledge in the Great Zawn, June 1976. Photos John Cleare.

Déjà Vu by Julian Lines

A branch of this primitive ape stock was forced by competition from life in the trees to feed on the sea shores and to hunt for food, shell fish; sea urchins etc, in the shallow waters off the coast... in time I see him becoming much more of an aquatic animal.

Professor Sir Alister Hardy. *New Scientist*, 17th March 1960

In the zawn, the sea had risen and calmed, and the sun had illuminated the smooth, granite folds. I shook with anxiety – I knew there just weren't any more excuses. I flicked through the guidebook for one last time to check the path of Déjà Vu. A few days ago, I would never have dreamt of soloing this route, but somehow the wicked parasite of desire had laid larvae in my mind. Déjà Vu was a brave undertaking when first climbed (in 1974) by Henry Barber – all brawn, self-confidence, white flares and a flat cap. He travelled the globe and left behind a trail of climbs that were both audacious and punishingly difficult.

From the viewing platform, I descended to a spacious ledge at the edge of the zawn. There, I was faced with reality: a small ledge known as the Green Cormorant Ledge, requiring a five-foot jump across a gulch. The penalty for mistake was a 30ft fall into the ocean. The jump was a point of no return and would seal my commitment to the route. The slope on the other side could just as easily tip me into the ocean.

A standing jump was the answer, but twice I tried, and twice fear glued me to the spot. I built up courage... then I was airborne... thud... committed... Pathetic now, hemmed in by warm, radiating granite walls, a mischievous ocean beneath my feet, and without a soul around to help if things went wrong. My pulse quickened.

Fifty feet up, my rhythm faltered because I had taken the wrong line; shadows started to be cast, and the blue waters vanished to black. I was shaking and started to lose focus as I regained the line. The odds began to stack up; everything felt wrong and fatalistic doubts crept in. It was my destiny to drown in the crushing swell, now 80ft below. The zawn baked like a kiln as I moved out onto a smooth wall with tiny holds. I started to wobble uncontrollably, froze, and then reversed back onto a guano-stained ledge to take stock of my limited options: either climb on and risk it all or jump off with similar consequences.

I was 'safe' on the ledge. I took off my sweaty climbing boots and lay down, using my boots as a pillow. I wiggled my toes in the warm air, closed my eyes, déjà vu again – juggling the dilemmas of being alone, crucial decisions to be made; trying to stay cool. I was in need of a hug, touch, or just someone to talk to more than anything.

After 20 minutes, the situation had not changed in the least. The abhorrent heat had made my mouth dry, and I could feel pain on my salt-cracked lips and see only demons dancing on the swell far below. I tied my bootlaces and once more began climbing towards those miniscule holds. Seeing it as the glass half full rather than half empty, I was only 50 vertical feet away from skin-tearing brambles and a clotted cream tea; life could not be simpler.

Facing page: Pete Livesey on Dream at the point where that route joins Déjà Vu. Left: Pete Livesey at work on the second pitch of Liberator. Photos John Cleare, June 1976.

Carn Gowla by Des Hannigan

From the tiny ledge at the base of the cliff, Carn Gowla seems a baleful place to be, even on this calm and shining day. The enormity of it all hangs in rocky clouds above your head. The sun casts a raw and sinister light; the ashen rock absorbs sunlight and deadens the blue of the sky. Only the Atlantic defines the cliff as being inescapably Cornwall. No other ocean breeds such restless power, even on a day as still and unthreatening as this one. You realise that from the ledge on which you stand, the Atlantic spreads outwards for thousands of unforgiving miles. On most days, where cliff and ocean meet, life would be unlivable.

The architecture of Carn Gowla is monumental. A line of gaunt cliffs sweeps away to either side of St Agnes Head. They extend for over a mile, are rarely less than 200ft in height and reach 300ft in places; a curtain-wall of dark grey hornfels shot through with quartz porphyry and 'reddened by impregnation with hematite' – as geologists explain with unintentional lasciviousness.

The name Gowla derives from the Cornish goel-va: 'the lookout place'. For most mortals, this is keep-well-back-from-the-edge territory, although sightseers park their cars with confidence on the flat cliff top. Steep slopes of heather and grass, patched with clumps of thrift, bladder campion and wild carrot, drop down for a short distance to end abruptly at the edge of the cliffs. The rest is thin air and a groaning sea far below. At the mid-point of the cliff top, a discreet memorial stone records the death of a man who went over the edge in his car; a tragic accident seemingly. His woman companion survived. At the last moment, she baled out onto those deceptive banks of scented wildflowers and heather just before car and driver tipped over the edge.

I got to know something of Carn Gowla in the company of a St Ives rock climber called Luke Pavey. Luke was more than capable. He started climbing in his teens, and it was immediately clear that he had exceptional talent. During the late 1980s, Luke spent much time at Carn Gowla. He may have worn out a few companions on the rope. Carn Gowla can be nerve-wracking. Why Luke decided that I might be useful on that daunting cliff, I do not know, but one early summer we made several outings to the cliff. It wore me out. The first route we climbed was called Journey to Ixtlan, named after Carlos Castaneda's book of the same name, although in this journey across the great cliff, Ixtlan is reached without sorcery or substances.

The route is a traverse of the southern end of the Carn Gowla cliffs. This is the kind of climbing in which both leader and second need to be confident and know what they are doing. The route is over 400ft long, a remarkable journey across the vertical plane, around corners and across rock of many colours and convolutions. It was early May. We started a touch late in the day and finished in the early dark.

Luke and I made other trips to Carn Gowla, during which we climbed several of the more popular routes. Luke then veered off to tackle challenges on other cliffs, although in later years he returned to Gowla to climb some of its hardest routes. Mild relief and fond nostalgia were my takeaways from our lonely sessions there. I thought I would never wish to return. And yet...

From any point along Journey to Ixtlan's wandering way there are inspiring views across a shallow bay to Carn Gowla's most famous climb, Mercury, a 300ft direct line at the heart of the highest section of cliff; a typical Pat Littlejohn and Steve Jones contribution. Mercury follows a corner to a small ledge halfway up the huge face. From here, the way curves upwards and to the right before making a final dash to the top.

A year or two after my visits with Luke Pavey to Carn Gowla, I climbed Mercury with Andy March, another seasoned climber whom you follow rather than lead. I remember, with undimmed intensity, paying out the leading ropes from the belay ledge 150t above the sea; more of a fabulous mirador than a ledge. 'Hidden Cornwall' at a genuine extreme; a viewpoint to beat all viewpoints.

Facing page: Max Dutson traversing at the base of Carn Gowla. Photo Grant Farquhar.
Right: Francis Ramsey on America, E4 5c, in April 1987. Photo John Cleare.

Journey to Ixtlan by Kev Howett

Jeremy Cole and I visited Carn Gowla on the tail end of a storm in November 1981 to find some spectacular waves pounding into the crag. It was obvious we could get nowhere near the base, but the traverse of A and B Buttress (the route Journey to Ixtlan – HVS 4c) takes a line above halfway and is described in guides as being 'safely above the reach of waves'; the crag ranging from 30m at the start to about 50m at the end. The climb starts on the cliff top (!) and takes a descending and ascending traverse leftwards, gaining height towards its end as the cliff increases in height. The sea was getting pretty close to us along the entire traverse, but when we reached a final stance somewhere on the exit line of the route Bohemian, a good 40m above the waves (we didn't have a guide with us and took belays where we felt safest), it seemed to come more alive.

Along the whole length of the cliff, the most perfect breakers were hitting head-on but were not gaining enough height to charge up the cliff face and reach us. However, for every couple of waves, the backwash would hit an incoming breaker, and the combined entity would rise vertically a few metres out from the cliff as a wall of malevolent brine edging ever higher towards me. This wall of sea stretched the length of the cliff and appeared to be nearing the top of the start of our route and seemed to be getting higher with each more powerful wave. Jeremy was leading the final pitch and stopped to take a photo as I perched, hanging from the stance, oblivious to the watery crescendo unfurling below. As I struck a nonchalant pose, I could sense it rising behind my back until it finally exploded in a mass of white tendrils as a final attempt to pluck me from the crag.

A Clockwork Orange by Des Hannigan

Sometime last century (circa 1985), the late great novelist, Anthony Burgess, came to Cornwall. The air turned blue (Burgess could swear with Shakespearean fluency). The bad language was due to Burgess's irritation with a television crew from Thames TV's South Bank Show that was filming a series about famous living writers talking about famous dead writers. They had lined Burgess up for the DH Lawrence slot. Where else but Penwith's north coast for location-location, given Lawrence's sojourn at Zennor during the First World War; given also the splendour of the Cornish land. I was a humble news reporter at the time. I was also writing an increasing amount of humble narrative about the rougher edges of the Cornish coast. Occasionally someone would give my name to big-time writers and filmmakers who were cobbling up pieces about West Cornwall and needed some local help. The South Bank Show (aka The Melvyn Bragg Show) hired me for the Burgess/Lawrence filming as a 'location fixer' and half-hearted 'safety adviser'. A gofer by any other name but with very good money to gofer all the same. The whole show was being run, on behalf of the mighty Melvyn, by a young woman who was hugely professional as well as hugely pregnant. She was hugely undaunted, frankly, even by Anthony Burgess who was in something of a grump from the word go.

I got lucky; Burgess and I took to each other immediately. Two grumps together. As soon as we met in the Queen's Hotel, Penzance – where else in the absence of a Ritz? Burgess dragged me aside and started to complain, in fluent Anglo-Saxon, about television, television crews, television programmes and, again, television. I had only heard such fabulously inventive cursing from certain trawler skippers. Burgess was, however, in a profane league of his own. In the foyer of the Queen's Hotel, Burgess's wife, Liana, the Italian translator and literary agent, and fierce protector of the Burgess Grail, fussed round Anthony with almost maternal affection. She took to me as well, I'm pleased to say. I was being doubly recruited beyond my brief. Mrs B had decided that since there was no way she was going to traipse about the Cornish cliffs, I was to be Anthony's guardia del corpo. 'You must watch his every moves, my dear young mans!' Mrs B said. 'Do not let him fall into the seas! And do not let him smokes too much!' This last was definitely beyond my brief. Burgess smouldered in every way. As we headed off on location, he gripped my arm. 'Stay close to me,' he muttered. 'You and I must at least remain fucking sane.'

I had suggested a spot on the Morvah coast just west of Rosemergy Towers as being a reasonably accessible cliff-top location for the Burgess-to-camera session. This

Above: A very young Des Hannigan in the Hole of Weems in Glen Clova. Photo Hannigan collection. Hailing from Dundee, long term Morvah resident Des Hannigan strayed to Cornwall in the 60s and was employed as a folk musician before infiltrating the tight-knit Newlyn fishing community where, despite admitting that he had no clue initially, he for many years was a successful professional fisherman.

A climber throughout his life he has climbed extensively throughout West Penwith, writing the CC guidebook.

He is very qualified to do this being not only a climber and a long term resident but also the writer of many books – a baton that he has passed onto his son Tim who he describes as 'the real writer' in the family.

Facing page: Kev Howett on Journey To Ixtlan, HVS 4c. Photo Howett collection.

was the rocky outcrop of Long Carn in the shadow of Watch Croft, Penwith's highest hill. To either side of Long Carn, steep vegetated slopes and granite cliffs tumble down for several hundred feet to the sea. I doubt that Lawrence ever crouched in gloomy reflection there. The coast path at Long Carn is quickly reached from the road, and although a short section of the path is seeded with boulders, the flat-topped rocks on the summit of Long Carn made an ideal location for the camera team – and for John Anthony Burgess Wilson, to give him his proper name. The day was perfect; the cliffs were at their sun-blitzed best beneath a blue sky. Lazy swells rolled across the sand in the bay below Long Carn and broke in ribbons of white against the granite shore.

We coaxed the camera team and equipment, including Burgess, down to Long Carn without mishap. Burgess was formidable when performing to camera. He perched himself on a convenient throne of rock with his back to sea and sky. The crew fluttered, fixed, focused; they became still as stone as Burgess launched forth. Unscripted; not a word out of place; not a pause; no retakes. Just a fabulous flow of mellifluous flummery about David Herbert Lawrence. Midway through filming, one of the area's resident peregrine falcons (celebrities in their own right in those days) zipped by and then curved back to hover directly behind Burgess's head from where it cast an imperious sideways glance at the interlopers. The camera crew were transfixed by the falcon but kept rolling. Burgess droned on, unaware that, for once, an even more formidable raptor was upstaging him.

When he finished, the entire team packed up and shot off back to the road, babbling about that 'amazing bird' and congratulating themselves on such a good wrap. I was left to escort a wheezing, cursing, panting, and blaspheming Burgess across the fang-like boulders and up the steepish path.

'You should stop smoking,' I said.

'I will when I'm fucking dead!' came the relatively robust reply.

At his peak, Burgess is said to have smoked 80 cigarettes a day. Sadly, he died, aged 76 in 1993, unsurprisingly perhaps, from lung cancer.

Postscript: Years later, the pioneering Devon rock climber Pete O'Sullivan and I tackled the high granite cliff in the zawn directly below Long Carn to produce a long and fairly challenging rock climb that is still a target of visiting climbers today. Peter had been reading Anthony Burgess's novel *Earthly Powers,* and so he called the climb Earthly Powers. Peter knew nothing of my brief shoulder rubbing on the coast path with the mighty Burgess himself all those years before, so there was a pleasing serendipity about the coincidence. I like to think that Burgess might have cherished the obscure detail of having an obscure rock climb on an obscure sea cliff in far Cornwall named after one of his great clockworks.

Above: America area. Photo Grant Farquhar.

Below: High over the Atlantic, Francis Ramsey leads America, E4 5c, on the intimidating America Buttress in April 1987. Photo John Cleare.

A Family Holiday by Graham Hoey

What could be nicer than a relaxing trip up a Cornish classic on a nice sunny day? Think again – Carn Gowla isn't for the faint-hearted; prior attendance at a lifesaving course may be advisable. For a climber, there are family holidays and 'family' holidays. The former are characterised by a complete lack of climbing, an obsessive fear of ensuing muscular atrophy and a perceived permanent loss of form: 'You'd better lead today. I've been on holiday with the family at Scarborough for a week, and I'm as weak as a kitten.' In the past, in an attempt to stay fit, I've stuffed a variety of training aids into an already overflowing boot, including finger exercisers, free weights and even a Bullworker (remember those?). 'Family' holidays, on the other hand, involve an arranged or 'chance' joint vacation with a fellow climber and his family: 'You'll never guess who just happens to be there when we are darling? I might just pop a bit of gear in the boot.'

The southwest, with its excellent beaches, walks, family attractions and (just coincidently) superb cliffs, was the ideal venue for Howard Lanchashire's family and mine. However, part of the deal was that because it was still a family holiday, opportunities for climbing would be occasional and brief. This proviso is responsible for one of the unique aspects of climbing on a 'family' holiday. Having only part of the day or evening to climb leaves no time for the casual warmup or two before the 'main event'; no chance to get used to the rock, style of climbing or protection, and no psyching up. One minute you are building sand castles or damming streams without a care in the world then, shortly afterwards, you are abbing into a tidal zawn, about to set off on some classic horror ahead of a rapidly encroaching tide, with just a few hours before sunset. It's all a bit surreal and at times you could be excused for imagining that you are in a dream (or more often a nightmare).

I did some brilliant adventure classics in the southwest with Howard, and most of these went without much fuss. However, the time we did America, the classic E4 on Carn Gowla in Cornwall, was an experience I'll never forget. America was first climbed in 1973 by the master sea cliff explorer Pat Littlejohn and ascends the daunting north-facing eponymous buttress by three excellent, exciting pitches. The route soon developed a reputation for seriousness, not least because of the approach to the climb; we were aware of a number of teams who had failed to get to the start. We'd also heard rumours of a possible abseil approach, but to us that would be missing out a major challenge of the route. We were determined to climb it 'the Littlejohn way'.

Knowing that a low tide approach was required, I arrived in good time at Howard's caravan. Howard refused to leave without a brew and his ritual wipe round the 'van (inside and out) with his 'signature' yellow cloth (the first sign of OMD[4], according to his wife). Eventually, we set off. After an 'interesting' V Diff solo descent above a disturbingly churning sea, we arrived at a ledge before a narrow zawn which required a Tyrolean crossing. Things were not looking good. For a start, the sea appeared to be a lot closer than it should have been for the time of day, and secondly, there was no sign of the spike you were supposed to lasso on the opposite side of the zawn (which probably explained the earlier failed attempts).

Undaunted, we tied a bight in the end of a rope and threw it across, perhaps hoping that the loop, in some way, would automatically 'home-in' on the (presumably hidden) spike. It didn't. Unfortunately, we then noticed that between the ledge and the wall behind was a narrow trough and had the bright idea of jamming a knot in this. Unbelievably, at the first attempt, the knot jammed, and for some reason that I can't imagine, I agreed to go first. The knot held, and we were soon assembled on the ledge with the Tyrolean dismantled before we noticed that, far from the expected simple 30m boulder-hop across to the start of the route, the way ahead was now a seething cauldron of white water with just a few exposed rocks.

Grabbing one end of the rope, I raced across between the gaps, bouncing from boulder to boulder to reach the sloping shelf beyond. This was by now underwater and necessitated a quickly assembled hanging belay 20ft up the first pitch. Safe at last, I took the rope in and beckoned Howard to come across. His words were lost

---

4   Old Man's Disease.

in the roar of the North Atlantic, now in full surge, but after a lot of screaming and gesticulations, the gist was: 'There's no effing way I'm coming across there; we'll have to set up another Tyrolean.'

Unfortunately, setting up a 30m Tyrolean with a 50m rope is not easy, and it became sickeningly clear I would have to go back for the second rope. Clipped into the single rope stretched across the gap, I scrabbled and hopped my way across, collected the second rope from Howard and eventually made it back to the belay somewhat gripped and completely saturated. I went under several times (carrying a full rack), and thinking of all those snagged runners on descent gullies still makes me shudder nearly 20 years later.

Howard set up the Tyrolean using natural threads in a high cave and eventually made it across to the belay, bone dry. By now, I was beginning to feel frozen and wishing to get moving, set off, shivering my way up the initial section of the route. As if the lack of protection wasn't enough, every time I stood on a hold, water oozed from my boots. I was filled with great relief on completing the pitch, as was Howard, who was on the verge of removing the belay and setting off to avoid being drowned. Fortunately, the rest of the route went without a hitch with two more superb, contrasting pitches. The second was a bold slab on the edge of the buttress, which Howard demolished quickly with his typical no-nonsense approach. I was lucky to bag the final one, taking a striking wall in a fantastic situation where memories of the earlier epic were almost forgotten.

Almost, but not quite, as we decided to abseil in later that week when we did the magnificent Guernica, Howard taking the honours of the crux pitch on that occasion. The following year, some friends of ours virtually walked across a dry beach, the sea an uncharacteristic millpond. You win some; you lose some. Sometime later, Howard was recounting our 'epic' in the pub with his usual enthusiasm and humour. Not known for splashing his money about, the audience was completely aghast on hearing he had left some slings behind for the Tyrolean. Suddenly turning uncharacteristically sombre, Howard replied: 'I tell you what; I'd have left my bloody car behind if I'd had to'.

Modern deep water soloing does exist in Cornwall but has few takers. There is a small and keen local crew, but they are reticent to publicise their activities so that who has done what first tends to get lost in forgetting and bullshitting with the only certainty being that whatever route it is... the Edwardses did it first.

From Tavy to Tazi by Nick Hancock

If you live long enough, you'll make mistakes. But if you learn from them, you'll be a better person. It's how you handle adversity, not how it affects you. The main thing is never quit.

Bill Clinton

Growing up in Devon, I suppose it was only natural to explore the amazing coastline of the county and its neighbour, Cornwall. From my early teens, I would coasteer along the sea cliffs and beaches near Plymouth. My first vivid memory of these early adventures was when my mum dropped me off on the way to her work at the chossy cliffs near St Austell. Things went from bad to worse as I traversed as far along at sea level as I could before being forced upwards by an impasse. I really thought the end had come when I was over 100ft above rocks, clutching desperately to soil and grass, as I topped out onto the coast path. Thankfully, this was an isolated episode, and at 14 years of age, in 1976, I fell into more a normal climbing apprenticeship on the more predictable granite tors of Dartmoor.

Not long afterwards, I began climbing with the late Dave Thomas (not the crazed soloist version), who set up the Kernow Climbing Club. With Dave and shortly afterwards Andy Grieve and Ken Palmer, I was introduced to the more sensible DWS area of Torbay. Initially, we treated the sea-level traverses as an extension of normal roped climbing although, in hindsight, it just made everything much harder. Rope drag, heavy gear and bad communications, not to mention the heightened risk of drowning due to entanglement, should have encouraged us to just get on with it and just solo a lot earlier. We didn't fully give up roped climbing above the sea until 1986. We first climbed Pinnacle Traverse and the Five Star Traverse with ropes, but in 1981 we soloed the Long Traverse.

The next year, 1982, I did my first real DWS in completely inappropriate conditions, luckily wearing a wetsuit. It was early spring with moderate seas when I set off along Magical Mystery Tour. At the Blue Grotto, I jumped in and swam across to the next promontory. What followed next was half an hour of desperate attempts to get out of the water as wave after wave knocked and dragged me back into the chilling seas. On my last legs, I just managed to get out and vowed to stick to warmer and calmer conditions in future.

By 1985, we were all climbing much stronger and began to polish off the safer and more technical traverses like The Watchtower and Kraken. That same year Andy, Ken and I climbed Rainbow Bridge using ropes. I remember the ropes being of limited use on the lower sections, but I guess our minds weren't ready for hard climbing 30ft above the sea just yet. It's hard to believe, now, that it was four more years before Crispin made the first solo.

By 1987 we were regularly soloing all of Magical Mystery Tour, but vertical deep water solos were still not really on our radar. In 1988, and I blame Chris Rees for all of this, he and I took the vector of DWS to some unfathomable places. Whilst mountain biking from Lands End, we thought it might be fun to mix things up a bit with some highball DWS at Carn Les Boel. First, we soloed Excalibur (HVS), which I'd led a few months before. I went first and did the thin section OK, but Chris struggled a bit when his turn came. Looking down at the shallow, boulder-filled zawn far below didn't fill me with too much confidence about his chances of making a successful splashdown if he didn't make it. Alarm bells should have rung, but after this sketch fest, we went down again for some more. We set out on Seal Passage at the more amenable grade of Hard Severe, and I topped out without any stress. Chris was finding things way too pedestrian and chose to attempt the Direct Finish up a nasty wide crack at E1 5b.

Things went OK until he got stuck about 10ft below the top. A pitiful voice wafted up from below as to his situation. I offered to sprint back on the bike to the car at Land's End and get gear for a rescue, but Chris dutifully informed me that it wasn't an option. The only solution was for me to reverse the top of the crack and offer my ankle to Chris as an additional handhold to negotiate the crux. Chris must have been somewhat relieved by the sight of an approaching ankle. But instead of gingerly pulling down on the offered limb he, somewhat surprisingly, laybacked up on me, causing my foot to pop out of the crack; we both ended up testing my wide crack technique a bit more savagely than during the average DWS. For a couple of seconds he was literally hanging off my leg, and our only means of attachment was a fist jam and a sidepull 200ft above sea-washed boulders. Unsurprisingly we have never been soloing together, DWS or otherwise, to this day.

In 1989, I continued to explore the possibilities of soloing established routes on the Cornish sea cliffs. First up, I soloed Amnesty at the Lizard, downgrading it from E4 to HVS. I'm guessing that the lack of ropes and gear, and the potentially false security of deep water below, made it a lot easier, as the climb has since gone back up to E3. Nare Head is another hidden DWS venue, and I also did Inner Secrets, E1.

Four years at university in Sheffield ended my DWS life for a while, but in 1995 I got back into it with solos of Astral Stroll and two on-sight first ascents at Cligga Head: Fear of a Black Planet, E2 5b, and Fingerin' the Dyke, E3 6a. Only the latter is a sensible DWS, the other being way too high above the ocean for it to offer much hope of salvation. Later that year, in company with Mike Wintroath, we DWS'd the first ascent of Basketball with the President, E3 5c, at Damehole Point. The name materialised when Mike told me that he used to go around to his neighbour to shoot hoops in Little Rock, Arkansas. That neighbour was none other than Bill Clinton.

In 1996, with Lee Earnshaw, we soloed from Long Rock to the Promontory at Baggy Point, producing the very unusual 3D climb of Through the Looking Glass at E3 6a. Next year I finally got adventure traversing out of my system with the low-level traverse of Gurnard's Head: Over the Ocean, E4 6a. Turning the arête crux below Tropospheric Scatter, knowing that falling would lead to almost certain death, made me realise that DWS should ideally be conducted above bottomless depths of calm, warm water on overhanging limestone at Torbay.

Shortly after, in 2000, I emigrated to Tasmania and was pleased to discover excellent deep water soloing venues there. So I was able to transfer those skills acquired in Devon and Cornwall to the other side of the world.

Nick Hancock on the first ascent of Wet Nurse, 23, Fortescue Bay, Tasmania. Facing page: Nick Hancock on the first ascent of Monty Python's Flying Circus, 25, Freycinet, Tasmania. Left below: Nick Hancock after the first ascent of Chaos Theory, 26, Friendly Point, Tasmania. Photos Hancock collection.

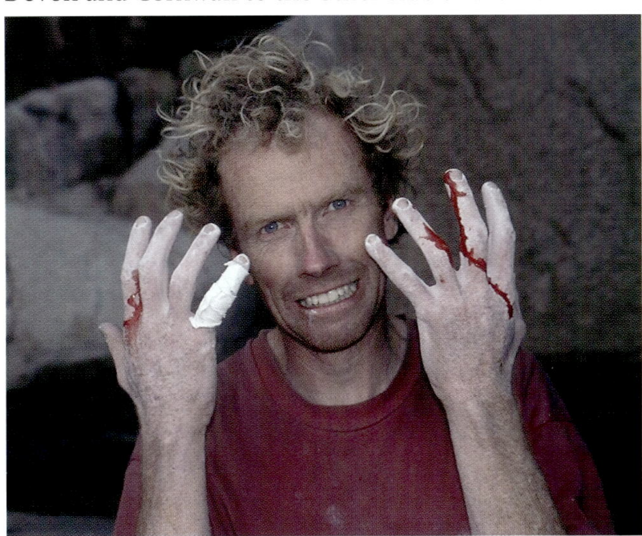

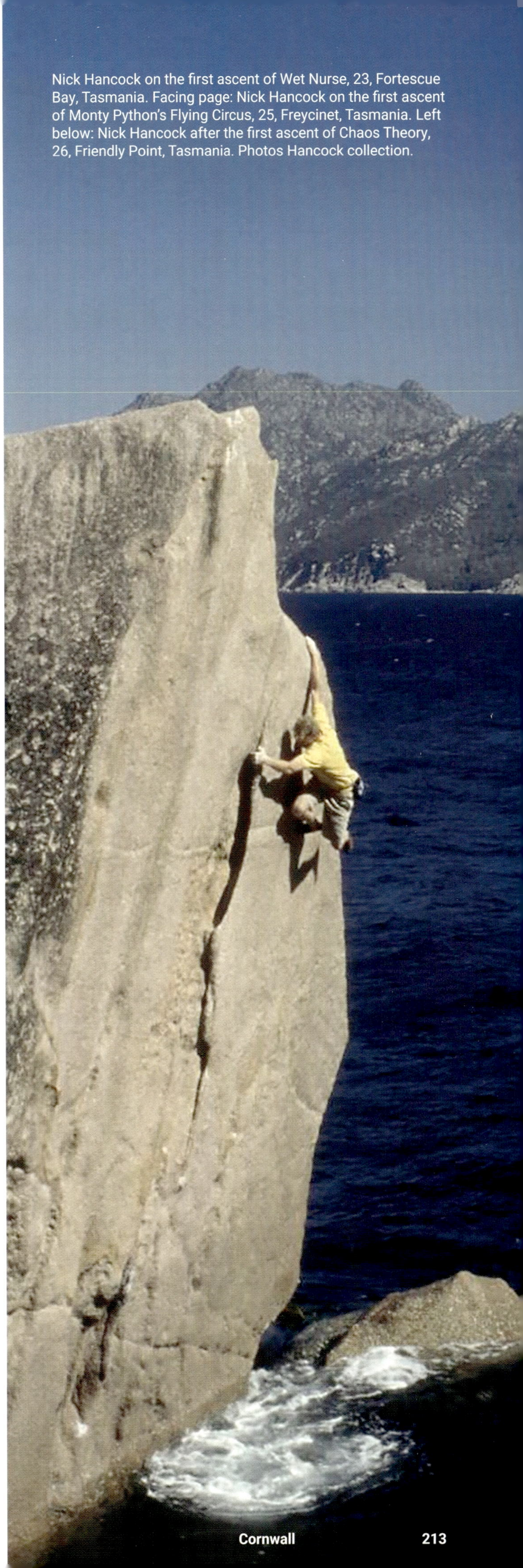

Dancing the Hidden Edge by Stu Bradbury

Dancing with Gravity by Stu Bradbury

The cycle flipped end-over-end and landed on its side, the knobbly tyre still spinning on its buckled front wheel. A nine-year-old me picked himself up off the ground, dusted himself down and inspected the grazes and gashes through torn clothes. The stunt had been destined to fail from the minute the wheels started rolling at the top of the steep hill. Gathering speed toward the ramp, a typically ill-thought-out escapade just for kicks, the bike hit the makeshift ramp halfway down, sending me high into the air. Unfortunately, gravity had its own plans, and instead of my idea of landing smoothly on the back wheel in control, the bike took a nosedive and planted its front wheel into the dirt, spitting me over the handlebars and sending me arse over tit down the hill to land in a heap at the bottom.

I'm sure, like many of you reading this book, my childhood was full of risk taking. It's a necessary part of growing up and finding out who you are and what you're capable of. For me, if there was a danger element then it always had a spell over me. As a youngster, I was always willing to attempt to give my hair-brained schemes a go; whether it was bike stunts, climbing up buildings or trees, jumping gaps or playing dare with cars, I was always in for the ride.

I am a big believer in the thinking that what comes before sets you up for the future, and the skills we gain through play in childhood can form a large part of the direction we take and the situations we can deal with as we evolve. I genuinely feel that is true for me.

I learnt to fine-tune my overactive imagination into the tool of what I later learned was called 'visualisation' at a pretty young age with the help of my role models: Spiderman and Evel Knievel. When I think back, bruises, grazes, and breaks were acquired like an extreme form of stamp collecting and worn with pride.

The whole idea and concept of active risk taking and the psychology behind it has always fascinated me. Why are some folks happy to take on risks and consequences which many find unacceptable? Science may have answers in the form of chemical reactions in the brain causing stimulation blah blah blah, but I am more interested in the individuals, their own personal motivations, and what makes them tick.

What really intrigues me are those that climb knowing they have the most to lose: soloists, those climbing bold hard routes, loose routes, new routes in isolated locations and so on. Those that actively go looking for that fine line and thrive on the journey, a journey that switches very quickly from the body physically taking the mind on a ride to the point where the tables turn, and the mind is taking the body. The point where, without mind control and acceptance of risk, you just can't get your body to perform.

Above: Stu Bradbury follows his nose on a new sea-level traverse. North Cornwall. Photo Lee Bartrop. Facing page: Andy Steinberg and Richard Pollard high on routes at Penally with Max Dutson and Lee Bartrop below. Photo Grant Farquhar. Below: Dancing with diamonds – Stu Bradbury soloing the classic Diamond Smiles, Lower Sharpnose Point, North Cornwall. Photo Ian Bryant.

I know from my own long-term ongoing experience that self-confidence and self-belief play a major part. From my personal angle, this has come from the fact that I have built a strong foundation over a lifetime from accepting and taking risks and getting used to what it feels like to be in that situation. One thing is certain: if it floats your boat, it's highly addictive.

The art of climbing is much more than its sum of parts. It's not just down to the physical act or the outcome – it's the overall process. I feel deeply grateful and privileged to have found my outlet through climbing with all its depth. I have enjoyed the process and journey to build the hard-won skills and judgement which allow me to take my personal climbing and operate with what the natural environment throws down, enabling me to become completely engrossed in its rugged splendour. This is why adventure trad climbing has always been my poison of choice. It has allowed me to follow my heart and find my own ideals. It has allowed me to create and has given me something which has totally consumed me above all else.

It's fair to say that if you play with fire for long enough, you are bound at some point to get burnt. Gravity is a harsh mistress and dishes out some strong lessons. If you are pushing the limits of what you feel capable of, things don't always go to plan. Those were my foundations: learning to play with fire and understanding the gravity of risk. Those mishaps and close shaves continued into the complex simplicity of my, often solo, climber's life.

***

I pulled the bike off the road, killed the rumble of the big single-cylinder motor and swung my leg off the seat. The void of silence was replaced with the familiar clicking and pinging of the cooling engine as it caught the gentle coastal breeze.

I lifted the open-faced helmet off my head, spat the insect debris from between my teeth and checked my face for bug splat in the bike's mirror. It's hard not to ride with a big, fly-catching grin on your face when you're trying to find the outer limits of the tyres on these twisty Cornish B-roads and, despite the tell-tale signs of age in the form of grey flecks in my beard and my thinning hairline, it still brings the same joy I got when my 10-year-old self first twisted a motorcycle throttle.

Lots of water has gone under the bridge since then, but the simple pleasures of this, combined with climbing, have been a constant throughout my life and still deliver the same thrill – even if the bike has been sidelined a little in my obsession to climb. However, on days when I am taken to seek isolation and total freedom, it provides an additional depth of pleasure to a day's soloing.

I threw the helmet and jacket into the backpack, double-checked I had packed the rock shoes and chalk bag and set off across the fields towards the coast path drawn by the ever-present noise and turbulence of the Atlantic Ocean as it went about nature's work.

Passing under the giant white radar dishes and skirting around the double perimeter fence on the way to Lower Sharpnose Point is always a surreal experience. It feels like a scene from a sci-fi movie – which seems a little out of place when surrounded by such rugged and wild natural beauty. Once the spy dishes are behind you, you can embrace the setting and take in some of the most stunning scenery that North Cornwall has to offer. Looking north on a clear day, you can feel the magnetic pull of the spectacular Lundy Island with its miles of granite and dream of its hidden zawns offering adventures to match any others.

As I break off the coast path heading down towards the crag, the roar of the ocean intensifies. I push through the waist-high gorse and am treated to a taste of its sharp thorns through my trousers, but almost as if in apology for its harshness the sting is softened by the sweet smell of coconut scent given off by its vibrant yellow flowers.

Looking down on the crag from above, I gulp in the updraft from the sea and feel the usual excitement as I view the steep walls and contemplate the movement and exposure of soloing in such an uncompromising setting. I am startled out of my trance as a family of peregrines screech past, twisting, stooping and playing in the updraft whilst giving their young fledglings a lesson in survival. How can such a sight taken as a whole not raise your spirits and swell your soul?

I like to spend time settling into a situation when I am out soloing – relaxing, breathing in the atmosphere, and starting to feel at one with the environment. I like to take my time, contemplate, slow my heart rate down with breathing and prepare myself for smooth, unrushed, precise movement so I can feel the joy and totally unrestricted flow – no ropes, no gear, no weight… simply pleasure.

Shoes on, chalk bag on, a few deep breaths… lightly and smoothly pull onto the wall; the joy of soloing lies in its freedom, its untethered movement. It requires total focus and commitment. No room for error or uncertainty. Its beauty is a complex simplicity, simple in theory but deeper in its undertaking. It requires you to really know yourself, to be at one with your inner thoughts and your ability to stay calm and focused. To keep bringing things back to the positive if they should drift, and the ability to shut out or deny the negatives so you can perform and survive.

I warm up on the crag's classic extreme. The start is steep but positive despite the holds feeling damp and greasy from the last high tide. The line follows a 30m diagonal crack with a sting in the tail but lives up to its name and always makes me 'Smile'. After three or four routes I feel sated, and I sit atop the narrow summit fin contemplating the thundering waves as the incoming tide cuts off the escape route from the zawn. As I relax, I am taken by the fulfilling glow which always follows a flurry of climbing action in such committing situations. My mind drifts off on the breeze and contemplates what I find so attractive about the whole game.

To say when soloing that you can always be in control would be misleading – unless you are climbing well within your limit. If you get drawn into the headspace of pushing yourself physically, it is natural that psychological baggage will surface, and it's a sure fact that you will be sailing close to the wind. The art is to find the simmer point and not let things boil over. If you are seriously searching for that inner buzz, you will inevitably end up finding that edge, and surprisingly it is often not at your physical limit.

It's not something you can encourage in others. Although many climbers will taste its delights, it is not for everyone; it's a very personal inner game – not a game of chance, but a game of both mental and physical control as often when soloing you find cruxes are not always in the same place as when you are climbing roped up. You can't take anything for granted. It always requires you to stay sharp and ready to adapt.

For me the act of soloing, with its commitment and simplicity, fits with my ideals. I have gone through the same steep learning curves in life that we all face, but my key aim has been to keep things simple, uncluttered; to live the life that I want. I have often lived on fumes so that I can climb, not driven by financial gain or possessions and moving to an area which feeds both my soul and my objectives. I can't say it has not been a selfish journey or a journey without casualties, but my biggest fear has always been the mundane – to live a life in beige.

During negative phases of my life, soloing has served as a strong medium for sharpening the spirit; for resetting the way forward. Its intensity brings clarity and focus in a meditative way. It lays the cards on the table; there is nothing like the feeling that you have something to lose to make you realise you have something worth saving. At other times it just served my want to be alone in a simple 'at one with nature' kind of way, and, if I'm honest, I can't say that playing with fear doesn't have a hand in it.

For a few years, I went through a period when in the process of searching to extend my personal journey by climbing bigger and harder objectives alone, I got totally embroiled in the art of roped soloing with a Silent Partner. I tramped around Europe in search of big walls to climb, but whilst I enjoyed the process and its challenges, in truth it is a full-on game which requires lots of effort and lots of gear. It can be an intense and exhausting process demanding total focus and your full attention. It is not without its merits, but in reality it's a million miles away from the freedom, joy and simplicity that is so attractive about free soloing.

As I ponder, my thoughts wander back to those early climbing days and my introduction to climbing. It is amazing I got drawn into it in the first place. My introduction was less than positive, and if my younger life had not been spent looking for risks and sticking my neck out, I may well not have got off the starting blocks. Climbing back then, pre-climbing gyms, was not as

Above: Richard Pollard traversing into the cave at Penally. Photo Lee Bartrop. Below: Max Dutson on TFM, F6c S1, at Penally. Photo Richard Pollard.

easily approachable as it is today. It was a chance meeting whilst working and getting my kicks as a tree surgeon in the mid-80s that I met an addicted and dedicated climber – a rare character who lived and took an 'off the beaten track' view of the world in general. I say 'oddball' in the nicest way, and we really meshed.

He would turn up for work in his Mk1 VW Beetle, which was hand-painted red and wax-oiled to within an inch of its life. His packed lunches looked like something from Peter Rabbit's larder consisting of raw veg, including carrots, radish, celery, garlic cloves and whole onions that he would eat like apples and stink out the cab of the wagon – much to his amusement and everyone else's annoyance.

His parents christened him Malcolm, but I later found out his friends had given him the nickname ALF (which stood for Animal Liberation Front) after he had got himself a job in the local abattoir to uncover and expose animal cruelty which ended up splashed all over the front pages of the local newspaper. Our planned first weekend's climbing was foiled when I got an out-of-the-blue phone call from his partner to say he would not be able to make it. ALF was lying in bed in Buxton Hospital after a 40ft fall whilst soloing. He had made a deep crater at the base of Ruby Tuesday, an E2 at The Roaches.

Having spoken to him later about the incident, he explained, in a matter-of-fact way, that he was partway through the crux when he realised that he had nothing left in the tank. He had to make a split-second decision whether to keep pushing and chance falling off out of control into the many boulders littering the base (which may well have been fatal) or choose his spot and jump, hoping that he could land between them and perhaps minimise the damage. He chose to jump – a decision that ended in a totally shattered and crushed ankle joint and a broken arm, but ALF survived.

So basically, my introduction to climbing included the pitfalls, consequences and aftermath of what happens when you screw up whilst soloing. To say he was lucky is an understatement, and after a long rehab, we formed a partnership climbing together for many years despite him having to modify his climbing style to work around his fused ankle.

I'm not sure when I think back now to that first year, whether it was inspiration or stupidity that found me soloing the roof of Wombat, an E2 only a few routes across from where Alf had impacted, but it just shows that you can't (or don't) learn from other people's mistakes. The practice of soloing had gotten under my skin from early on.

Left top: Andy Steinberg on the first ascent of King of the Castle, F7a S1, at Penally. Photo Lee Bartrop. Left bottom: Luke Holmes on Cold Rush, F6c S1, at Penally. Photo Richard Pollard.

It goes without saying that soloing is a risky game with little or no margin for error, but those close shaves come in some strange forms. Over the years, my soloing has been peppered with a few spicy moments. All have left lasting impressions and taught me to never take anything for granted. I have learned to always be on my guard for a multitude of possibilities and pitfalls which have rarely been down to lack of preparation.

On numerous occasions, I have literally found my life in the balance when startled out of focus by an unforeseen incident, such as being hit in the face by a startled Tawny Owl on my then local crag of the Roaches. How it didn't knock me off is a miracle, and it was only by instinct and fast reactions that I managed to hang on by the skin of my teeth. Needless to say, I was a little startled myself.

Another day while soloing in Lawrencefield Quarry, I reached up to mantle onto a ledge only to be met with an army of wood ants who drew weapons and sprayed me in both eyes with formic acid. I hung there blind for some time whilst my eyes worked overtime to water away the sting.

Once, to my horror, I dislocated a shoulder when a foothold snapped unexpectedly, leaving me hanging from a hand jam. Luckily for me, it clunked back into place as I gently tried to move it. I will never forget the sickening pain or the fear of the consequences. At that point, I was committed so had no option but to continue upwards. It had trapped the nerves and left my arm in a weakened state, so it was a tenuous and careful exit, merely holding with the left arm and pulling with the right.

As well as those lessons to always expect the unexpected, I've learned over the years to listen to the omens. For example, if I trip over my shoelaces on the way to the crag, I try to accept that I'm not in the right place at the right time to climb and just sit back and enjoy the view. When I've ignored those signs and allowed myself to be forced by a ticking clock and then got away with things by luck rather than judgement. Well, those experiences are truly frightening.

As a self-confessed sea cliff addict and a compulsive soloist, it was inevitable at some point that these mediums would collide and that meeting would naturally morph into deep water soloing, but for me it was not a straightforward or easy progression as I had a few barriers to break down first.

My climbing came from an old-school trad background, and my drive was for adventurous on-sight ascents of hard or new routes, so with that approach I was blueprinted to dig in and keep fighting.

Likewise, my soloist mentality was hardwired: no grey areas, you just DON'T fall. I was happy with that and lived and played by those rules, so it felt totally alien to be soloing and accepting falling off. Add to that a dislike of cold water, plus an overactive imagination of what lies in wait under the surface, and I had to reset my approach and mentality.

Once the spell was broken and I started pushing myself, I found it added a totally new dimension to my climbing, allowing the freedom of movement I found in my soloing, but with the addition of a fallout zone. This offered the opportunity to push to, or even past, my technical on-sight limit and enjoy it in relative safety.

The north coast of Cornwall is not renowned for its DWS potential as it has a reputation as a shipwreck coast which is force-fed more than its fair share of big swells and undiluted Atlantic storms, so conditions are fickle at best, and the stars need to align for suitable conditions at its limited venues.

My interest was piqued when the ever-inquisitive Lee Bartrop, in his quest to uncover new route potential, discovered a stunning overhanging amphitheatre of rock on North Cornwall's Atlantic Coast. He instantly saw its deep water solo potential, and the hushed word went out to the inner circle which was made up of a small but dedicated bunch of north coast new route activists who, although known to each other, had up to that point all been operating in isolation and over that season the likes of Rich Pollard, Jess Carr, Andy Steinberg, Luke Holmes, Max Dutson, Maca McManus and myself went down to check it out.

It ticked all the boxes: a huge sea-level traverse, a sea cave and heaps of overhanging options over deep water. Over that summer a buzz of enthusiasm was ignited, and after Lee's first ascent of the complete (and very conditions-dependent) traverse, the gauntlet was laid down and picked up with gusto by the dynamo that is Rich Pollard and the cool-headed Andy Steinberg; light-hearted competitiveness ensued.

The rock offers flat, sloping holds which remain greasy until it's blessed with a breeze from the right direction, ideally in conjunction with late afternoon/early evening sun. That summer, round-robin text messages followed whenever the conditions were right and banter-filled evenings of successes and splashdowns were followed by some fine local ale.

The north coast had finally got its very own quality deep water solo venue, and for me, after years of sea cliff climbing in isolation, it was a novel experience to spend time with the bunch of inspiring and colourful characters that I have the pleasure of calling friends.

Topping Up The Soul by Richard Pollard

I'll try to explain what's important to me in terms of enjoyment and experience; most aspects of my life are planned around climbing above the sea, so as it stands, I'm all in, all the time.

So many of my friends have helped or inspired me. It can be a real eye-opener to see how we all react at the sharp end or in those moments of absolute joy. I've been lucky enough to witness greatness in so many of them, regardless of age, sex or ability. I don't feel like I've ever had much talent or skill myself, but I do have a will to keep on trying.

Rich Pollard

From the start of my climbing, the shared memories, friendships, fun, and piss-taking have been as important as the movement over the rock. I can't help but feel a little sorry for those who don't understand the level of trust and sometimes intensity that binds us as climbers.

Deep Water soloing plays a significant part in my life. The rock and the ocean can be a comforting space to rid yourself of any problems and focus the mind.

Growing up in Cornwall, the ocean has been my escape. In fact, I can't remember a time before the sea. I was lucky; my lovely parents would take me, my brother, and our mates to the beach whenever the weather was good.

Work, home life, travel, and even writing this all come second to my overhanging addictions. Relationships, time and money all pay the cost for my devotion, and I wouldn't have it any other way.

No one owns a location. The sea cliffs are for everyone, but on occasion, a place can really grab you; for me it's Boscastle.

When conditions are perfect, it's as beautiful a place as any other I've climbed at. The afternoon and evening sun can warm not only the rock but something else too. Spring 2018 was the first year I climbed there, but years before, Lee Bartrop and Rich Hudson had been putting up some new trad lines. Later he took another addict, Max Dutson, and they traversed in. Lee had seen the potential of the place as a DWS venue. It was early days, and we'd been, sort of, invited to check it out.

Now, it's important to say that some parts of the north coast can have a reputation for being a little loose, so I was slightly sceptical about the rock quality, but in the excitement of seeing somewhere new, I got going first. I headed all the way over to the far pillar next to the entrance to the main cave only to plop off before the rest. The line was without chalk, and I was hooked in those first moments.

In that first season, Andy Steinberg and Lee put up some brilliant deep water solos. Lee, Max, myself, Jess, and Andy all had projects, and some were done fairly quickly, but it took me a few seasons of falling in to climb Cold Rush.

Friends arrived to check this place out, too: Stu, Macca, Ken, Crispin, Solly and Luke, to name a few. I believe we all knew something special was happening here.

The lines went up, and our great crew grew.

Left top: Richard Pollard. Left middle: Chalk handprint. Left bottom: Richard Pollard and Jess Carr on the traverse into the Penally cave. Photos Pollard collection. Facing page: Andy Steinberg on Penally Dreaming, F6c+ S1, at Penally. Photo Lee Bartrop.

Some places are to share, and some are to inspire; maybe not all experiences are better described first!

Even with all of its history, the north coast still feels fresh and new to me. We are blessed with so many coastal locations, but conditions are fickle, and alignment is key. Lee's Penally cave traverse is like this. It took seasons of work, but what an achievement. His smile and happiness on completion were unforgettable. I had followed him most of the way, but after a fall, I swam to the end to watch him complete this stunning line. The climbing is varied and exciting; the situations and rests are improbable. This cave is seldom experienced so being in this realm feels sort of precious and rare. Such places are special, refilling my soul. Connection, happiness and satisfaction are the result.

I'm not alone in my love for this place. All my friends have something to give, and without naming everyone, it feels pretty bloody amazing – together, for me, we are like family.

It's funny how climbing can inspire such focus and intensity on a particular place or even just a few holds. The on-sight still has its part to play for me, but sometimes familiarity can help the head: learning, progressing, and getting further.

Embracing the immersion.

Breathe in, cool down.

Swim back, looking up at the rock.

Rest and reset.

Try again.

We had just got back from Mallorca; the conditions hadn't been great for DWS, but the sun had been out, and spending time with friends is priceless. We didn't want the holiday vibe to finish, so myself and Lee headed straight for the north coast. He is one of the most positive, easy-going people I have ever met. For your abundant enthusiasm and kindness, Lee, I thank you.

We'd been chatting about a possible DWS line over a cave roof, and as he'd climbed some routes there previously, he also knew where the best abseil point would be. I don't like abseils over sharp rock; they have always scared me, but the juice looked worth the squeeze, and Lee seemed happy for me to go first. Once down at the ledge, I chalked up and set off, heading up and right, feeling the solidity of the rock as I went. The line sort of formed without pressure leading in the direction I envisaged it would take.

The climbing wasn't particularly hard, but it completely absorbed me. An absence of other thoughts was refreshing, and even now, the singularity of those moments stays with me. Traversing the cave roof with all senses turned up to 11, I spotted some fool's gold, or what looked like it in the rock. Cool route name, I thought for just a second, then back to now. Carrying on, I made it to a resting ledge brimming with joy and exhilaration. I sat there contemplating but focused.

I didn't want to drop it from that point. I was already fairly high and to fall off would have been a shame; I needed to do the line justice by topping it out. Anyway the next section above was too high to fall from, so I just went for it steady, checking the holds until a walk-off was possible. The congratulating handshake and a hug from Lee finished off a very memorable experience I'll always treasure.

This, for me, was the perfect return home and showed how good it can be climbing in Cornwall.

***

I swam over with my dry bag for a traverse on the start to Pleasures of The Fresh, and it didn't disappoint. I was soon resting on a ledge at the entrance to the cave, waiting for the pump to ease. Placing my hand in my chalk bag – even though I'm not ready to carry on helped me to relax – a comforting habit I expect I share with many.

I've never seen an orange glow anywhere else so encircling than at Penally; it really is magnificent. Swapping hands, I watch my dust in the sunlight slowly fall; it moves first left and then back right. The cave was breathing.

Now, for me, the expression, 'Getting a place wired,' has come from my surfing roots. Catching and then sliding down waves is made easier if you understand the factors that influence conditions.

Tide, wind direction, swell direction, swell size, sandbanks, and rocks can all influence and affect how waves break. So, I was dead set on getting this place wired. In this moment of clarity, I felt contentment. I had realised and understood at the same time what was happening way before the chalk dust hit the sea.

This was why, on certain days, it dried out in here and not always when the sea was flat calm. The gentle rise and fall of the swell was causing air movement in and out: it was a billows effect.

The flip side was that when the swell got too aggressive, the back wall became the source of spray, which made the internal walls damp. A small swell, a rising afternoon tide, dry conditions and still a fair amount of luck gave perfect conditions.

It all fitted.

Facing page: Richard Pollard on Cold Rush, F6c S1, at Penally. Photo Arron West.

Above left: Silver Studded Blue. Photo Jess Carr. Above right: Jess Carr after the first ascent of Silver Studded Blue, F7a S1. Photo Richard Pollard. Facing page: Jess Carr and Lee Bartrop on the first ascent of Amber Glow, E2 5b, Boscastle. Below: Jess Carr on Amber Glow. Photos Mike Hutton.

Silver Studded Blue by Jess Carr

Back again in this familiar place where our small clan gathers on hot summer evenings. The sun has moved around to its furthest north position in the sky, illuminating the orange streaks which look like flaming lava flowing down the black cliffs above the cave. The dark rock ledges absorb the heat of the sun and make for the perfect sunset spot. Sometimes you even see the elusive green flash from here; sun meeting sea.

This evening it's only me and Rich. It was late in the day, and the sea had already moved from turquoise to inky blue. I wonder how deep the water is in the centre of this cave – it seems bottomless like the universe turned upside down.

I have been here so many times, moving across the traverse line to gain entry to the cave; I know it so well that my body flows with little thought. Before I head in today, I notice a butterfly on the rocks. We share these cliffs, this ocean; we are part of it. Jellies who seem to have no particular mind as where to be, just happy floating with the current. Pigeons who dwell in the small cave high up above the reach of the waves. Guillemots, cormorants, oystercatchers, gulls pass by. This butterfly sat alone, set strikingly against the matte black rock, stood out to me. I wondered what they called it.

Super-high tide tonight, so high it laps your feet on the way in. Like satin this evening; calm. A rarity on this coast. So magical when it's like this. The sea glugs gently against the rocks and reflects the mood – the ambience – back and forth, inside and out. As I reach the resting ledge, Rich calls to me and points to the horizon. A pod of porpoises breaches the otherwise unbroken skin of the water. They are beautiful to watch as they move eastwards.

I set off on the line I have been attempting, vertically unfinished, reach the semi-rest position, arms outstretched, glance down and out. I have an audience: three seals curiously watching with their knowing eyes. I admire these creatures, but when I'm in the water with them, I get a little nervous that they will come and nibble my toes.

I turn to the rock, take a breath, and move on. I know where to put my feet, know which damp holds to avoid. Moving around onto the headwall, pumped but calm, I reach the top. It's amazing – those moments when the stars align.

Little Did I Know by Andy Steinberg

A thousand mile journey must start with a single step.

Lao Tzu

Above: Andy Steinberg.
Photo Steinberg collection.

Seven years old, I climb into the footwell of my dad's work van so I'm not seen by any other council employees, and we head off to the beach. Dad is the head lifeguard of the Newquay beaches, so my school summer holidays are spent mostly at Fistral Beach while he works. I quickly realised that the traditional beach activities of building sand castles, rock pooling and catching waves, although fun, just didn't cut it.

I wanted adventure, to push the boundaries, to take a chance. I found myself exploring the immediate area of the coast, looking for different areas of rock that I could jump from with a suitable deep water cushion. I loved the feeling of falling, the excitement and the rush, then trying to pick my way back up the cliff again, weaving back up the path of least resistance, at some points above the safety of the water and sometimes not. I loved the uncertainty that went with it.

As the years went by, the jumps inevitably got bigger, the landing zones got tighter, and the water shallower, always looking to add a touch of danger – always looking to push it a little further.

At 12 years old, I found myself with a couple of friends standing at the top of a huge cliff, known by us as 'Gully'. Rumour had it that previous generations had jumped from up there, but we had never witnessed it. Had they jumped from up here, or was it just hearsay? My friends had no intentions of jumping that day. They were there for the spectacle or to raise the alarm.

I offered my feet up to the edge as I had done many times before, but today was different. Today, I intended to take that step. Everything was telling me to walk away, but I wanted to jump. I'm going... I'm not going... I'm going... I'm not going. With the tide dropping, every minute of hesitation, the fall gets a little higher, and the sun's beginning to set, adding to the atmosphere and increasing tension. My stomach is churning, the adrenaline is pumping, and I haven't even jumped yet. I'm going... I'm going; I took that step.

The second my feet left the cliff edge, the drop became real, and everything turned silent. The empty space below immediately filled with the nesting gulls as I dropped through the air. I kept falling and falling. Boom! I hit the water harder and went deeper than I ever had before. Everything turned dark for a split second until I pushed up through the bubbles out of the darkness towards the light. The second my face breached the surface, my ears were met by the deafening noise of the panicked seabirds above and a shower of guano. I pulled myself up onto dry land. I was covered in bird shit, but I was buzzing. Little did I know how the skills honed in all those years of throwing myself off hundreds of cliff jumps would come into play later.

Fast forward six years: the endless time and simplicity of my childhood years were behind me. The cliff jumping had faded out, and there was a distinctly empty space in my life. I'm sat at home watching TV, and I stumble across a rock climbing documentary: My Right Foot. Instantly, I knew I needed to try climbing. That's what was missing: the adventure, the exploration, the uncertainty, the absence of rules, nobody to answer to, back out around the ocean and the cliffs that rise above it.

That weekend I walked out of a local outdoors shop with rock shoes, a chalk bag and an offer from the shop assistant (who happened to be a climber) for a day's climbing lesson. We spent the following Saturday at the magical Cornish sea cliff of Trewavas Head, climbing a few of the routes that scaled its granite walls. From the moment I took that first step, and my feet left the ground. The void had been filled. This ticked all the boxes.

I came back to that crag where it all started later that summer armed with rock boots and a chalk bag – no safety gear, no climbing friends, mechanical or human, and no instructor. I set off and soloed three of the four routes that we had climbed on that first day. Next was the harder, taller, more exposed of the four routes, Williams Chimney, with a grade of HVS 5a, which at the time meant nothing to me. I set off, slowly edging up one move at a time; the anxiety began to creep up the higher I got: 'Breathe, don't look down, just relax and keep moving up towards the security of the chimney where I knew I would feel comfortable again. I knew the technique that was required. I had shimmied up between two opposing walls onto building roofs many times whilst out on those childhood adventures, but I arrived only to find it was wet. Shit! I was back up on the edge of that cliff jump again, stomach-churning, adrenaline pumping and legs shaking!

Little did I know that during the next 30ft of climbing, I would experience for the first time what would later become my main attribute in the years to come. I'm not the strongest, not the most talented, but trying hard and not giving up; not taking no for an answer is what seems to be my strength and is what was required in that chimney section.

That experience should have put me off, but it didn't. It was an early lesson. I was out of my depth, but it showed me things will not always go as expected. Nothing's a given, but moreover, it cemented the try-hard attitude in me and a love for free solo, losing yourself, going at it alone without the hiccups of stopping to place a wire, without giving gravity a helping hand by weighing myself down with equipment and without needing to rely on others. Just the pure joy of the unencumbered movement and the commitment that goes with it, the commitment required is not for everyone, which is what makes this practice all the more special.

The years that followed were full of excitement and adventure: magical days out soloing amongst the glowing architecture of West Cornwall, meeting the many colourful characters that bless the same sea cliffs and share the passion for climbing and who would ultimately become lifelong friends. We would go on to adventures together amongst the golden granite of West Penwith and Lundy, the remote and intimidating walls of Carn Gowla and the brooding giants of the North Coast, to name just a few.

The penny had to drop at some point: to bring together all the ingredients, the skills learnt, and the freedom achieved from soloing blended together with the protection that can be offered from the ocean. The resulting combination provides perhaps the ultimate style.

It's mid-summer, and the sun bakes the granite, so we head to the shaded cliffs of Penwith's North Coast to enjoy a day of trad on the secluded and quiet crag of Pedn Kei. As the day nears its end, we pack up our gear and start walking the dry and dusty coast path that weaves its way back towards Gurnard's Head and our car, but at the junction, we don't turn right back up to the vehicle. Instead, we branch off left towards the headland and a route that I noticed on a previous visit that rises straight out of the ocean: Tropospheric Scatter, E4 5c, a 30m vertical face rising straight up from the looming ocean below. The tide's not great as it's already started to retreat, and I'm not particularly well-prepared, dressed in jeans and a shirt, but it's off the cuff. It wasn't planned, and just like that cliff jump, I want to go for it. I don't want to wait or leave it for another day.

I abseil down as low as the sea permits and pull onto the wall with already tired arms. Rich reluctantly pulls up the rope with a look of concern on his face, but I know this has got him thinking: soloing is not for him. He quite rightfully isn't willing to accept that level of risk, but this form of climbing offers a bridge, a bridge that can be crossed to sample the delights and the freedom that soloing has to offer.

I watch the knotted rope bounce back up the face and disappear over the lip. I'm back here again, the uncertainty, taking a chance, where I want to be. I start to move up the wall, following the cracks and edges that lead up the face, passing the small roofs and features the wall has to offer. Abruptly, the holds seem to come to an end. I take the opportunity to stop and shake out tired arms, take in the position, soak up the atmosphere and scan the rock above, where I spot a distant hold a little higher. I look down. I'm not totally convinced there's enough water to catch my fall at this point. I'm not sure if I'm deep water soloing or just free soloing, but I'm here, and the decision has been made.

I'm going for it. The next 15m pass by smoothly; I'm totally absorbed in the moment, not thinking about the landing and not thinking about what's above, just steadily moving up across the rock, enjoying the climbing. The volume has been turned down; this style of soloing offers a second chance should you need it. I reach the top and pull over onto the platform to join Rich again: 'Well done you crazy bastard!'

Both of us are full of excitement for that moment and for the lifetime of adventures that is to come in this beautiful style of climbing. The Cornish coast is a special place. We don't have the concentration and convenience of routes like the neighbouring areas of Devon, or where it all began on the cliffs of Dorset, when it comes to deep water soloing. But what we do have is special: gems scattered along our stunning coast.

Nothing will be handed to you on a plate; effort will be required before your hands even touch the first holds. The fierce Atlantic will need to be sleeping; complicated access overcome, humidity and wind direction playing ball and sometimes a combination of all of the above. But when it all comes together, you will be rewarded with memories and experiences that won't ever be forgotten.

Dreaming My Way Past Trenarren by Tom Last

The undulating, seldom-visited coastline that bounds the west of St Austell Bay is only middling in height relative to the great cliffs of the Southwest. However, it is cut with tiny coves, ruined buildings and the flotsam and jetsam of centuries of lives lived along the shore. The names on the map echo like adventures from Robert Louis Stevenson, heading out from lonely beaches on a following sea – Phoebe's Point, Silvermines, Ropehaven, The Bite – but you are back in time for tea.

I'd stared from my window at this coast for years: parallel lines drawn south to The Gribben and to The Dodman, their features sharpening then fading from dawn to dusk with each swing of the Sun across the bay. I'd map the coast's lines to tidy my mind, picturing its ways as I had as a child mapped toy soldiers onto battlefields, weaving regiments through my parents' kitchen.

Some way down the bay's western shore is Black Head, the muscular bully of the bay, poised like a model tank, ready to charge off the breakfast table. Steep greenstone walls crane over calm waters, humming with potential energy. Things changed, and I moved away from my window – just a few miles away.

Covid came and went and hovered around our lives. Under quiet skies and empty hours, I visited the headland again. Time and again through Polgooth, Tregorrick, Lobb's Shop. Past the old rifle range and the tiny carved compass where someone spent similar ages occupied; driven by my own distraction through the blackthorn that cornices the zawn and down to its sanctuary.

Some days, smacks of jellyfish were a relief, and I would rather watch than climb. Laughing in November as a southerly swell engulfed the once sunny gearing up ledges where I had dozed above.

One evening, I arrived to find chalk on nearly every hold. Blood spatter at the scene of the crime. I knew the culprits for sure. Friends had found the head, too, and joining forces we mapped the wall: an easy traverse, high feet.

'No, higher still…'

'Crimp the slot, don't jam it…'

Wild dynos gave way to static beta. Waters which grew deeper with every splashdown and every subsequent reimagining. Five metres… eight.

'I've never seen the bottom; it has to be at least 10…'

My route ended in a staccato fight, as things do. Limpet like on the crux, unshakable. Then, stabbing at the easier crack. Now, wobbling on the ledge. A ring of faces appears above and talks me through the final moves – soup up.

The world opened up, the moment passed, and we moved on and away, but there on the map in my mind is a pair of crossed swords where we drew our battle lines on that quiet coast.

Above: Tom Last. Photo Last collection. Facing page top: Andy Steinberg on Release the Hounds, F7a S0, at Black Head. Photo Richard Pollard. Facing page bottom: Andy Steinberg climbing Ring of Bright Water, F7a+ S0, at Black Head with Jess Carr and Rich Pollard looking on. Photo Tom Last.

Above: Max Dutson on Magic Fly, E2 5c, at Carn Gowla.

Facing page top: Max Dutson traversing from the abseil along the base of Carn Gowla.

Facing page bottom: Looking along the north section of Carn Gowla from America Buttress to the Coastguard Point.

Below: Doctors Farquhar and Dutson on duty with Festival Medical Services at the 2002 Glastonbury Festival.

Photos Grant Farquhar.

How Cold is too Cold? by Grant Farquhar

As Max Dutson and I traversed in from the abseil rope under the Baptist Cliff to the Sewage Pipe area of Carn Gowla, I remarked to him that this would make an excellent sea-level traverse. Shortly after these words were uttered, we were both soaked and scrambling to avoid being washed into the sea carrying full racks of trad gear.

So, on the following Monday after work, I was fully psyched to attempt a complete sea-level traverse of Carn Gowla. It was a beautiful blue-sky day with a light breeze and only a small swell running – the surf at Perranporth was forecast to only be in the 1–2ft range, and it rarely gets less than this on the north shore. I didn't have a wetsuit at the farm but figured that boardshorts would be OK as I would, I reasoned, only be in the water for quick dips between sections of traversing. After all, it was the height of summer, and I regularly drove past old ladies 'cold-water swimming' in Penzance harbour. It couldn't be that cold, could it?

I got to the coastguard lookout car-park high above the sea at 5pm around mid-tide on the ebb. All the way in the car from Penzance I had been deliberating whether to start from south or north but decided on north as it had a more defined starting point – America buttress – even though the guidebook advised that low tide was required to negotiate this area which would coincide with my arrival if coming from the south. If things went according to plan, that is. Which they didn't.

So I parked up and started trotting north towards America buttress. Glancing down and left, it looked an awful long way to (what I thought) was the other end of the cliff, and thoughts like benightment flitted through my brain. I quickly countered them by reminding myself that I had a torch and there were, most likely, some escape routes, and I could always just have a quick look and bail if it looked too gnarly.

The way down to America was not that easy to find, and I don't think I went the right way, having to do a lot of downclimbing in serious positions, but I just got on with it rather than scouting around for an alternative way and before long I was peering into the gash guarding the base of America where the guidebook recommends lassoing a spike to facilitate a tyrolean traverse. It was only about 10ft across but tapered into barnacle-encrusted and greasy depths. I could see a possible crossing point at this state of tide, but the rock on the other side was very wet, and falling in was a distinct possibility. The ledge on the other side was large, so my plan was to chuck my rucksack across onto that and then climb over.

1,2,3, heave. The rucksack hit the ledge, teetered on the edge for an agonising moment and then plopped into the agitated sea below. I immediately jumped in to grab the bag and swam across to the other side. I sorted myself out and then started traversing but before long reached an impasse of slimy rock and took to the sea again. It went on like that for a while, with a combination of climbing and swimming to make progress when the rock was too wet or slimy, which was often. The rock architecture there under the black walls is very impressive, and there weren't many sea-level ledges with the rock mostly continuing steeply into the water.

After about half an hour, I realised that when I was in the water the current of the ebbing tide was helping me along but would have been a bastard to swim against given the sea state which seemed much rougher than the forecast 1–2ft swell at nearby Perranporth would suggest. Surging white-capped waves continuously broke against the cliff causing significant backwash which made swimming and, especially, exiting the water tricky – and tiring.

To begin with I didn't feel that cold in the water, even though I was simply wearing board shorts and climbing shoes. Cold water swimming? No problem, I thought. I had a T-shirt and a windproof in a dry bag inside my rucksack, but they wouldn't have helped me in the water. Before long I realised that I was committed – my 'look' had turned into a one-way trip, and there was a long way to go to the point with an island that I could see in the distance. The climbing interludes became less and less until I decided, finally, to just swim for it.

At this point, I had been going for over two hours and was very cold and shivering. If there had been an option to bail up an easy section of cliff then I would have gladly

Max Dutson taking flight at Carn Barra.
Photo Dutson collection.

taken it, but there was nothing: just massive cliffs of black rock dropping into the water. So I cracked on, alternating between swimming on my back to check out the future new-route potential of the cliffs and more determined swimming on my front to make progress.

The point and, I was hoping, an exit was nearing, but it was still a long way off. After a few goes I exited the water onto a ledge and realised that along with uncontrollable shivering, I was losing dexterity – my arms and legs were turning into rubber and not obeying the signals from my brain. I had to visually scan where they were before I could rely on them on the hand and foot holds. It belatedly dawned on me that I was in quite a serious situation at this point. Nobody knew where I was – which was in the swell and current below a massive, inescapable north-facing cliff, in boardshorts with hypothermia.

But there was nothing else to do but keep on going. The next section of cliff featured a massive arch which I went underneath. The scenery was incredibly spectacular, but I was not in a mood to appreciate it and carried on to find a little ledge in the sun that was out of the wind. I needed to stop and try to warm up although, in retrospect, this was a mistake as I've since read that your body temperature continues to drop for 20 minutes after exiting the water, so I should have just carried on and swam the final 100yds from the island I was on to the next point where I could, finally, see an exit.

I stopped, emptied the water from my rucksack and got the drybag out. The first order was to put on my T-shirt and windproof and then extract my cigars from their waterproof case. As I sat in the sun it felt like I was warming up, but I couldn't stop shivering. Then, with relief, I noticed a guy fishing on the next point so knew there was definitely an escape from there. God knows what he thought of my sudden appearance on the rocks offshore.

There was no point in putting off the final swim although my body was rebelling at the thought of it. I packed everything away again and then, without hesitation, plunged into the sea and struck out for the sloping ledges on the other side of the channel. With my rubber limbs exiting was not easy. At one point I got washed off a slimy slab from 20ft up falling uncontrollably into the water which, fortunately, was deep. This was not a place to break a leg. Next time I made it, dried off as best as I could and put on my clothes again.

As I mechanically trudged up the path to the coastguard station, I came across an elderly couple sitting in the sun who I guessed were Dutch. I don't know why; they just looked Dutch. They both made an involuntary jump when I appeared before them on the path. 'Where did you come from?' they asked. I tried to reply, but my mouth wasn't working. 'The sea,' I eventually croaked, gesturing behind me. They looked at each other in puzzlement and, not in a position to elaborate, I kept on going to my car patiently awaiting me in the car park. I had the heater turned up to full blast all the way back. In fact, the metal-framed sunglasses that I was wearing heated up so much they started burning my face, but I

was still shaking like I was having a seizure. The bearded guy behind the counter at the petrol station looked at me with concern as I repeatedly tried and failed to insert the vibrating credit card in my hand into the slot of his card machine. He probably thought I had a bad case of the DTs, this being Cornwall after all.

'How was the climbing?' my wife enquired when I got to the farm. 'Interesting,' I responded. The visible shaking had subsided, and she didn't ask me any more questions, so I just left it at that.

This personal epic occurred in August when the water temperature in Cornwall is at its annual maximum, which, according to the internet, was 16 °C at St Ives and 17°C at Perranporth that day, so it would have been somewhere around there at Carn Gowla, and I wasn't expecting to get hypothermia on that particular summer's evening after work.

Aware that I should have considered this before, I belatedly wondered: How cold is too cold when it comes to water temperature? Researchers from the University of Portsmouth set out to determine exactly this in a study they published in the *British Journal of Sports Medicine* in 2019[5]. In the study, 12 competitive swimmers swam three times, for up to two hours, in different water temperatures between 14 and 20°C wearing swimming costumes. Their temperature was measured (rectally) to determine the onset of hypothermia which they defined as a deep body temperature of below 35°C.

They found that the majority of their subjects, who were athletic swimmers, could not complete the two-hour swim in 16°C water without becoming hypothermic. They concluded that humans vary widely in their ability to maintain body temperature in cold water and that heat loss in water is increased by exercise. The main factor in maintaining deep body temperature when swimming is the amount of body fat, with thinner swimmers showing more rapid falls in temperature due to lower levels of insulating subcutaneous fat. This would generally describe the body shape of most climbers. They also noted that swimming performance deteriorates before the rectal temperature reaches 35°C and that swimmers could not be relied upon to assess their own deep body temperature while swimming.

It may only be one study with a fairly small sample size, but their methodology appears robust. After being provided with the data from this paper, FINA, the international swimming organisation, were sufficiently persuaded to amend their rules for racing in low water temperatures and made wetsuit use compulsory in water temperatures below 18°C. On the basis of my own, n=1, experiment I can only agree. I wasn't expecting to get hypothermia on my 'look' at the Carn Gowla traverse. I didn't have Dr Dutson handy to take my rectal temperature, but I had all the symptoms of it: shivering, loss of dexterity, and slurred speech (and I hadn't been drinking). Several hours in the water in boardshorts in that temperature can do that (I now know). Of course, it would have been a completely different, but no doubt less interesting, story if I'd worn a wetsuit.

5   Saycell et al. *British Journal of Sports Medicine* 53(17), 2019.

The Sheltering Sea by Grant Farquhar

His 'help' or 'save me' like a bird
Bore back only the stretched, white fingered
Hand of the swelling sea and found him sheltered,

Sheltered in soon all of us to be
That memory against the scuppering rocks,
The spilling aprons of the sea.

At Whose Sheltering Shall the Day Sea by WS Graham

It was high summer; we were staying at our farm near Penzance, and I was keen to climb. I hadn't done any DWS in Cornwall at that point, but Gurnard's Head and Carn Gloose, on the north coast of Penwith, looked like the most obvious candidates for exploration. So, I decided to do Astral Stroll as my first route – it looked ideal for DWS, and it was an *Extreme Rock* tick, after all.

Bikini-clad beachgoers wandered past as I descended – on the wrong track – from the Gurnard's Head Hotel. Realising my error, I was soon walking past Gurnard's Head itself and found the ridge of Carn Gloose. Scrambling down this, I arrived at an obvious gearing-up spot and a fine, lofty perch from which to contemplate the ever-changing moods of the sea. Often hostile, that day the sea was welcoming – offering respite and shelter from the strenuous effort promised on the cliff. But shelter too long, and you will not be leaving that cold embrace.

The recommended approach is by abseil. Well, I had neither a guidebook nor a rope but found downclimbing the line of the abseil to be pretty steady. It's above a platform and therefore death if you fall, but the rock is sound, and there were no hard moves although the downward sloping low-friction ledges were disconcerting.

When I got to sea level, the crag looked amazing, and there was no one on it. The tide was out, and boulders were visible underneath pitches 2 and 3. I could see that the tidal range was massive – at least 10ft – so while Astral Stroll should be safe at high tide it was not safe at that time, I reckoned. I decided to solo the first pitch anyway 'to have a look'. Above the belay at the end of the first pitch is a groove, so I climbed up this and stepped right into another groove which had a clean fall into the deep water below. It was all fairly easy climbing apart from one tricky 5a-ish move at about 20m. I stood and contemplated it, revelling in the situation – this is what you came for, I thought to myself. Feeling the holds, I figured out the easiest sequence and then executed it. Looking at UKC later, it appeared that I'd soloed High Frontier, first climbed by Rowland and Mark Edwards in 1980 and, appropriately enough, often used as a bail-out from Astral Stroll if conditions are bad or, in my case, if the tide is out.

I was heading to Carn Gloose again except this time I didn't take the wrong path and get lost, and it was high tide. With a light north wind but substantial north swell, the low version of the first pitch that I'd done the last time was inaccessible. I knew my way to the high variation from my previous visit, so I was soon climbing around the arête and into safe DWS territory. There was no sign of the boulders below Astral Stroll this time; they were safely covered by the tide. The crux of the second pitch is an awkward, blind step down from the belay that it looked like you could easily bypass from the lower variation, but I did it anyway and carried on to 'the bidet' belay which is a small ledge in a crack near sea level. This feature is well-named – if a big sea was running into the crag you would be getting soaked by spray coming up the crack between your legs. But today it was hot, out of the wind, and the swell was running perpendicular to the crag.

Right and facing page: Chris Hall on Leviathan, F6c S1. Photos Grant Farquhar.

My feet were swelling in the heat, so I sat happily on the bidet and took my shoes off for a bit. Like last time there was nobody on the crag, and it felt like a very special place with overhanging rock all around. I marvelled at the line of Astral Stroll which takes an inescapable line through very impressive territory; a bit like an E1 version of The Moon at Gogarth except even better because you can DWS it. The crag was baking in the sun with the sunlight sparkling off the water. I still didn't fancy falling in the 16°C water, mind you, and so was not climbing with the usual degree of relaxation that I would above the warmer waters of Bermuda.

The crack at the back of the groove was a bit damp, but the holds were massive, so it didn't matter. The traverse right was obviously going to be a long step onto a big smear. That this was probably another crux was reinforced by a stuck cam in the crack above. Around the corner, chalk disappeared down a groove where someone had gone off route, I mused, or maybe regained the rock after a fall into space?

Astral Stroll continued above on good holds and in-balance to a steepening with some well-chalked crimps that I didn't bother to use as a low sidepull permitted a long reach to massive jugs. Stepping right to the belay below the final pitch, I looked down and still appeared to be above water although I knew the base of the slanting gangway below the north face must be lurking down there, somewhere.

One steep pull on massive holds led to the easy final section up a groove with good foot ledges on the wall to the right and the finishing ledges. I scampered back up and left to the ledge and basked in the sunshine, admiring the ever-changing ocean. Often hostile, today it looked welcoming. At high tide, Astral Stroll has got to be one of the best DWS in the UK, comparable to The Conger. Not many moderate DWS take the easiest line through such overhanging rock.

My next visit was to Gurnard's Head where I met up with Chris Hall, and we sooloed the superb arête of Leviathan. Sitting on ledges in the sun afterwards, we bullshitted and admired the line of Tropospheric Scatter, another Rowland Edwards route, but it clearly needed a spring high tide to DWS as there are ledges underneath. So with that in mind, I was back there on a boiling hot Friday evening at 6pm in time for the full-moon high. As seemed usual on the Cornish crags there was nobody else there, but there were signs that people had been there, namely a streak of turds down the Severe on the right side of the wall. It looked like someone had stuck their arse over the edge and shat down the route. Dirty bastard! I thought as I down-climbed while trying to avoid the shit and holding my breath against the stench.

The sea-level traverse to the base of Tropospheric Scatter looked hard. There were no inviting sea-level ledges with the slate-wall dropping sheer into the brine and the only seemingly available foothold being lapped by waves.

Facing page: Dave Pickford making the first DWS ascent of The Silence of a Lamb (F7b S2 or E6 6b) – an example of a route that is far better as a DWS than a trad route. Photo Pickford Collection.

Reaching out from the corner brought a good sidepull within reach. I only needed to use the foothold briefly in order to get established in a layback and bring my feet up onto high smears, but it was long enough for both shoes to get soaked. I only had one pair with me, so changing them wasn't an option anyway. I continued to the base of the crack of Tropospheric Scatter and made a couple of moves up to where there was a decision to be made about which hand to use in a finger jam. I already felt pumped and dithered to the point where I thought: best climb down and continue traversing into the corner of Shark where I thought there might be resting ledges. This proved to be a mistake because the corner dropped sheer into the water with only a leg-pumping bridging rest available.

I traversed back to the start of Tropospheric Scatter feeling more, not less, pumped when I got there. A couple of determined pulls, no dithering this time, brought a large hold within reach, and then I continued to motor onwards. Good handholds beckoned me on, but they always seemed to come at full stretch from the footholds, so it was difficult to recover, but soon I was able to traverse onto a ledge and an à cheval no-hands rest on the arête. This was a very spectacular position to be in and a good rest, but my arms already felt far gone, and I did not recover as much as I hoped, even after 10 minutes sitting there. Nice as it was, I wasn't going to sit there all day.

I duly set out into the crux which the guidebook had said required a 'long reach'. I set out in search of this, transferring as much weight as I could onto my damp shoes and using all my old-climber cunning to avoid having to pull hard on the handholds. The long reach done, I hung off a small but incut hold and fingered the holds above which were good but, again, at full stretch. Looking down, I could see a good foothold at my waist, and all I needed to do was smear with one foot and rock up onto it, but every time I tried to do this my arms didn't have the beans to make the move.

Rapidly pumping out, I was running out of options, but I could see another large hold and potential rest on the arête. A precarious foot swap with 1% power in my arm batteries put this hold within reach, but as I stretched for it, I started to barn-door off, and it was game over. My arms wilted, the wall tilted and then rushed past as I pushed off to make the leap outwards to clear the underwater ledges at the base. Lying on my back in the sheltering sea, I gazed up at the towering walls of slate and marvelled at what a fantastic crag Gurnard's Head is.

The next essay is dear to my heart and typical of the unsung exploits of Max Dutson. He first recounted this story to me as I was driving him from my house in Fachwen to the pub in Llanberis. It is the story of how he wasn't 'officially' late to report for his first day at work in Treliske Hospital. The problem is that he arrived by helicopter, as a patient, with hypothermia. It was just another day in the life of Max.

Lucky February by Max Dutson

Sitting on the cliff top at Boscastle after another beautiful evening deep water soloing. For once, there is no wind. The sun is setting, and the noisy surge of the waves down below penetrates through the eerie calmness of an otherwise still North Cornwall coast. A darkening sky and the booming below evoke powerful memories of an experience years ago when I first moved down to Cornwall. Despite the passage of time – over 27 years – they cause me to shudder and consciously push them away in a mindful attempt to enjoy the here and now.

I'd just moved down to start my second intern job at Treliske Hospital in Truro. I'd had a good six months prior to this, spending more time on the grit than in the hospital where I had my first placement. The grit soloing was delectable, and I yearned for more of the same. Luckily, for February, there had been a few dry days, and I felt there was no reason why the North Cornwall cliffs couldn't offer a similar experience. I finished work early and found myself cycling to Cligga Head. I had, maybe, an hour of daylight. I stashed my bike near the clifftop and scrambled down a fisherman's path to a sloping granite promontory. To the north was a wide-mouthed zawn with 120ft cliffs on either side. Already, it was gloomy as the night approached, but the guidebook's description of Queen Jane beckoned me. Purportedly, it was a 'pleasant' two-star Diff up a solid granite slab on the edge of the zawn. I quickly changed into my climbing boots. My trainers joined my jumper, Swiss Army Knife and camera in my daysack. The sea was relatively calm as I traversed above it with only a small swell pushing up the walls of the zawn, lucky for February.

The guidebook mentioned climbing up 'a narrow slab', but there were a number of narrow-looking slabs above me. I started up one that looked relatively amenable, although the granite was surprisingly rotten for such a local classic. As I pulled around a small overlap into the base of a flared, slabby corner, I realised how dark it was getting. I needed to make a decision about whether to go up or down. It seemed to be a bit too dark to start climbing down again, especially as I would have to reverse-mantle a few narrow, loose ledges. So I pushed on. I carefully stemmed up the corner thinking how hard it was for a Diff. By this time, the profile of the clifftop had merged into the darkness although the top must have been close, surely?

I made a cautious mantle onto a large clod of thrift growing in the back of the corner and searched upwards for my escape. The rock had become very loose, the holds requiring gentle downward pressure rather than outward pulling. Also, the corner seemed to steepen a bit, and everything had become somewhat difficult to see. Why hadn't I started earlier? It was half past five and dark. I realised – too late– daylight hours in February are in short supply. The decision to remain where I was seemed easy at the time. It was too dark and too loose to climb on; it was too dark to descend. And, having been climbing, I was reasonably warm. Also, the thrift that I was standing on just accommodated both feet and even allowed me to, very carefully, change into my trainers and don my jumper. I also found a hat at the bottom of my bag. Once I had sorted myself out I started to contemplate my predicament.

On the plus side: I was warm enough, at that point, and it wasn't particularly windy (indeed the dew had started to collect on the thrift plant). In the dark, however, the waves sounded larger than I remembered them at sea level, and the holds that helped prevent me from teetering off my precarious perch, 100ft above the sea, were small. A crimp was what I had for my left hand, and my Swiss Army Knife levered into a hairline crack would have to do for my right hand. I was very grateful that I'd packed that penknife. My camera also came in useful for taking a couple of selfies – one of my feet and one of me looking somewhat solemn in my bobble hat: to help me remember my night out, if I survived.

I checked my watch – quarter to seven. How odd. It already felt as if I'd been there for hours. My eyrie of thrift had also started to concern me. Damp, it was now slippery. I stood closer into the corner and focused on putting my weight through my toes. I was starting to feel a tad frightened. I started to shout. Initially a forceful, but gradually a more feeble: 'Help!' The words seemed to float off out to sea where only the seals could hear them.

Above: Max Dutson during his epic described on these pages. Photo Max Dutson. Below: Max Dutson DWS on Cligga Head. Photo Mike Hutton.

A cunning thought passed through my mind: six SOS flashes – dots and dashes – could summon help, I hoped. Holding the knife in my teeth I reached over my shoulder into my bag and pulled out my camera. Then, pointing it out to sea over my shoulder, I pressed the shutter six times. I hoped it flashed. A few minutes passed, and not much happened. I did it again and then realised how futile it was.

Time passed; I was more frightened. Swiss Army Knife back in hand, toes digging deep where the thrift anchored into the rock, I checked my watch again. Ten o'clock: better get a couple more pints in before last orders. Time was passing slowly; I was cold. The sea sounded rougher. My watch was going too slow – half 11... midnight... 20 past – I was shivering, cold and scared. With a slight start, I suddenly woke. I must have very briefly dozed off. How had I not fallen off? Christ! Keep awake.

Two-ish. Frigging cold. Talking about Christ, praying didn't help. I didn't know whether I'd see my family again; things started to become interminable. My mental torture reflected the state of the thrift beneath me. It was on its last legs. The covering of green had all but gone leaving a sorry plug of mud still somehow anchored to the back of the corner. At least, I thought it was.

Armageddon came at about three am. Suddenly, the remnants of my beautiful eyrie parted contact with the rock. I guess 'suddenly' is not quite correct. I had all of a couple of seconds to realise that I would no longer have anything to stand on and jumped into a bridging position with one foot on either side of the flared, slabby corner.

But, this position couldn't last – I was 100ft up; the waves were crashing below; it was pitch black. It was three am for fucks sake! Then I started sliding. This was the end.

I was expecting to accelerate over the overlap and into the dark abyss below, but somehow I managed to brace my feet against the corner, so the slide was, almost, more terrifying. Slower and more protracted – but with the same inevitable consequence. Every few seconds, I would slip a foot closer to my doom. I didn't get any flashbacks of my life – I was too busy being consumed with terror.

But then my atheist beliefs were obviously turned turtle. I stopped. The plug of mud that had disintegrated under my feet had partially accumulated on the lip of the overlap, 20ft below, and, before I started my final descent, my trainers embedded in this meagre collection of earth. I scrabbled for handholds, grabbing at the rock for a renewed purchase. A lifeline had been thrown to me. Christ doesn't mind a few blasphemies, and I let rip.

The sea was still surging below, however. The passage of time remained interminable. The darkness was almost overwhelming: I couldn't stand the mental torture. And, on top of this, my mud lifeline was again starting to wither.

I made the decision to jump off. I didn't know what I was jumping into, but as, at long last, shards of dawn light filtered through the devilish black, I could vaguely make out the zawn below. I resolved to jump off at nine am. I decided that by this time my mental anguish would have become too much. I thought I might survive with maybe only a broken leg or similar. If I ran sideways as I was falling, across the 80-degree slab, I might just land in deep-enough water and avoid the projecting rocks.

Of course, this didn't take into account the waves now spraying 30ft up the zawn walls. Also I was probably going to be forced into the decision to jump off as bit by bit, my foothold was shrinking as the soil crumbled away.

All was not lost, however. Despite the increasing swell a small fishing boat was out on a morning foray. It was 20 minutes to nine, so just in time. As it chugged closer, I yelled... and yelled. It went past and headed off. The one and only chance gone. My resolve reflected my foothold – tenuous and marginal.

The nine o'clock watershed came and went. The mental turmoil was less powerful than my absolute desire to cling to the rock, though I didn't know how this would end.

Another boat. A smaller one. Two people. I shouted again, hopeful yet realistic. And then, amazingly, a response:

'Where are you?'

'I'm up here' ...stupid.

They didn't know that I couldn't wave as I would have fallen off. Eventually, they spotted me.

'I need a helicopter.'

A vague response, and then they disappeared out of view. What? Bloomin' fishermen!

Time slowed down again. I waited, not knowing what to expect. I waited, knowing not to expect anything. But the turtle turned again. The distinct noise of an approaching helicopter almost caused me to discard my atheism. A winchman was lowered a few metres away from me. The downdraft almost nailed my coffin, and he could see the petrified look on my face as he prepared the underarm harness:

'Don't worry!'

'I'm slipping!'

The downward force from the rotor blades caused the last remaining clod of earth that was once my comfy thrift perch to part beneath me. I fell just as he reached and put the harness under my arms. Lucky for February.

Compounding Errors – memories of a near disaster by Lee Bartrop

The Atlantic Ocean. It's May, around 2005. If that's correct, then I was 32, and Matt, my brother-in-law, was 20-something at the time. We are off to climb a cliff long since departed from the mainland. This sea stack in Bossiney Bay, North Cornwall, has been calling for some time. A tall, jagged edge standing out in the sea. There is a cairn atop, but we know not how many have been. We want to go there. Though visible from many places, it has a hidden side facing the sea. Reasoning that it's not that far, it's not that cold, and the sea's not too rough, our intention is to swim to the stack with our rucksacks in bin bags. We have no phone, no head torches, no dry bags, and no wetsuits. That is our plan. I have an idea of where to start – scramble down a grassy slope and downclimb the rocky ledges. From there we can swim, climb out onto a ledge, climb it, abseil down and swim back in time for a six pm family dinner. A lovely famous sunday roast with roast potatoes, a roast beef joint, gravy, fresh veg and wine. Lovely people, loved people, loving people: warmth, love and a cosy home.

It's not that you just don't care. It's not that you just want to get away. It's not that you just are not prepared (hmm... perhaps we could have been more prepared). It's just that you've been in the sea in May in your pants for a short swim like that. You've clambered out onto barnacled rocks. You've climbed a climb like that. You wanna climb that cliff? Go. Have an adventure and then come back. It's just that you've always not really had any spare cash, so everything is basic needs. In moments of capability, basic needs are the focus. When things are going right, one might not need to think of what might go wrong. We left late. The tide was not really a problem – it was the light and how long we needed to get back in time for dinner. It's a four-pitch climb and a 100 yard swim. The climb was really good. It's not an XS; relies on a peg on one pitch, but it's OK. Great positions, good moves, good rock, and good fun. We got to the top, added a stone to the cairn and set up the ab. We're pleased with the ascent. It's a great adventure route. Thank you Mr Fowler. By now it's about nine pm, so light is fading. I ab off to a ledge with one more 90ft ab after that to the bottom. Our plan is to swim, get back to the car, go home and apologise for being late, eat dinner and drink wine. Lovely.

Matt duly abs down. I go to pull the ropes, but the rope is stuck. It stretches like a worm still in the ground to the limit, except it won't snap, and it's not a worm that I'll let go of because I don't want to hurt it. It is a climbing rope, and it is stuck. Our problem is we need it to be not-stuck so we can get down and back. Matt is strong, lean and fit. He pulls on the rope. The ledge is quite big, so he can vary the angle a bit – that sometimes helps. He pulls hard; the worm gets longer. I guess we could sleep on this ledge. We would survive, but we wouldn't be home for tea. Thankfully Matt is strong and the rope, eventually, moves. At the base we pull the ropes, easily this time, and put them in our rucksacks which go in the ripped bin bags.

Stripped to our pants we contemplate the swim back. It is raining and getting dark. The sea has picked up; it is now a lot different to when we left. I'm not sure we should risk going in it. We can see the ledge where we went in, which is where we were planning to get out. But the sea is very rough with waves rushing up and washing back down the rocks. The exit is covered in mussels, and we are nearly naked. We would probably be washed over it and could be sliced and diced. The rest of the cliff is wave washed; smoothed over and over for thousands of years by the sea. The swim is not far, but we would find it hard. We might not get out; we might not make it. But if we swam over there, to where that slope is – yes, the sea is rough, but we could get washed up onto it, and if not swept back out, we could land there and get out. Yeah, sliced and cut, but we could make it. I don't think we should go; the sea is too rough. Oh, but wait a minute now, it seems kind of OK? Nah, it's too rough to swim, it's dark, it's raining. We should stay on the stack and swim back tomorrow when it's light. I am slightly lower than Matt and sat on a barnacled ledge. The sea is rising, falling, and churning just below my feet. I agonise: should we stay or go? Then a larger swell engulfs me and sweeps me off the ledge. I am in the sea; Matt jumps in. We are going.

I have my bin bag under one arm and scooping the sea with the other. Matt has his bin bag on his chest, kicking backwards. We are bobbing, and the sea is chopping. We are going, but not very well. We don't seem to be going to the right spot. We are not making progress. Matt is moved away from me. I've been concentrating on getting

Above: Lee Bartrop.
Photo Rich Pollard.

to the other side, but we are not getting anywhere; we can't get to the landing spot. If we don't get out of here we will drown. But we are trying. We keep going. For a moment Matt is out of sight. The sea is salty, rising, falling, black. I cannot get out of here without Matt. We are being held in place by the standing waves made between the open sea and the backwash from the cliff. The sea doesn't care if we are lost land mammals in the wrong place or dead driftwood dallying until oblivion.

I still can't see Matt. My bin liner is heavy; I'm not making progress. We have to get out of the sea. I see Matt – my wife's brother, my friend, my children's uncle, my inlaw's son, my friend's boyfriend, my friend's friend – and shout over to him. We have to change direction; go sideways into an area of sea which seems calmer, in the lee of the sea stack. We have to go there. We are back together. It is dark. We are near the cliff. It is black. The sea rises; I must grab something. I know when the sea falls I will be heavy, the sea will leave me, and I must hold on. I am lifted with the sea, and I reach with my left hand. I feel for a hold. It is wave washed, but I have something: an edge. I crimp down – I MUST hold it. As the sea falls, I get heavy. My rucksack is in my other arm. The rucksack is falling. I pull on my left, and then the bag is washing away. I pull it up and hold on. The sea falls. I hang in the balance and then up, up. I hold on and get my feet on. I am on; I am getting out. I am standing on a ledge in the dark, out of the sea. I remember thinking everything that could happen was 'no', and what happened was 'yes'. I shouted 'YES!' out loud, yes like I meant it more than anything. What could have happened was 'no'. Instead I shouted a roar, a deep animal guttural adrenaline surge: 'YEAHSSSSS!'

I was shivering from head to toe; my teeth were chattering. I could hardly tear open the bin liner. The rucksack was full of water. I had to get the rope out for Matt who was still in the sea being washed up and back. I couldn't risk going back in the water. I pulled out a length and threw the end. He said he couldn't hold it. Make a loop, he said. I made the loop and threw it again. He put it over his head and under his arms. I let the rope slide as he was pulled back and waited for the next surge. As he was washed up towards me, I reeled him in. He was out. We were out. On the ledge I held his shoulders and looked towards him. It was primal. I stood there and released another animal cry: 'WE ARE AALLIIVVEE!' Matt was silent and still, completely still. I was shaking all over. He didn't speak. He didn't move. He just stood there with his hands by his side not touching anything. I said, 'Matt, we have to put our clothes on, even though they are wet. We have to tie ourselves to the rock. We cannot go back in that sea.'

It is pitch black, so we are feeling our way around on some kind of ledge. I get my clothes on, my harness on, and find placements for gear to attach ourselves to the cliff. The sea is still rising, and it may get rougher. We cannot get back. Matt puts his clothes on, gets his harness on, and we are tied in; we will not die. It is very windy, so we huddle together in an alcove, sit on our ropes, and sleep.

Above: Climbers return from an ascent of Grower Rock near Boscastle, North Cornwall, on 27th Oct 1973. Below: Un-named sea stack, some 200ft in height, off Gunver Head, near Padstow. Johnny Fowler makes a crossing by Tyrolean traverse on 4th Sep 1971. Photos John Cleare.

The climber is climbing as long as his nose is above water.

AW Andrews

North Devon

Very worthy coasteering area between Clovelly and Widemouth… It is useless to judge this as if it were mountain rock, and in general routes which look easy to a mountain-trained climber turn out to be fairly difficult, while those which look difficult often prove to be impossible.

EC Pyatt

Described by his GP, Tom Patey, as 'the greatest mountain explorer of his time,' Dr Tom George Longstaff (1875–1964) traversed parts of the North Devon coast from the age of 12 years during visits to his cousins who lived in the area. These explorations did not come to light until the publication of his 1950 autobiography, *This My Voyage*, in which he described the sea-level traverse of Baggy Point incorporating the Very Difficult Scrattling Crack, which he climbed in 1898 and is now regarded as the South West's first seacliff climb:

'My cousins lived on the North Devon coast, facing Lundy Island. The cliffs are only slate but finely fretted by tumultuous seas. We began climbing in a small way in 1887, and by 1892 we were using rope. Such climbs were nearly all horizontal traverses. The rules were to get around the headlands between the top of the cliffs and the sea below, keeping above the high water mark if possible. There was a good climb from Thrift Cove around Bull Point to the buzzard-haunted cliffs of Rockham Bay. Some of the passages were difficult. The sharp slates of 'Gory Corner', the crux of the traverse, drew blood.

'But the finest climbing of all was on bluff Baggy Point, facing Lundy. The complete traverse is hardly possible in one day, for two sections some considerable distance apart can only be passed at low tide and during a calm. At the southwest corner of the headland is a deep overhung inlet which must be crossed by an awkward descent followed by a tricky jump onto a spike of rock which is submerged except at low tide. We did not complete this section till 1898. The northwest corner of the traverse of Baggy Point is done by Scrattling Crack, a curious freak at an otherwise impassable corner. The crack is about 130ft long, but not difficult since it is not really as vertical as it looks, and a foot or knee can always be squeezed in. It is quite exciting to race for this corner on rising tide with a good sea running.

'But climbing on sea cliffs does not grant us that freedom of spirit which we find on mountain tops. The surf confines us with elemental restraint: thus far and no farther. Yet it is this very intimacy with the sea's infinite variety of moods which gives to cliff climbing its unique fascination.'

Exmoor

Exmoor remains an esoteric climbing destination. It has some of the most wild, remote and adventurous coastal terrain in the entire British Isles. Exploration here is extensively covered in the highly recommended *The Hidden Edge of Exmoor* by David Kester Webb and Elizabeth Webb.

James Hannington (1847–1885) became curate at Martinhoe church in 1870 and was delighted with the discovery of remarkable and inaccessible caves in the cliffs below the hamlet. He resolved to construct a path down to them and, with the help of parishioners, completed this perilous task. Terry Cheek writes of a near-drowning

Facing page: The Exmoor Coastal Traverse looking west from the Red Cleave. Photo Grant Farquhar.

incident when Hannington became stuck between two levels of a cave which he named 'The Eyes' and had to wait for the tide to rise and pop him out as if he were a cork into the above level. Cheek believes that Hannington's escapades required rock climbing at a grade of Severe. Hannington was later ordained as a bishop and died a Christian martyr's death on missionary work in Uganda in 1885. The memorial stone at the site of his death reads: 'The Blood of the Martyrs is the Seed of the Gospel'.

Edward Alexander Newell Arber (1870–1918) was a Cambridge University professor of palaeobotany who was described in his obituary in the *Geological Magazine* as 'no armchair geologist', which certainly applies to his 12 expeditions to North Devon where he extensively traversed sections of the coastline from 1903 onwards. Among many academic works he produced *The Coastal Scenery of North Devon* in 1911, in which he wrote:

'No words can convey the wildness and grandeur of these cliffs... Rock climbers and those fond of scrambling in unfrequented places will find in the beaches of the roughest portion of this coast a new paradise, which is certain to meet with their approval... It would indeed be a proud accomplishment to have traversed the whole of the coastline from Porlock to Bocastle. Whoever manages to accomplish this feat in the future will have seen wonders in the way of coast scenery and can also boast of a remarkable record... Certainly, it will not be perfected without special studies of the difficulties and opportunities, studies as serious, perhaps, as those which were necessary when, in former days, some untrodden Alpine peak was to be attacked.'

Arber also commented that 'the progress is slow, a uniform rate of about a mile an hour being reckoned as the average', which remains a useful rule of thumb when estimating times traversing the rough terrain of the inter-tidal zone of the North Devon coast.

Clement Archer (1902–1967) was a meteorologist and a member of HM Diplomatic Service in the Far East. During his time there, he climbed extensively in the mountains of Japan and Korea, even said to employ grappling-hook techniques for climbing zones of extensive root systems on the Korean granite domes. In the 1950s he retired to West Somerset. His longstanding climbing partner Cecil Agar had also retired and was living near Ilfracombe. The magnificent coastline nearby offered a 'natural meeting place' for days out together. The pair started with walks and scrambles before it dawned on them that a continuous low-level traverse of the entire coast was possible. The pair started investigating the traverse of the coast from Porlock to Saunton Sands from 1954 onwards. They first explored the section of Exmoor coast between Woody Bay and Heddon's Mouth, expanding their efforts from 1958 onwards to other sections, including Baggy Point. Archer employed unorthodox techniques such as swimming while roped up and the use of nailed boots and ice axes to ascend steep, vegetated cliffs. Archer recruited the teenage Kes and Tim Webb as 'sherpas' on his expeditions. Terry Cheek, who was his neighbour from 11 years

of age, describes him as 'an eccentric, pedantic and very impatient man. He had been known to throw his tangled rope into the sea. But he was driven to complete his quest and in doing so has given us a route that takes us back to the basic sport of man against the mountain'.

Archer's explorations lasted 10 years, during which he privately published his *Coastal Climbs in North Devon* (1961 with supplements in 1963 and 1965) giving 'an account of climbs on and along the coastal cliffs of North Devon, undertaken by my friend, Cecil Agar, and myself in the years from 1954 onwards'. John Cleare:

'Coasteering in the 'grand style', as exemplified by the coastline of North Devon, involves maritime knowledge and matters of judgement similar to but quite distinct from mountaineering. Technical difficulty on steep rock was generally avoided. Yet such was the scale and intricacies that faced Archer and his companions that a new scale of problems peculiar to coasteering arose. The cliff structures and beach terrain are enormously varied. In the absence of frequent descent routes, escape upwards was rarely easy. The party learned, a few hundred yards at a time, the area of cliff and shore that was swept by the changing tides. The ground to be crossed and obstacles to be overcome varied accordingly. Every yard of this intricate coastline was fought for in route planning and route finding until all the pieces fitted like jigsaw... The general climbing public did not become aware of this great adventure until it was almost complete.'[1]

Archer contacted retired Admiral Keith Lawder for help with the Baggy Point section of his project. Lawder commented in the *CCJ*: 'The idea is, of course, directly in the tradition of AW Andrews, who made similar routes in West Penwith 50 years ago, but they knew nothing of this until they finally met us.' Lawder's grandson, Iain Peters, remembers:

'Archer kept meticulous records, which he had privately printed. I have a copy, as my grandfather and I were invited to join the team for the final section around Baggy itself, which involved more technical climbing. I remember Archer as a crusty old man with very fixed opinions and a contempt for modern climbing equipment and methods. My grandfather was appalled at his rope work and taught him how to set up and use a Prusik. Unfortunately, this led to his downfall as he lost his footing whilst on a rope and instinctively grabbed the knot, thus preventing it from engaging on the rope. I believe he broke his leg badly in the fall.'

In 1965, Clement Archer made the headlines of the *North Devon Journal Herald* with: 'Somerset Man Saved in Cliff Drama. Six hour North Devon rescue'. The accident is described by Terry Cheek:

'Archer fell on the 11 July 1965. He devotes two and a half pages to the accident in his supplement; he slipped on wet grass whilst descending the east side of West Lymcove Point, protected by a Prusik knot, which failed when he grabbed the rope above the knot. He landed on

---

1   *Sea Cliff Climbing in Britain.*

the beach, breaking his leg. It took 11 hours and 1,100ft of rope and fire hose to extract him.

'His leg repaired well enough, but he could not trust it, so he gave up climbing. He was 69 years old at the time of the accident. He didn't survive much longer than a year after. My father believes that he died of a stroke or some form of brain failure. He was a character who made life interesting. He was not a good climber, but he excelled in pioneering escape routes on mixed rock and grass. If he had not existed it is possible that the traverse and the many buttresses that support modern routes along the Exmoor Coast would not have been discovered for many years, or if at all.'

Archer had fractured his femur in the fall. His climbing partner Cyril Manning pulled him above the high-water mark and then prusiked back up the ropes to raise the alarm. A helicopter was called but couldn't reach him, so it required the coordinated efforts of the fire brigade, police and coastguard to raise him to the top of the cliff. As Martin Crocker said, 'It goes without saying that this is no place for an accident – but, there, I've said it.'

Equipment for Amphibious Climbing
by CH Archer

It stands to reason that where the time factor is so vital, a watch must invariably be carried. It is perfectly safe from seawater if first it is put in a small polythene bag and then the bag is put in a screw-top tobacco tin.

We have found chocolate admirable as an iron ration; its restorative value is high, it is compact and easy to manipulate, and its taste suffers little from immersion in seawater.

Rocks covered with acorn barnacles give admirable footholds but are too sharp for bare hands. Efficiency and comfort alike are promoted by the use of lightweight leather gardening gloves. These are also invaluable when one meets brambles on the cliff routes.

One of the great difficulties on this coast is that even at the end of a prolonged calm, the water is always cloudy. Visibility underwater rarely exceeds a foot, and so the climber loses the use of his eyes just when he needs them most. An ice axe makes a useful probe, and if this be not carried, any kind of a pole will often find an invisible rock which saves the need for deep immersion.

If one has to swim heavily laden among unknown currents, a cord is certainly advisable, but the cord itself can become a problem by getting wound round the leg in the process of swimming. I have not tried, but I suspect that the nuisance could be abated by fixing corks or some other kind of float two to four feet behind the body, so as to prevent the cord sinking until it is well clear of the legs. As words are often distorted and hard to distinguish even at ranges of 30 yards, it is well to agree on a few visual signals before becoming separated on an amphibious section of any length.

Above: Sheila Nicholls leading Scrattling Crack at Baggy Point on 30th Sep 1974 Below: Pat Littlejohn and Keith Darbyshire on the first pitch of The Archtempter, E3 5b, on Blackchurch Main Cliff, North Devon on 6th April 1975. Photos John Cleare.

The Hidden Edge of Exmoor by Martin Crocker

'We asked climbers to recite four things about Exmoor; our survey said: deer/fox hunting; Exmoor ponies; Lorna Doone and… ' Awkward silence… and… too late… 'Uu-uugh' goes the buzzer as the clock runs out. Inevitably, a few wider-travelled climbers tuned in to Family Fortunes will have got the fourth answer. What about some of Exmoor's more celebrated features? For example, how about the bizarre pinnacles of The Valley of the Rocks or the dramatic coastal incision of Heddon's Mouth or Exmoor's highest point, Dunkery Beacon (519m high, presumably excluding what must be one of the world's biggest cairns).

Older climbers will recall the devastating floods of 1952 in which 34 people in Lynmouth lost their lives, while the more cultured may know that these parts were a favoured retreat of Coleridge and Wordsworth and that the coast was a haven for 17th-century and 18th-century smugglers. And – best of all, maybe one or two will even have heard tales of the dastardly and enigmatic Beast of Exmoor that likes to maul sheep in its spare time. Idiosyncrasies aside, the chances are that for many visitors, the enduring impression of Exmoor will be one of rustic colour and light, of sparkling streams and the warm autumnal richness of its rolling heather moors, all invigorated by Exmoor's moody history and mysterious folklore. You might think that there would be little scope amongst its gentleness for the climber. And you would be right. Instead, you have to think laterally and take an outrageous trip down to where this mountain of sandstone meets the sea – to 'The Hidden Edge Of Exmoor'.

Anybody who has rambled the wonderful Somerset and North Devon Coast Path will realise what a formidable prospect the Exmoor Coastal traverse is. Huge stomach-churningly steep slopes with dense scrub and gorse plummet an unfathomable distance to God-only-knows what sort of crumbling crud overhanging the sea. And with the ferocious tide and swell of the Bristol Channel, the sea seems so rarely satisfied or restful, baring only slim crescents of boulders or pebbles for a few precarious hours, minutes, or sometimes never at all.

'What took you so long?' a bemused Kes Webb had enquired of me during my first trip to Hurlstone Point. I struggled for an answer and, even now, I can only reason that I had not read sufficiently between the glib lines of the guidebook or that I'd got too preoccupied to picture what might happen when 1,000ft of Old Red Sandstone does battle with the sea. My ignorance was not so much a measure of how well the few local climbers could keep a secret but more one of how marginalised the mainstream had become from the maverick, hard-core activities that represent the essence of the mountaineering spirit. The proof is that Exmoor is one of the few areas regionally to have escaped the South West classics megalomania of the 70s. I was on the scent and eager to find out more. But only by probing, literally, into the very guts of the place, alone, would the locals consider me worthy enough to be taken into their confidence. Talk about getting blood out of a stone.

Kes Webb now whets the appetites of local groups with his awesome slideshow 'The Hidden Edge Of Exmoor' and runs a boat to give tourists a hands-off flavour of the real thing. However, it is Terry Cheek, a community bobby, who bridges the time warp between the wild and wacky days of Cyril 'The Squirrel' Manning – whom he idolises – and modern climbing. In the delightful narrative of Cheek's guidebook, neither style is in disharmony, as he unlocks the imagination of the reader to a wonderland of geomorphology, local history and daredevilling the tide. You'll journey with him along the Exmoor Traverse upon which, still to this day, he seems to spend every free moment, refining his route, testing his tidal science, surveying, guiding; all 20 years after leading the first continuous unsupported traverse in 1978.

Today, The Exmoor Traverse represents the blood of Exmoor climbing, and it is the thread that links and feeds access to the many promontories, buttresses and slabs that occasion its route.

Left: Peter Biven and US climber Leo Le Bon on Heart of the Sun, E3, on Baggy Point on 21st April 1974. Photo John Cleare.

The 1978 Exmoor Coast Traverse by Terry Cheek

I was Clemmy Archer's neighbour at Torre in West Somerset and 11 years old when he started his quest. I was allowed to follow him around Exmoor reading his weather instruments, but he would not take me climbing. He was grumpy, impatient and eccentric. If anything did not work as expected, he would throw it around. I witnessed many of his antics as he played with climbing equipment, including rope ladders, on the cliff above his house. One of his more common faults was to lose his temper and throw his tangled rope into the sea and then expect others to retrieve it. He was inspired by the challenge laid down in the book written by EAN Arber, who explored the coast from Porlock in Somerset to Boscastle in Cornwall, taking 15 weeks over a seven-year period. Archer found the Somerset section of the Exmoor Coast boring and did not bother with it; his route starts at Foreland Point below the lighthouse and finishes at Combe Martin. His guide also includes a few places west, and there is a large section devoted to Baggy Point. There is little mention today of his climbing partner Cecil Agar, who appears to have led many of the difficult parts. Maybe this is because only Archer put pen to paper. He does not hide the fact that they swam past many of the difficulties. They were climbing in nails.

Also during the same period, Cyril Manning and his future wife Pat Harris were climbing on the coast. Not to establish routes but to gain access to the nesting colonies. They had already climbed through sections where Archer and Agar had been forced to swim. Both parties finally met up by chance at a talk given by Archer. In the supplement to his guide, Archer includes the areas which Cyril and Pat later led him through. The route finally came out of the water, only two short sections, one below The Valley of Rocks; the other below Martinhoe remained unclimbed. These were finally done by the 1978 expedition, although a considerable amount of verbal encouragement from Cyril Manning was required at The Flying Buttress below Martinhoe. Archer writes in the foreword: 'The tidal limitations mean, in fact, that the expeditions must be so scattered in time as to be impracticable except for local residents.' This has turned out to be correct. I know of a few local climbers who chose to do the route in sections; they are near completion in their fifth year. This year has only produced two tidal periods on weekends when the weather and sea conditions were suitable.

Clemmy was a retired colonel and provided me with an impressive reference when I decided to join the army at 15 as a boy soldier. I quickly joined the battalion's climbing club and learnt to climb properly, although anyone who had watched army climbing parties in the early 60s might challenge that. I returned home in 1968 and found that Clemmy Archer had fallen and broken his leg at West Lymcove Point. The rescue had taken 11 hours up 1,100ft of cliff, and he had given up climbing. In 1974, I became an instructor at The Police Cadet Training Centre at Taunton. I met and commenced climbing with Cyril Manning on The Exmoor Coast and quickly realised that I would have to adjust and ignore quite a lot of what I had been taught if I was going to keep up with this man who was older than my father. It dawned on me that Cyril's technique was not bred out of ignorance of basic safety measures but of necessity. The tide waits for no man. Here was a place of awe-inspiring scenery with huge cliffs and massive waves and, at best, only a six-hour window in which to achieve your goal.

Speed and knowledge was the key. I quickly got up to speed, but the development of my nerve when the tide had turned and was threatening to cut us off was taking a little longer. With the knowledge of more and more escape routes committed to memory, I began to feel at ease. Cyril told me of his adventures with Archer from 1963 and how the route had been pieced together over a 10-year period. He lent me one of the rare copies of Archer's guidebook. Cyril was a little more conventional than Archer and Agar and employed modern climbing skills. Cyril and his future wife Pat Harris were responsible for opening up much of the inaccessible section. In pursuit of their bird watching, they had already climbed through sections which Archer and Agar were forced to swim. Eventually, only the section below The Valley of Rocks and the route onto The Flying Buttress under Martinhoe remained unclimbed. I climbed with Cyril during the 1970s and found a man free of the constraints of modern climbers. Often solo, he climbed in big, bendy boots across walls that many would consider to be HVS+ today. He would not have even considered buying rock boots that were

Above: Terry Cheek on The Exmoor Traverse in 1998. Photo Martin Crocker.

In 1978, Terry Cheek led a team of three police cadets, using traditional rope techniques, on a five-day expedition along the ECT, the first time it had been completed in a single push. The team had Cyril Manning to advise and lead them on pre-expedition climbs across some of the difficult sections. They had a whole week of very low spring tides plus good weather. And, as Terry said, they got paid to do it.

on the market. He graded his routes as Easy, Moderate, Difficult and Hard. His way of descending the cliff was to place a stake or piton, then tie one end of the rope to the anchor and the other end to himself. Should he fall then he would stop after 120ft, hopefully short of the water, where he would sort the problem out. Normally he would descend the length of the rope, untie and solo the remainder of the descent. He was an inspiration to those who, at the time, felt that the direction of climbing was being dictated by trendy magazines, equipment manufacturers and armchair climbers obsessed with grades and ethics.

Climbing on The Exmoor Coast has as much history attached to it as any other climbing area in the British Isles, but it remains a largely unknown area because of access problems due to the tidal range and high cliffs. Strangers to the area find that whole days are wasted trying to locate the routes down the convex hog-back cliffs, and many turn up on unsuitable tides. Some state that getting down to the beach and back is enough for one day. At 40ft, the tidal range along the length of The Exmoor Coast is the second highest in the world, crashing against the highest sea cliffs in England. It is being between these two heavyweights of nature with no obvious avenue of escape that, in my opinion, makes this traverse the ultimate adventure experience. Although the route does not require a high degree of climbing skill, the extreme tidal conditions, route finding, changeable Exmoor weather and tides, and overall length make this route a very serious undertaking. The lack of escape routes causes a great deal of concern. It should not be undertaken without reference to the tide tables and weather forecast, both of which are predictions, not a certainty. No matter how skilful you are, it will be the natural elements which decide if you will be successful or not.

The Exmoor Coast Traverse starts at Goat Rock, Foreland Point and finishes on the seafront at Combe Martin. Only the section under The Valley of Rocks and a short distance from Red Slide Buttress to The Claw beneath Martinhoe remained unclimbed. Archer swam past both locations. In early 1978, a very small intake of five 18-year-old recruits arrived at the centre. A special training program had to be developed for these older-than-normal cadets who would join the force within months.

The traverse was chosen as a project which would have all the usual fitness, adventure training and planning exercises involved. The aim was to complete the traverse in one push carrying all the provisions and equipment required without re-supply or leaving the route. The route was to be any beach or rock between the water and the grass line above the cliff. This would mean not entering the water and solving the two unclimbed problems without climbing out onto the grass slopes. The training went well, and Cyril led us all around Little Hangman and across the face of Yes Tor. The cadets nicknamed him 'Kill Me Quick'.

Left: Ian Parnell soloing Shangri-La, HS,
at Baggy Point. Photo Dave Pickford.

At 4am on the 6th of April 1978, we awoke and travelled down to Foreland Point. The weight of our packs had been reduced to 30 pounds, and the cadets were able to climb with the packs in bendy boots at a high standard. Three had been selected to climb; two were going to attempt to monitor our progress with a small VHF radio. No one, most of all myself, expected us to be successful. The aim had been reduced to: 'Let's see how far we can get.' We were allocated a week. I expected us to be somewhere in the Inner Sanctuary at the end of the week having escaped and bypassed at least a couple of the difficult sections.

At the start, Goat Rock on Foreland Point, we were already in trouble. There was a large swell, and we needed to be under the Valley of Rocks in time to take advantage of low water that day in order to cross the unclimbed area. Foreland Point has strata leaning to the west; many of the reefs were undercut and flooded on the west side. With the climbing around the point behind us, we were able to walk along the beach with only the problems of Higher and Lower Blackhead Points to slow us. Cyril was waiting for us at Lynmouth and wished us the best of luck under The Valley of Rocks.

Our late start had resulted in missing the low water under the Valley of Rocks. We had arranged to meet our comms group just around the corner at Wringcliff Bay that afternoon, where we intended to camp. We were forced into serious climbing on an incoming tide. We overcame the flooded east inlet but came to a halt at the large west inlet of Mother Meldrum's Gut (inlets or zawns are known as 'guts' in North Devon).

We sat on the ledges above the high water mark for seven hours and waited for high water to pass and the tide to recede back towards low water. The radio did not work. Because of the need to reduce weight, we could not carry water; we were relying on streams and springs. It was bitterly cold. We could not sleep or cook. Darkness came, and at 10:30pm we saw the first boulders in the bottom of the gut. We knew that this section from the gut to Dog Hole on the other side was just over one rope length, but it was hard. We decided to go for it in the dark. With torches clamped between our teeth, we made our way across the last 200ft of cliff at about 40ft above the water. It took us four hours. At 2:30am we walked into Wringcliff Bay and woke our comms group who were asleep in the boulders. We had been awake for nearly 24 hours.

The following morning we had a conference on the beach at Wringcliff Bay. Our plans were in shreds. We had climbed for one day and were exhausted. Yes, we had cracked an unclimbed section, but we were heavy and moving far too slowly. We were using modern rope handling techniques with 3 x 150ft ropes, which were wet and heavy and a nightmare to sort out when the four of us were on a cramped ledge. Reluctantly we had to accept that we would have to climb on the low water at night and ditch two of the ropes, one of which we had already left behind in Dog Hole. This was later retrieved by Duncan Massey, a local climber who was with the comms party. He was also trying to keep our training

inspector calm. We had to adopt Cyril Manning's tactics or fail. We devised a system of having the leader and last man climbing conventionally with belays and runners. The other two made their way along the rope via-ferrata style with two cow tails and had to sort themselves out if they fell. There were times when even this was abandoned for what later became known as deep water soloing.

Our plan for the second day was to traverse Duty Point, Crock Point, Woody Bay, Wringapeak, and The Three Bluffs. An inlet called Duty Creek splits Duty Point. Clemmy had experienced difficulty on the slabs on the east side of the creek because his nailed boots would not grip on the smooth rock. We found it difficult on the steep west side, where the weight of our packs put unbearable strain on the arms and fingers. We made it to Big Bluff, the first of three, without incident. Cyril had told us that at low water we could descend down into a cave that ran through the bluff. We had arrived a little after low water and could not find the cave entrance. Years later, I discovered an easy line across the face of the bluff which did not require the tide to be so low. Again, water was the problem. We had failed to reach Hollowbrook Waterfall. We had to collect drips from a spring above to cook the evening meal. We eventually found the cave entrance partway up the cliff, but the sea could be heard 30ft down in the darkness. There was no way through. We were about 200m short of our planned destination, Hollowbrook Waterfall, and knew that we would have to make up that distance on the next low water at midnight. We settled down in the boulders and slept.

At midnight we dropped into the cave through Big Bluff. We could hear the sea in the darkness, but we landed on firm ground. We ran through the length of the cave, and to our relief, we found ourselves on the dry beach between Big Bluff and Double Bluff. This was it; we were committed to 1,500m of traverse which had the reputation of being hard, and at that time only two known escape routes. Luck was with us; we were able to walk through the cave which divides Double Bluff and along the reef to the beach between the bluff and Great Bastion which has to be climbed. It is about 220ft high and has a frontage of about 75m. We had been advised to cross it at about 30ft. As we found to our cost, the rising strata tends to take you up far too high, especially in the dark. Having descended and located the correct line, we found it to be an agreeable climb apart from one small bottomless recess in the middle of the face. Here the swell tended to be thrown much higher than normal. We passed through The Yogi Hole at the NW corner and immediately heard the roar of Hollowbrook Waterfall above the noise of the swell. Having already eaten on Big Bluff, we settled down for the night. It had taken us less than an hour.

At first light we awoke from our sleeping bags and gazed up at the source of the noise that had been dominant throughout our shallow sleep. It was huge. Archer put its height at 186ft. The volume of water pouring over the top of the cliff seemed way out of proportion to the small stream that rose a short distance away at

Martinhoe. We walked out and climbed the broad back of Cormorant Rock and immediately had a view of the difficulties ahead. The ominous Black Wall totally blocked the route about 300m ahead. We were now on our third day and much further ahead than I had thought possible. We were only using the rope when there was the possibility of a serious fall. Even then, the whole party would tend to be mobile while roped providing we had at least a couple of good runners between the leader and the last man. It was now well over a day since we had any contact with the others, and we knew that they must be concerned for our welfare. But there was nothing we could do; the coast path was not in view, and the radio required line of sight in order to function. We made it to the foot of the Black Wall one hour before low water. We climbed the strenuous 60ft overhanging corner-crack-come-chimney on the left side of the wall at Severe grade. We later discovered that this was a first ascent. Easy scrambling brought us to the unmistakable A Cave with its rock bridge joining the main cliff to a flying buttress.

It was low water, and the sea was calm. We were able to scramble under the bridge and climb onto the wall of the main cliff. We continued west unroped across the gap of what is now known as Pharoah's Chimney and down an easy slab to the beach on the east side of Red Slide Buttress. Here we paused and stared at the east wall of the buttress. It was unclimbable, but Cyril had told us that only The Black Wall, The A Cave and the climb onto The Claw would give us problems. One of the cadets wandered into a cave in the back corner of the inlet and came running out, shouting that he had found a way through. We passed easily through the cave behind the buttress and came out facing Archer's Corridor Route – a swimming route through a cave under The Flying Buttress.

We were in a place where only a few people had stood when we heard 'Hello' from above. Nonchalantly perched on a narrow ledge 40ft above was Cyril, tied on the end of his rope which disappeared up over the grass line. While he was explaining that people were getting anxious, Pete Hopkinson, an instructor from the same centre, came alongside in an inflatable. We bid farewell to Cyril and Pete informing them that we would camp that night at Heddon's Mouth, one of the few places that can be reached by walking in from the road. The tide had turned. I knew that we had about two hours to get there over 650m of boulders followed by a climb around Highveer Point. Cyril informed us that he would see us in Combe Martin. He appeared to believe that we were going to make it. We had a quick look around the famous Hannington's Cave then made off across the boulders of Bloody Beach, as Archer had named it. Here and there, Hannington's paths could be seen on the slopes above.

At the west end of Bloody Beach is Highveer Point. It can be walked on the low water of some spring tides, but we were now two hours into flood tide, and the point presented a strenuous, almost out-of-reach mantelshelf move followed by a climb to the top of Pimple Rock. Once there, it was a gentle stroll down the back of the rock onto the beach and a reunion with the others. It was still cold as it had been throughout the previous three days. We discussed communications and decided that being above us was of no use. The comms group would have to drive to some place where they could almost achieve a line of sight into the shoreline and give the radio a better chance of working.

After they had left, presumably to the Hunter's Inn Hotel a short distance upstream, we set about cooking a meal and drying out our sleeping bags. These were damp from the condensation caused by the plastic survival bags. We were extremely tired and not talking much. One of the cadets said, 'You know we are going to crack this.' Ahead lay The Swim, North Cleave Gut, Yes Tor and Little Hangman. They had already been around the last two with Cyril Manning. Pete Hopkinson and I had taken them down into North Cleave Gut which had not gone well. They had not seen the swim at West Lymcove where Clemmy had his accident. Quietly I thought, yes it's possible, we'd got further than I thought we would. It was cold, but it was dry, and the sea was calm. If it stayed that way, we were in with a chance of success. As I was placing a piton to support the washing line, I thought that it could be something small that would defeat us like a twisted ankle or some other injury; we were vulnerable because we were tired. It was at this point that I struck my thumb with the peg hammer.

At Heddon's Mouth we had a leisurely start to the day. High water that morning was 9.8m. We could not move west until about 10:30am. We moved quickly, un-roped along the tops of the reefs past the Tuesday Route and were soon getting our gear on for The Swim. The climb of this went off without incident, and we were soon motoring towards Bosley Gut where a short climb around the corner brought us onto the beach leading to North Cleave Gut. We had been here before and knew what to expect. We were surprised to find that the tide was exceptionally low, and we were able to walk into the gut. It is an awesome place. Archer describes it as being an inlet of the sea, which at high tide runs up a channel for 120 yards, and most of that distance is less than 10 yards wide. Sheer walls of 250ft on the east side and nearly 400ft on the west side dominate the inlet. At the back of the gut in the east corner, a waterfall drops 350ft in a horizontal distance of 50ft. I was 30 and would not have believed that 25 years later, I would be still climbing and Martin Crocker would be dragging me up routes on the massive west wall.

By mid-afternoon we had reached Sherrycombe Waterfall and drank heartily. I had expected this to be our fourth camp, but the tide was still way out. Was there a chance we could force another 1,500m to Yes Tor? We decided to go for it. We could always spend the night safely in the bottom of the Great Hangman Gut, which was about halfway. We passed the bottom of The Great Hangman and out onto Blackstone Beach without climbing. The tide had been coming in for an hour, yet the beach was still clear. I had never seen it so low. Within 20 minutes, we were at the east side of Yes Tor. The stepping stones leading onto the face were still above water. We stepped casually onto the face where, before, we had to wait and judge the time to run and jump between the waves. I was getting crazy ideas about finishing the route in Combe Martin that night. We reversed the route across the face that Cyril had shown us a couple of months earlier and came out on the ramp at the foot of Yes Tor Ridge. We saw the swell hitting the reef ahead at the foot of Little Hangman and decided we had done enough that day. When we left Sherrycombe Waterfall, the question of the availability of water did not cross our minds. Yes, we had made it a little further than planned, but the cost was going to be another cold night without food or drink. We tried the radio and could faintly hear the others calling, but they could not hear us.

The morning was grim. It was cold, and we wanted to end the trip ASAP. The sea had become angry with white horses visible. We got moving far too early, and the water had not receded far enough. We made our way easily across the beach below Little Hangman Gut but ran into difficulty as we edged along the cliff towards The East/West Cave. The swell was striking the cliff and climbing up towards us. The climb onto The Ramp did not look too good. It took us well over half an hour to climb over The Ramp, down into the gully behind and up onto the large ledges leading to The Black Chimney. Here the gully into the chimney was flooded. At low water, the corner on the west side of the gully is difficult to climb with a pack. We stood around waiting for the tide to ebb. It began to snow. Balaclavas and hoods were pulled over our heads as we stood and stared at the Scotchstone Reef a few yards offshore. Eventually, we were able to cross, but we had not considered the problem of barnacles. Our hands were scratched and cut. They stung from the wet and cold. A couple of weeks later, we all noticed that the skin on our hands began to flake and peel.

We arrived at Combe Martin on the morning of the 10th April 1978. It had taken us four and a half days. There was no sign of our comms group; the local bobby took our photo. We had stayed on our intended route between the water and the grass line, and no one had taken a fall. As far as I know, the route has never been repeated in one push and never done west-to-east in one go. I have no doubt that if Cyril Manning wanted, he would have soloed the entire route in two to three days, counting the birds and collecting minerals as he did so. If asked what grade is The Traverse? Cyril Manning would have replied that it varies between easy and hard. I don't know; we found some VS sections, but we had good weather and stayed reasonably low. If you are forced up higher into exposed positions where the tide has not cleaned loose material from the cliff, I cannot say what grades you may encounter. You may have it easier than us and be able to walk past some of the difficulties and wonder what all the fuss is about. The weather and tide on the day dictate the line. Very few climbers have enquired about The Traverse. The main interest today is from those who go coasteering. It would seem that the clock is being turned back 40 years to Clemmy's aquatic era.

Right: Ian Howell leading The Almighty, E1 5a, on Blackchurch Main Cliff, North Devon on 6th April 1975. Photo John Cleare.

Aqua Alpinism – an attempt to complete Britain's longest climb in a single push by Dave Pickford

The pioneers from that golden era called their breakneck lateral pursuit 'Coasteering', a sort of bucket-and-spade rival to mountaineering and every bit as exacting.

Martin Crocker

The heaving swell slaps at my feet, licking along the narrow sandstone shelf like a dragon's tongue before swirling into a through-cave. Reflected light glints on the water on the other side. Swimming across the cave, I point out to the psychiatrist, will cut off a few hundred metres of climbing, saving us precious time.

The sun has started to sink into the sea with the coming of evening. Just then, a massive swell set slams into the cavern, and the constricted entrance takes on the form of a cataract on the upper Brahmaputra. The psychiatrist looks at me quizzically.

'Not such a good idea, perhaps?' he suggests. I look back at him, shiver, and laugh. This is the point where we decide to abandon ship.

In case you're wondering, I'm not recounting a flashback from rehab. The psychiatrist is, in fact, my partner in crime, Dr Grant Farquhar. He is also largely responsible for the fact that we're traversing towards the base of the highest cliff in England, the Great Hangman, three hours before sunset. It's just after high tide, and we both have mild hypothermia. In early June 2013, when this mission took place, the waters of the Bristol Channel were still only 11°C after one of the coldest winters in a decade.

We've been on the move for 11 hours already on the Exmoor Coast Traverse, Britain's longest climb. We're attempting a one-day ascent of the most demanding section of the route: Lynmouth to Combe Martin, a distance of some 17,000m: around twice the height of Everest from sea level. It's a mammoth undertaking and has much of the atmosphere and seriousness of a big alpine climb on an unexplored peak. It had never previously been considered feasible – or indeed attempted – in a single, one-day push.

The history of the Exmoor Coast Traverse spans more than a century, and is as complex and convoluted as the route itself. Following the lead of Victorian pioneers James Hannington and Edward Arber, the route was first completed by Clement Archer and Cecil Agar in 1954, but only in different sections climbed on separate days.

A couple of decades later, in 1978, Terry Cheek, Trevor Simpson, Graham Rogers and Robert Simmons made the first continuous ascent over four and a half days, using ropes and an expedition-style approach with a support team. The route involves serious and, in places, difficult climbing on wet rock, with only a handful of safe exits in its entire length. Our approach, by contrast, was fast, light, and totally unsupported. We climbed solo, wearing 2/3mm wetsuits and approach shoes, and carried compact dry bags for our food and water.

David Kester Webb and Elizabeth Webb explain the nature of this route in their superb book *The Hidden Edge of Exmoor*: '[This] is a serious mountaineering venture that is compounded by a tide that can rise vertically over six feet an hour and by cliffs that tower over 600ft in places. Out of sight of civilization, it is an awe-inspiring wilderness, boasting the highest cliff in England, a waterfall as high as Niagara, and a colony of ancient stunted yew trees that may prove to be the largest in Britain.'

The seriousness of the undertaking is multiplied by the fact there is no phone reception at the cliff base anywhere along this part of the Exmoor coast. In this respect, as in others, the traverse is actually more serious than a big climb in the Alps, where helicopter rescue is just a speed dial away. If you had a bad accident anywhere along this desolate shore, you'd likely die there long before help arrived.

We made good progress, covering three-quarters of the distance between Lynmouth and Combe Martin in 11 hours. The psychiatrist is a proficient surfer, and I'm an experienced open-water paddler. But with mutual respect for the power of the sea, we called the 'Brahmaputra Cave' our full-time whistle and made our escape up the steep incline of Red Cleave, a fine 1,100ft, 50-degree bramble-festooned couloir.

Lynmouth to Combe Martin in a day, or 'The Exmoor Coast Integral' as Dr Farquhar and I have called it, is undoubtedly feasible by a highly proficient team of two or three and under more favourable conditions than those in which we attempted it.

This alpine-style adventure, completed without boat support, is undoubtedly one of the biggest challenges of total physical endurance, logistics, commitment, and route-finding skill in Britain, and presents the possibility of a whole new sub-genre of climbing: the psychiatrist and I call it 'aqua-alpinism'.

In the same way that para-alpinism links climbing or mountaineering with BASE jumping, 'aqua-alpinism' links rock climbing with extreme coasteering to produce challenges that are beyond the scope of either activity alone. In order to complete the Exmoor Coast Integral, you will need to be a strong, experienced climber and an equally strong and experienced coastaleer and swimmer. As Daft Punk said, you'll also need to stay up all night and get lucky.

When the late, great Swiss alpinist Erhard Loretan pioneered his 'night naked' approach for climbing hard routes with maximum efficiency in the Himalaya, it's unlikely he imagined his theory might be used on a sea-level traverse of an obscure section of the southwest coast of England. The beauty of climbing in general, and alpinism in particular, is its stubborn refusal to conform to a single definition of what it might be. This is surely one of the many reasons I'm so drawn to these strange, dangerous, and exhausting activities.

Facing page and right: Dave Pickford and Grant Farquhar attempted the section of the Exmoor Coastal Traverse from Lynmouth, in 2013 in a day, using modern DWS techniques and wetsuits but were forced to escape at Red Cleave, three-quarters of the way to Combe Martin, due to encroaching darkness and hypothermia. Photos Grant Farquhar.

ECT by Grant Farquhar

Think of it: perhaps the more insane a man is, the more powerful he could become.

One Flew Over the Cuckoo's Nest by Ken Kesey

07:00: 'I'M A NINJA. MY LIFE IS LIKE A VIDEOGAME' blares the iPhone on my bedside table. I don't feel like a ninja, but it's time to get up. Leaving the slumbering wife and child, I stumble in the pre-dawn light through to the kitchen and get the coffee going.

'Meow! Meow!' I trip over and then feed the cat. Swinging in the hammock seat outside the front door, the sun rises north of east as I read the UK newspapers on my iPad and caffeinate my brain. Time to shower and dress.

07:45: After a few kicks, the moped farts reluctantly into life. I'm sweating because it's already 35°C and 80% humidity in Bermuda. The road leads past the usual in situ ducks through the Botanical Gardens to the hospital doctors' car park, already almost full. Two flights of stairs lead to the theatres. The head OR nurse informs me which theatre I am in.

08:10: The anaesthetist gives me the nod, and the room goes quiet as all eyes turn to me. I lean over the supine patient and apply the paddles to either side of his head. I press the button. Electric current passes through his brain, causing him to stiffen and grimace. I know that he is unconscious, and paralysed, under the general anaesthetic. This contraction of his facial muscles does not reflect pain but the initial tonic phase of the induced seizure. Nevertheless, it's always disconcerting.

This treatment modality, Electro-Convulsive Therapy, is something that we still use - albeit rarely - in psychiatry to treat severe and life-threatening mental disorders such as the severe catatonia that my patient is suffering from. It remains controversial mainly because of its portrayal in the film *One Flew Over the Cuckoo's Nest*, where it is depicted unmodified – without anaesthetic – as a punishment, not a treatment.

The initials ECT also apply to something that may, rarely, be recommended for individuals suffering from severe forms of climbing addiction: the Exmoor Coastal Traverse. Are those people also mentally ill? I'm not sure, but a dose of ECT certainly results in a temporary cure of that particular insanity.

05:00: 'I'M A NINJA. MY LIFE IS LIKE A VIDEOGAME' blares the iPhone next to my head. It's still dark as Dave Pickford and I drink coffee at his mum's house on Exmoor. We are intent on an in-a-day 'ascent' of the Exmoor Coastal Traverse, or at least the 12-mile section from Lynmouth to Combe Martin. The weather forecast is good with sunny blue skies, although there is a 10 knot east wind. The internet further informs us that the air temperature is 15°C and the water 11°C. We plan to start traversing at first light on a high, ebbing tide. Apart from the swell forecast from the west of two feet at 12 seconds, it's an ideal day for it. We are equipped with wetsuits – full-length surfing summer suits. Mine is new but won't be by the end of the traverse. We have sticky-rubber climbing approach shoes on our feet, and we have rucksacks with drinking water and our cell phones in waterproof cases, although we will later discover that there is no reception at the base of the cliffs.

06.30: The sun has risen, and the tide is ebbing as Dave slots his latest sports car into a handy, but obviously dodgy, parking place next to the pub at Lynmouth. Having bigger fish to fry, we'll worry about any parking fines later. We walk to the end of the short road and hop over the wall onto the jumbled boulders littering the foreshore. Only just uncovered by the tide, these are slippery and treacherous, but with around 60,000ft of sea-level traversing looming in front of us, we make as rapid speed over them as we can.

The impressive cliffs below the Valley of the Rocks are the first major climbing obstacle, but these are quickly passed with a combination of climbing and swimming aided and abetted by the tide ebbing with considerable force westwards down the Bristol Channel.

09:30: After three hours we arrive at Woody Bay and continue across the beach past some bemused walkers. 'Where have you two come from?' one asks as we trot past, intent on the next section. I can't help but look into the distance. Fuck! It's a long way to the end of this behemoth.

10:30: Low water occurs not long after leaving Woody Bay. For the next six hours, the tide will be working against us along with the west swell which is not massive but is enough to give us pause on the sections, typically rounding points, where we are exposed to breaking waves. This section is well described by Martin Crocker:

'This five-mile stretch of spectacular coastline from Wringapeak to Great Hangman really belongs to The Exmoor Traverse and to the pens of its originators. The first kilometre is as demanding a piece of coasteering as any since its character and one's chances of success are so closely intertwined with the synergy of sea and weather. It is complex, thrilling and spookily overshadowed by mountainous slopes of choss and often impenetrable gorse or scrub.'[2]

The A Cave is one of the spectacular highlights of the traverse and doesn't disappoint. We pass this on the right with some tricky downclimbing above boulders.

12:00: Heddon's Mouth is the halfway point and accessible by footpath from a nearby road, although there is nobody there when we pass through. Beyond that, there is more complex route-finding with lots of climbing in the HVS/E1 range, which is never that hard but is often committing and serious.

14:00: Beneath Black Cleave, we rest up for the three-hour mid- to high-tide interval, planning to start up again after the turn of the tide. We lounge for three hours in the sun in our wetsuits. It's a welcome rest from the exertion of the first half of the day, but I never really warm up, even wearing my wetsuit in the sun.

17:00: We leave Black Cleave and resume traversing. I follow Dave as he frantically traverses a wall while a set of waves approaches. He nips around the corner as it hits. Following behind, I'm immediately washed off the wall into the rinse cycle of a washing machine with boulders added into the mix. I emerge unscathed, and then we contemplate a shortcut through a lagoon.

'We can get in down that ramp,' suggests Dave. Not sure about this, I advise waiting for a bit to observe, and within a minute, the lagoon is filled with heaving waves that toss the boulders around like they were made of polystyrene. If we'd been in there, we would have been minced. We take the long way round.

18:30: Dave and I stop before an obvious impasse to discuss tactics. Even with the wetsuit, I'm feeling very cold and reveal to Dave that I don't think I can face another swim. We estimate that there are still three hours of traversing to go, and the risk of benightment

Facing page and above: Dave Pickford on the Exmoor Coastal Traverse. Photos Grant Farquhar.

and hypothermia is high. Not knowing where the next escape will be, we decide to reverse back to the Red Cleave and bail. It's steep and brambly and feels like a Scottish winter grade III gully. My legs are rubber, but it's an escape. Eventually, we get to the top and look back to where we started from – it's a long way away.

The question that will always haunt me is: would we have made it that day to Combe Martin? Red Cleave is about halfway between Heddon's Mouth and there, so we completed three-quarters of the traverse that we set out to do. It turns out that our 'high point' beyond Red Cleave was just before Sherrycombe Waterfall which has a neighbouring path, so we could have, at least, pressed on and escaped from there rather than retreating to the Red Cleave. From Sherrycombe to the end, it probably would have been about another three hours. We would definitely have been finishing in the dark with hypothermia. We had torches, but I think it would have been too close for comfort, and we made the right decision to bail.

Sometimes, you have days out climbing with good mates that turn into epics and you will remember forever – this day out with Dave was one of those, but on this occasion the ECT definitely felt like a punishment rather than a treat(ment).

Another question, as Terry dealt with above, is: What is the climbing grade of the ECT? On our DWS attempt, Dave and I soloed a lot of serious ground in the HVS/E1 range, but that was within our comfort zone. We could have taken more time to find easier passages, but I doubt if you could find a line that did not require soloing at the grade of at least HVS in very serious situations.

<hr>

2   *High* Magazine #210, May 2000.

Martin Crocker (above) describes some aquatic exploits during the first ascent of Fish or Man? with Terry Cheek on 28/9/03.

Facing page: Terry Cheek following Fish or Man? (E4), North Cleave Gut, during the epic first ascent. Photos Martin Crocker.

Below: Martin Crocker on the first ascent, solo, of Extreme Lives, F7b+ S3+ or E7 6b, Portland. Photo Black Planet Photography.

The King of Exmoor by Martin Crocker

Opportunities to climb in North Cleave Gut are precious, yet today a window of elusive co-conditions appeared to be opening for us. A gentle northerly breeze – a welcome friend in the wake of a cold front – had nudged away yesterday's moist air and plucked any trace of sea damp from the gut's towering sandstone walls. And one of the highest spring tides of the year was coaxing the sea away from our climbs; with good fortune, we would have sufficient time to climb the three new lines we'd been saving up for four years.

For starters, an overhanging groove was dispatched in a flash of grunts and yelps at E5, as did an easier two-pitch extreme on-sighted near the mouth of the gut. It was pleasing to find the rock in perfect nick: rough, juggy, and the afternoon sunshine was beginning to bounce off the east wall, almost within reach. Today North Cleave Gut scarcely felt like one of Exmoor's most intimidating cliffs.

Terry's watch chimed 3:30pm as we geared up for our third route – a huge diagonal line of juggy breaks that cuts 30m across a sheer wall to the top of our E5 groove. High tide wasn't expected until 6.30pm. 'How much time have we got down here – an hour do you think?'

'About that,' Terry confirmed.

Now, I used to reckon I could sense the point at which the tide turns. It is a wave, a little more boisterous than its contemporaries, that rolls from the sea and teases into the consciousness – a subliminal warning that the clock has started ticking.

'Don't you think you ought to uncoil the ropes on that boulder?' I proposed to the guru, having observed how the sea was stealing our floor space minute by minute.

'Nah, we'll be alright,' came the nonchalant reply.

Are you sure?' I insisted, hoping more for the air of an executive decision.

Nod. Without further ado, I set off along the breaks strategically fixing runners for me and for him. Twenty metres along, I was surprised to note that all I could see under my feet was sea; the boulder bed of the gut was flooding – spring-tide style. I leant out from jugs, craning my neck; Terry was only just in view beyond my left foot, the ropes having been hastily thrown into a heap on the boulder.

'Shall I take a belay?' I asked.

'Nah, you're almost there', he directed. So I was, and so I accelerated to the stance, fleeting evaluations – like three-star E4 5c, one of the finest on Exmoor – barely taking form in these circumstances.

'Taking in!' I bellowed.

'Wait a minute', he reacted, sounding flustered for the first time that day. I peered over the capping roof and down along the break to see what was up. The sea appeared to be everywhere, the boulder submerged, and Terry was fighting with the mother of all knots on the few remaining pebbles at the back of the gut. 'Told you so,' might have been a deserved knuckle rap, but I kept quiet; after all, he was the ranking officer here.

'Take in, take in!' Terry shouted with a new-found sense of urgency as he started splashing back to the base of the route. Ten valuable minutes had been lost in unravelling the ropes; I hope this turns out alright, I thought.

Now, climbing smooth sea-washed sandstone with dry shoes is one thing, but doing so from wading in the sea is quite another. Hats off, Terry somehow climbed up to the first decent hold – a little spike just before the break where it yielded good jams and its first runner at five metres. A remarkable effort. Then suddenly: 'Hold!'

I did, but nothing registered as Terry's body weight was absorbed pound by pound along the torturous line of the ropes. Looking down, I was flabbergasted to discover that he was once more in the sea, but up to his knees this time. It wasn't easy disguising my chuckles as I marvelled at how legends got their kicks on the crags. Ten minutes later, the sea had risen to his waist. 'Are you alright?' I asked.

'Yeah, I'm alright…' reassuring me also that he'd rigged up a prusiking system with two thin tapes. Twenty minutes later, after a lot of huffing and puffing, Terry managed to gain the cam runner again. Problem solved? More like wishful thinking – the sea was intent on outwitting us.

'Hold!' he shouted once more. Concluding that this might be a long show, I carefully extended my belay and moved to a ringside seat on the lip of the roof. That was just as well, otherwise I'd have missed Terry flipping upside down and diving head-first like bait cast into the sea. What great theatre! How did he do that? He didn't seem to know either but quickly wriggled around into a more conventional posture on the rock. After a quick breather, he climbed to the cam runner for the third time and then onto the next, managing to ease himself beyond the immediate reach of the hungry sea. But the sea was voracious today: in the last hour it had risen by three metres.

A further hour passed, and then another, but the distance Terry had ascended could be measured by three loops of rope at my feet. The sea was at his heels again, snapping, and the light was beginning to fade… could an epic be brewing? 'Where are your jumars?' would have been a flippant question from me, so I tried better with: 'Can't you just yank from runner to runner?'

'They're too spaced, and I'm exhausted. Can you do anything?'

Now it is all too easy to underestimate the difficulty in jumaring overhanging ground while keeping in contact with the rock. You have to learn the art on the job, not wait 40 years to start practising when it was the only option to get out.

'I'll reverse-jumar the abseil rope and get it to you,' was my offer of assistance. It really was the only solution since, at this rate, he'd be underwater in 30 minutes. And the thicker rope would be easier for him to jumar with his Cloggers that he'd left on my stance when abseiling in. However reverse-jumaring a 30m overhanging diagonal crack comes with its own challenges, and today was a good day to apply all the short-cutting tricks of the trade that are a middle finger to health and safety. But I got to him in five, having clipped some of the runners into the jumar rope so he wouldn't catapult out into the middle of the gut when he weighted it. So back to the stance for me, and Terry transferred his sodden self onto the jumar rope.

'What's happening!?' But it was too late by the time circumstances had caused us to remember that the rope was not a static line. It stretched magnificently, and –

with the vaguest expression of dissatisfaction – Terry was dunked into the sea again with a 'sploosh'. Nothing less than a full immersion this time, he bobbed up with a float of knotted ropes like entrails around him. Once more, I shouted to see if he was alright.

'I'm alright,' he said. 'I'll get out.'

At least there was sufficient tension from the jumar rope now to keep Terry's head above water. But the minutes passed, and I grew worried as I wasn't able to take in any more slack. Again: 'Are you alright?'… and, persisting, 'Are you sure?' Jumaring underwater can't be easy. Then came the crunch:

'I've had it.'

The mood in the gut, and my gut, changed instantly, the blow delivered by an unsettling calm and resignation in his voice. A chill swept through my soul and, physiologically, I was adrenalined-up for a big effort in a second. I decide what to do, or maybe it's instinct. I can't get down on the abseil rope again because Terry's on it, so I secure him, untie from the lead ropes and re-belay them. Reverse-jumaring on the lead ropes, I'm quickly by his side and discover he has managed to haul himself out of the water.

Perhaps I overreacted. We all yearn to be heroes if given half the chance. But progress halts; he's trying to use a pair of shunts that belong to the Iron Age, and there's gear everywhere – slings, nuts, and karabiners cocooning him. Together, we sort out the chaos, and I loan him one of my jumars in place of one of his shunts. He's fixed up real neat now and starts to jumar towards the stance on the remaining reserves of his strength. It's painfully slow work. It's pitch black. It's a long haul to the top – this is Exmoor.

Half-awake on the coast path, at long last I can make out the rustling in the bracken below. I pick up my camera and take a picture of the dark shape that could be Terry as it swaggers drunkenly onto the path, yet still smiling. At times like these, you might wait for a punchline; sometimes they come, sometimes they don't. Today one did, and I say it to him before he can say it to me:

'Don't call me. I'll call you.'

But I called him the next day, and we start out all over again; that's because Terry's King of Exmoor, and the cliffs are his throne.

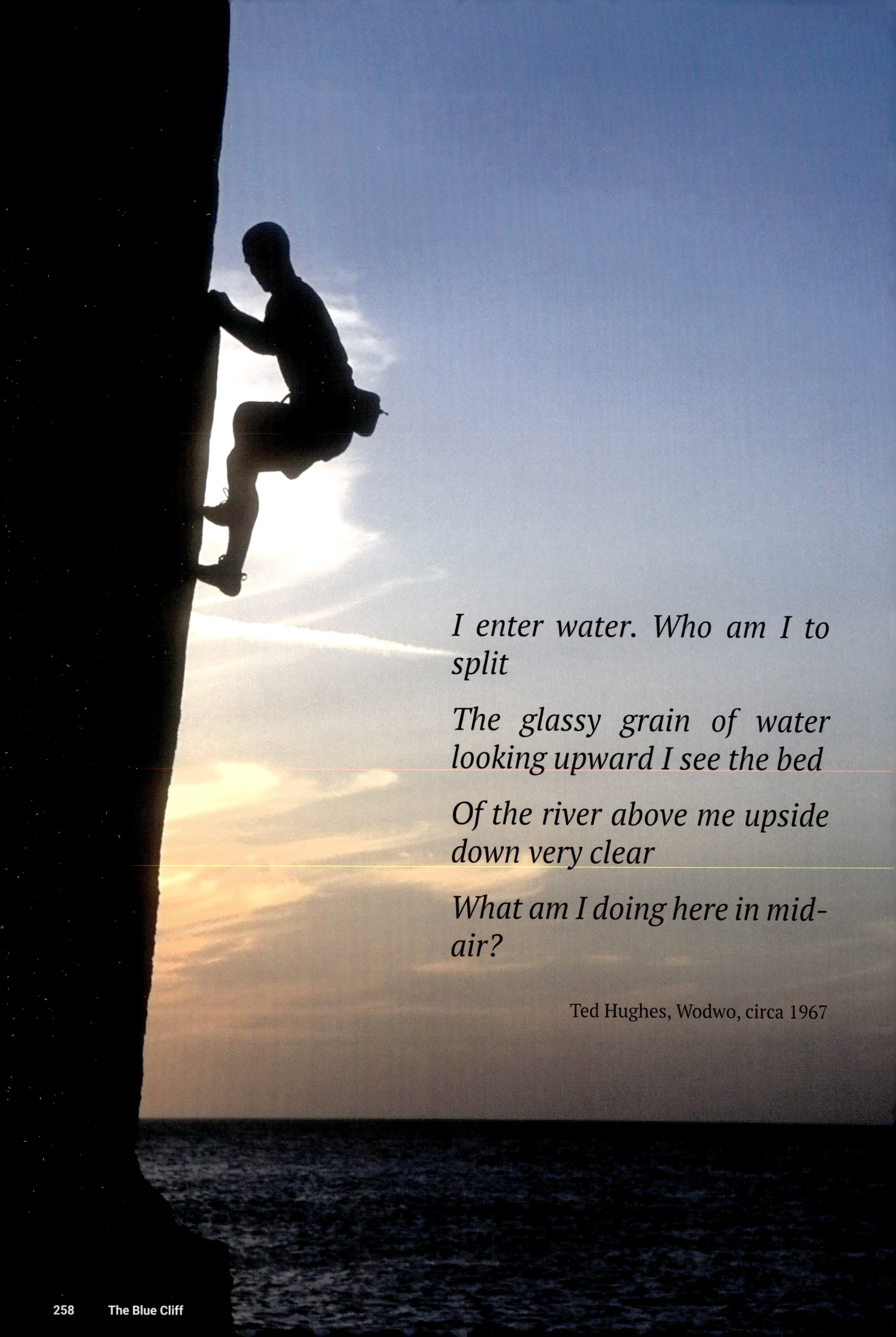

I enter water. Who am I to split

The glassy grain of water looking upward I see the bed

Of the river above me upside down very clear

What am I doing here in mid-air?

Ted Hughes, Wodwo, circa 1967

Lundy

Equipped was the prison of Gweir[1] in the Mound Fortress, throughout the account of Pwyll and Pryderi. No one before him went into it, into the heavy blue/grey chain; a faithful servant it held. And before the spoils of Annwfyn bitterly he sang.

Preiddeu Annwn in *The Book of Taliesin*

In the second century, Ptolemy (90–168AD) wrote his *Geographia* – a compilation based on earlier sources of what was known of the geography of the second-century Roman Empire. He names Lundy as Heraklea – The Island of Hercules – and Hartland Point (the nearest point of land to Lundy) is named as Herakles Promontory – The Headland of Hercules.

Lundy has been inhabited for at least 3,000 years – archaeological investigations have discovered considerable traces of Bronze and Iron Age settlements. The Dark Ages, following the fall of the Roman Empire, left Lundy shrouded in myth and legend. Marauding Vikings around the ninth century AD coined the name 'Lund-ey', meaning 'Puffin Island'. The name 'Lundy' is first attested in 1189 in the Records of the Knights Templar in England, where it appears as (Insula de) Lundeia.

The Templars were a major international maritime force at that time, and Lundy was granted to them by Henry II in 1160. This may have been a response to the increasing threat posed by the Vikings; however, it is unclear whether they ever took possession of the island.

For well over a hundred years, Lundy was home to the 'troublesome' de Marisco family whose favour with the reigning monarchs waxed and waned, the low point being when William de Marisco was hung, drawn and quartered for treason in 1242. For the next 600 years the island was variously a base for marauders, a fortified outpost loyal to King Charles I, a retreat for disgraced nobility, and the centre of an ingenious smuggling operation.

In 1627 a group known as the Salé Rovers, from the Republic of Salé (now in Morocco), occupied Lundy for five years. These Barbary Pirates, under the command of a Dutch renegade named Jan Janszoon, flew an Ottoman flag over the island. Slaving raids were made embarking from Lundy by the Barbary Pirates, and captured Europeans were held on Lundy before being sent to Algiers to be sold as slaves.

In 1833 the estimated population of Lundy was 10 people, a single family living in a cottage and the four keepers of the lighthouse which had been built in 1819 by Trinity House. In 1969, Lundy was purchased by British millionaire Jack Hayward, who donated it to the National Trust. It is now managed by the Landmark Trust.

---

1   Gweir ap Geirioed was one of three famous island-fortress prisoners in *The Mabinogion*. Lundy was thought to be Ynys Weir or 'Gweir's Island'.

Facing page: Neil Gresham on The Flying Dutchman, F7b+ S2. Photo Dave Pickford.

Rock Climbing on Lundy by EC Pyatt and KM Lawder

Above: Keith Lawder.
Photo EC Pyatt,
courtesy David Medcalf.

The whole aspect is reminiscent of the breeding places of giant petrels in some of the islands of the Southern Ocean. Off the combe, Gannet Rock rises sheer out of the sea. One day we rowed out and climbed it: a reputed first ascent. Guillemots and puffins eyed us curiously, and staring seals poised themselves upright in the water. I was astonished at the size of their eyes. The pillar of the Constable at the north end of the island always defeated us, though the lighthouse keepers said it had once been climbed by a sailor. The Shutter Rock where Amyas Leigh cast his sword into the sea, and where the cruiser Montague was wrecked, was a problem dependent on the great tides of the Atlantic.

This My Voyage by Tom Longstaff

Though mountaineers have been climbing on the cliffs of Western Cornwall for more than 50 years, the systematic exploration of the rocks of the West Country only began about five years ago. A peep at Lundy during a day trip encouraged us to return for a thorough investigation and to attempt climbs on the apparently promising crags. The chief aim of the mountaineer is the ascent of new, untrodden summits. He accomplishes it, first by the easiest route, then progressively by harder ways as his proficiency advances. In Britain where the mountains themselves present no problems, steep crags on the mountainsides are made to yield rock climbs, which become of progressively harder standards as one generation succeeds another.

The aims of the cliff climber are similar and twofold. He tries first to climb all the isolated rock masses or pinnacles – the untrodden summits of the mountaineer, and second to make rock climbs on steep cliffs similar to those on mountain crags. Pinnacles can be divided into three groups: cliff pinnacles above high water mark, beach pinnacles between high and low water marks, and sea pinnacles (islands) below low water mark. Frequently, the problem of ascent is one of access rather than of technique. Sea pinnacles have obviously to be approached by boat or by swimming; beach pinnacles need routes down the cliffs and close attention to the tide, and so on.

The climbing crags, too, present problems of access. Ideally, there should be an easy way down the cliff and a traverse just above the sea to the foot of the climb. It is seldom this easy in practice; the traverse may be long and arduous, while in the extreme rope, ladder or boats may be needed before a start can be made on the rocks. Let us see how these principles apply on Lundy. The granite crags are familiarly similar to those of Western Cornwall. In addition, there are a few walls and faces reminiscent of Dartmoor tors with rounded ledges, few holds, and offering very little to the climber. In places the rocks are studded with sharp crystals – sharper than anything in our experience – and cuts on the hands were easily and casually acquired.

Later on, we found that Dr Tom Longstaff had been there in the 1890s and had climbed on St James's Stone in 1903 and again in 1927. The biggest climb is the Devil's Slide. This is a smooth granite slab, about 400ft high and 50ft wide, rising from the sea to the top of the cliff, mostly at an angle of 50 degrees but with the upper part approaching 60 degrees. Holds are nearly all frictional, and in the upper part it was necessary to lead out 80ft or so before coming to a crack that would take a peg for belaying. To reach the start, we got onto the slab at about half-height and roped down to the foot, as we didn't seriously try to find a way to cross the zawn and climb the retaining wall of the slab. Right at the top, wild goats traverse across on very small holds and jump the last six or eight feet – we had to do it more cautiously. Presumably, there is an exit at the top right-hand corner where the goats come in, but we didn't use this and traversed across the slab to the left, which is a better finish. We graded it as Hard Very Difficult with the last 100ft or so Hard Severe.

We scrambled around a great deal and saw enough to convince us that Lundy is an area with great possibilities and an important climbing discovery. We also felt that it has an atmosphere of its own, and we can strongly recommend others to go there – it is indeed something to be on really good rock with no one else in sight and no other climbers within a hundred miles.

Facing page: Jeremy Cole on Satan's Slip E1 5a, Lundy in April 1981. Photo Kev Howett.

The Greatest Unclimbed Route in Britain by Andrew Walker

Everyone delights to spend their summer's holiday
Down beside the side of the silvery sea
I'm no exception to the rule, in fact, if I'd my way
I'd reside by the side of the silvery sea.

I Do Like to Be Beside the Seaside by John H. Glover-Kind, 1907

August 1992: a fifth trip to the Lundy Island granite in a long run of epidermis-shredding shenanigans. Quite a lot of climbing took place, but my strongest memories relate to an amphibious attempt to rescue a goat from Gannet's Rock and someone seeking midnight sanctuary in St Helena's Church to escape adulterous wrath. St Helena discovered the True Cross, and the aggrieved party attacking the sanctuary-seeker in the church was Truly Cross. We'd kayaked and climbed all along the west coast over the previous years, often combining the activities – including a surreal post-pub paddle around the island in the full moonlight. From these amblings emerged the germ of an idea to put up a 5,000m granite route to challenge the length and reputation of El Capitan: the complete sea-level traverse of the west coast. And way before Alex Honnold milked the idea, we were also to do it with no ropes. A false sense of familiarity bred the attempt. It was clearly the greatest unclimbed route in Britain.

Unusually, we had a relatively strong team and were not on drugs. Four of us had trained by collectively spending hours in the 'green room' of a hundred mountain river hydraulics, and I had a 25-yard swimming certificate. We took it very seriously: I had a dry bag for the fags, and a hip flask. One of us even had a waterproof camera. We were also suitably rubberised, with buoyancy aids and harnesses. Seeing the smeared streaks on the sea, we also took a couple of 20m kayaking throw lines and a very light rack. To save weight, we'd share the helmet. Although, I'd already learned the flaw of this approach after an ice-axe sharing scheme on a very snowy Walker Spur.

We opted for a clockwise traverse (south to north), starting at high-water slack tide from Great Shutter Rock, the exposed southwest point of the island. The previous year, we'd been paddling into the relatively inaccessible seaward face of the otherwise disintegrating Shutter Rock. We called it Kayak Wall and did a few routes there including A Localized Puddle of Indifference and Sink the Admiral – named after the local maritime vicar told us of a burial at sea and the subsequent burst of naval gunfire when the strangely buoyant cadaver wouldn't sink. Our route descriptions were subsequently torn from the Marisco Tavern new routes book, possibly for ecological or scatological reasons. I don't know why, but I had to start it somewhere, so it started there – and we were very common people.

Orientated north to south, the island is like a fist in a faucet. The tides ebb west and flood east creating 'a puzzling and unpredictable set of currents'. The only sensible boat mooring is on the more sheltered east coast. The tidal range on Lundy is around 10–15m, and the stream on full bore reaches around 15 knots. For an attempt on the full traverse, there seems to be no ideal solution – inevitably, it will tug at your toes somewhere. Where you need to get it right is at the south and north ends, where huge bitch-like offshore tide races can form. Even trim on the cliffs, these can be too strong to swim, or even kayak, against. Our plan was to start at high water slack so there would be no current at the exposed southern tip, with the gathering prevailing current gently stroking us up the coast in a northerly direction. It would also get progressively easier with the ebb tide dropping, exposing more traversing options. But as 20th-century philosopher Mike Tyson noted: 'Everybody has a plan until they get punched in the face.'

We scrambled down the gully by Focal Buttress, under Ulysses Factor, to meet the foam and surf at the seaward end of the tunnel into the Devil's cavernous blowhole. The section between here and Montague Steps is relatively straightforward but quite committing at high tide, with the many low-water boulder beaches largely submerged and kicking up spume. The only 'easy' escape route between here and the steps at Pilot's Quay has some Severe climbing followed by a considerable state of choss, which is normally abseiled to gain access to routes. The long, narrow tunnel

This page and facing page: Sea-level traversing on Lundy. Photos Andrew Walker.

Above: Sea-level traversing on Lundy. Below: Grant Farquhar and Adam Wainwright on the first ascent of Adolf Gibson and the Lunar Tuna of Caned Furniture, E4 6a. Taken whilst sea-level traversing beneath. Photos Andrew Walker.

under Leaning Buttress, although very sheltered, was an ominous mass of beer-like froth. There is another northward tunnel here, starting from the entrance to the Limekiln. Having moistened our gussets at the end of the tunnel, the coast from here traverses a large sea cave to the dimensional 'otherness' of Weird Wall – even more Wodwo-esque than usual as we got our first battering in the strong surf.

We'd once paddled into here to climb on a scorching millpond day. Now, Goat Island and its petit Aiguilles du Montagu provided a temporary breakwater from the battering waves. Where we took to the water, the outgoing tide was, counter-intuitively, going against us. Various kayaking cockups had failed to inform our understanding in relation to the formation of large tidal eddies at both ends of the island and in Jenny's Cove. While the Bristol Channel empties its bowels westwards around the two points, it creates powerful back-eddies and contrary currents along the west coast. Fortunately, we were distracted from these tidal subtleties by the wind, which was around six on the Beaufort Scale. I'd seen worse on Lundy, including one winter with the waves going over (not just up) Flying Buttress and when a Christmas Day attempt to abseil out of the Old Light tower resulted in going upwards and outwards – kite-like. On that trip, the *MS Oldenburg* didn't sail for 28 days, and we needed a helicopter to make an urgent New Year's Eve party in North Wales.

Traversing on crozzle to the steps at Pilot's Quay over Force 6 seas, it was easy to understand the fate of the HMS Montague in 1906, particularly when you've built a lighthouse that is invisible from the sea. In addition to the steps, there is another potential escape just south of the Old Light cliff which I'd previously descended. Described in a subsequent guide as 'one of the most harrowing descents on the island'. It is. The coast from here to Sunset Promontory has some deeply munched zawns, deep caves, promontories and impressive cliffs, notably Black Crag, Two Legged Zawn, and Wolfman Jack Wall. In serener sea conditions, they are separated by slithery tidal boulder beaches. For us, pressed at their roots with a big sea and dented eyeballs, it was wild, intense and isolated. The throw lines were used to negotiate several zawns – although little 'swimming' was undertaken, or even possible. Substantial rollers were breaking, and it was with relief that the slathering familiarity of the boulders in Landing Craft Bay was reached, with an option of an easy escape to the pub. This was even mentioned by the weaker members of the party.

Crossing Landing Craft Bay to Battery Point was a slippery horror for numb and wobbly legs. Although battered, bruised and tired, there remained only the last 3,500m of our new route left to do. I made my case as forcibly as I could over the booming of the pounding seas but was narrowly out-voted by four to one (for these were democratic times). We escaped from the chaos up to the Battery by a hideous gully on its south side. Having clearly done the hardest part, and all of 1,500m, we reluctantly left it for someone else to finish the final easy part. It might need a bivi. And a calm day.

An Introduction to Salt Water by Kevin Howett

During my first year (1977) at Exeter University I joined the climbing club, although initially I was more interested in orienteering and hunt sabbing. But by the first Christmas, I was enjoying the climbing more than the running, and the inbred followers of the hunts were beginning to get more and more violent towards us, so climbing seemed the safer option.

There was a 'thing' at the time in the southwest climbing world called 'Glubbing'. It was discussed in the Ewe Bar at the university with some of the locals and described as a graded categorisation of how badly a climber got wet on a sea cliff. I remember it thus:

G1: washed by the spray from a wave
G2: the wave extended over your head
G3: washing-machined and thrown around violently
G4: washed out to sea

'Glubbing' is defined in the online Urban Dictionary as the act of going to a gay club, or more relevant to climbing as 'an inarticulate strangled sound (as of someone attempting to speak while underwater)'. Some advocates intentionally searched out big waves to increase their G points. I was fascinated by the idea and soon found out that it was inevitable when climbing around Devon and Cornwall.

It seemed that it should've been almost obligatory that the club should visit the island of Lundy in the Bristol Channel on an annual basis, but my first trip there didn't materialise until the year after I had graduated, in Easter 1981, spurred on by the publication of Pat Littlejohn's guide and Bob Moulton's third edition of *Lundy Rock Climbs*. We booked passage on *MS Oldenburg* from Ilfracombe and were geared up to camp for a week. The weather had been appalling for weeks, and the boat – the only means of supply to those living on the island – had been held up; the islanders were running out of basics. But the boat was scheduled to try and leave very early in the morning, so we motored up to Ilfracombe and dossed on the quayside. We woke to a miserable day with high winds, and it was unclear if the sailing would happen. Finally, the message came over that they were going to try as there was a lull in the storm before another was due, but they warned it would be a rough crossing.

Even just out of the bay the seas were rough, but once into open water the waves were huge, and the inevitable sickness very quickly overcame me first. Whilst the others languished in the galley desperately trying to stave off their own malady, they needed me out of range or else the entire team would succumb to a collective subconsciousness. Now almost delirious in my own private hell, they duly wrapped me in two duvets, a huge oversized waterproof jacket and trousers and lashed me to the rails on the port side, where I spent the entire journey. The horizon was nowhere to be seen as the boat repeatedly plummeted into huge watery-grey crevasses, its redundant propellers shaking the bones of the ship as they exited the water before a slide into an explosion as the boat hit the floor and waves broke over the entire length turning the grey to pitch black. In no time at all, I was shaken to the bone along with it. The vomiting continued throughout a journey that should have taken less than two hours but took nearly five.

Luckily, the landing bay on the island was sheltered and the waves substantially less frightening. We, and our kit, were unloaded into a very marginal and battered-looking landing craft that came out to meet us suitably helmed by a man whose trousers were held up by string. It accelerated through the surf to strand itself at speed on the shore and brought an end to the worst day of my life. We established basecamp in the shelter of high walls and went for a walk in the continuing storm to recover our land legs. That night it went from bad to worse. Alone in my Vango Force Ten tent, I woke feeling strangely warm and weightless, like you do floating in a wetsuit. Extricating my head and an arm from the sleeping bag, I sat up to find I was lying in three inches of water; the wind had forced the rain through the canvas to fill the built-in ground sheet to the brim. Everything I had was either drowned or floating. There was no sound from the other tents, and the time of day was not discernible, but once out of the tent, I was greeted by an empty field. A search revealed how every other tent

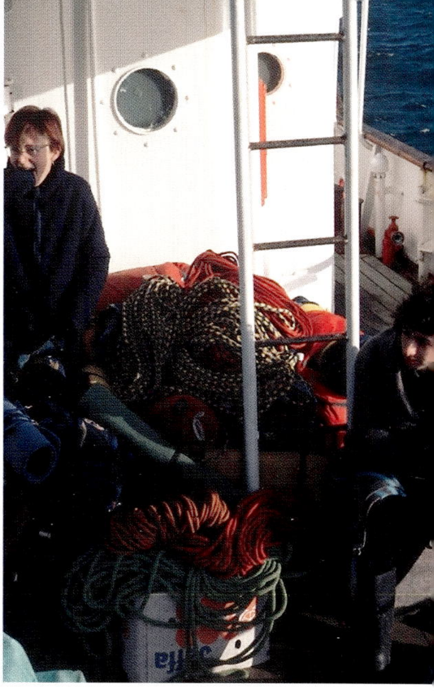

Above: Exeter University team en route to Lundy in April 1981. Below: Arch Zawn. Bottom: Weird Wall. Photos Kev Howett.

Above: Lundy Arch Zawn Purple People Eater base.
Below: The Devil's Slide, April 1981.
Photos Kev Howett.

had been demolished by the storm, and everyone had evacuated to the comfort of a nearby house leaving me happily snoring.

I was climbing with Jeremy Cole, who seemed to relish the sometimes loose and scary north coast adventures as much as I did, but this trip we had our sights firmly on some classics. The eye-of-the-storm weather had turned to sunshine, but the winds and the seas had other ideas, and we were only able to access the upper two pitches of Satan's Slip, and even that was touch and go as the sea seemed to defy gravity and reach the mid-way fault. Searching for something less affected by the high seas, we investigated Weird Wall with a view to doing Navigator which was described as on 'superb rock'. On reflection, this was not the best choice as the crag is described as 'large and rather strange' with a long and torturous access that is 'complicated and adds considerably to the interest and character of the climbing'.

We thought we'd get acclimatised by climbing Wodwo, which was given a sensible HVS in the guide. The descent requires a low tide and involves a steep exposed grass slope leading down a headland, downclimbing a buttress, and traversing back to a ledge beside a cave. Wodwo then starts beyond by jumping a sea-filled gap which was being washed every now and again. There, we dithered and waited, but it wasn't getting any better, so we went for it just as the seventh wave struck. Wetted by a Grade 1 glubbing, we had been lucky and continued to complete the route in good time with just a little further concern on the rather loose rock at the top.

The day was still relatively young, so we had time for Navigator, but there were large thunderheads approaching from the west. We huddled in a depression on the cliff top, watching an ever-increasing swell start to batter the descent and the traverse as the first of several thunderheads rained down on us. Our dilemma was whether to chance the rain but get along the traverse while we could or risk waiting for a dry interlude and lose the chance to do the route. As it climbs the wall beyond the gap, we had to make a quick decision so, reasoning set aside, we headed down.

Descending the now soggy grass slope in rock shoes was scary enough, but looking at the traverse being swept occasionally was more so. We should have roped up, of course, but we both felt speed would be our friend on this occasion, and if we got glubbed again we reasoned we could cling easily to the large jugs – it was, after all, only Severe. There followed a staccato scuttle to try and gain the sanctuary of the ledge, with us both crabbing between sections of the traverse which were being brushed with rivulets of sea froth. It was a relief to gain the ledge which remained safely dry, and we quickly belayed and relaxed.

We were confident we could climb Navigator as it was not graded much harder than Wodwo, and there was now no way to get back anyway. Unfortunately, anything beyond the ledge was now being sprayed. Happily ensconced and dry, it was an easy decision to change our plans; Umbgozi started looking like THE route of the crag! Another HVS with an attractive description promising 'spectacular climbing'. As it started from our ledge and climbed above the sea cave, it was spared the gap jump, and we would be well away from any more froth.

The route traces a line up the slabs on the left side of the sea cave, then traverses the lip into a hanging slabby ramp from where the wall above leads to the top. The pitch grades were each given 5a, so I headed off and easily gained the edge of the lip. The rock was not too bad, if a little snappy, but halfway across progress was barred by a strange stalactite hanging from the roof above and pointing dagger-like into space below. Fatter at the top, it was essentially very overhanging, and it was just more than an arm span in girth, add to which turning it was upping the exposure substantially. More worryingly, it was shattered and loose, and I didn't have a lot of confidence in my last bit of protection. I clawed at it in the hope of finding something solid. Shards of rock fell into the abyss like leaden confetti, and it seemed there was nothing that would hold my weight.

I reasoned that normal climbing techniques would not work on this, but by hugging it the pressure of my body would hold all the looseness in place long enough to swing to the far side. Getting onto it had to be done quickly to maximise friction, so I laybacked off the most solid hold on the left-hand side and simultaneously threw my right arm and both legs around it to get a full-body embrace. I was a little relieved that my theory had held together (literally), but there remained the problem of extrication. At this point, bits of it started to fail which forced my hand, so to speak. Keeping the hug going with my legs, I leaned my upper body rightwards to enable my left hand to cross above the right hand whilst performing a side-ways jump (parkour long before such a thing had even been invented). I'm not sure how, but it worked. I shakily threw in some protection and rushed quickly to the start of the hanging ramp to set up the belay.

Now Jeremy is a bit heavier than me, and I considered what would happen if he fell off, so I dug out some better runners. I also realised that should he fall on the traverse, he would be in space and wondered if it would be possible to lower him into the sea. My harness had a belay loop on the front, as normal, and a loop on the back whose function had eluded me until now – I realised I could clip the belay runners into the back loop and sit on the very edge of the ramp (directly above the cave roof) and be much more comfortable and better to see what would unfold below. Jeremy sauntered up to the far side of the stalactite and stalled. He shouted above the wind: 'How did this go?' It was hard to explain, but I thought he got the gist of it. However, suddenly, he was airborne along with quite a bit of the stalactite. All the runners between us ripped, and he swung in a huge arc under the roof below me with a diminishing exclamation whose volume increased and decreased as he reappeared and disappeared and reappeared and finally disappeared amongst the sounds of the crashing waves and whistling wind.

Now I realised the error of my belay system. With Jeremy pulled tight against the front loop and the belay pulling on the back loop, I was now suspended between the two and getting squished about the waist: a severe Victorian lady's corset. I didn't really need my bosom accentuated, and asphyxia seemed a very real possibility. I tied the live ropes off but couldn't move an inch in any direction to relieve the pain. Then it got worse. Jeremy was now prusiking up the rope. I retreated into myself to try and ignore his bouncing weight by staring at the waves which were now displaying impressive white horses a long way out of the cove. It was then that I noticed what looked like two figures braving the waves on the traverse to the base of the crag. I was transfixed because I couldn't believe anyone would consider going down there now. It seemed like an age before Jeremy finally appeared at my feet, slightly frayed about the edges. I rearranged my belay and my innards. I pointed to the folk below. Sure enough, there were two climbers roped and pitching the traverse, and they were getting a thorough soaking. We surmised that they were heading to do Umbgozi because, by now, the traverse past the gap jump would surely be impossible. It was getting late, and a dour sunset was struggling to light the western sky. More thunderheads appeared on the horizon, and the atmosphere was foreboding.

We couldn't wait to see what the two heroes below were going to do, so Jeremy quickly sorted the gear for his lead of the next pitch whilst keeping one eye on our neighbours. We couldn't believe it when they didn't stop at the ledge but continued the traverse under the cave. 'Have they got a death wish?' we both thought out loud. There was nothing we could do, so Jeremy set off at a steady pace up more solid terrain to find the peg belay. He never found it, and I hoped he would opt to continue up what was described as a 'loose finishing wall' as I had had my injection of excitement for the day. But he managed to get a belay in and handed the lead of the unprotected rubble to me. Dusk was gathering in the last rays of light as I clawed at the grass on the top.

We walked down the grass slope again to see if the climbers were still alive, and there they were, battling along the traverse towards the start of Navigator. The lead climber was actually in a dry patch, but the poor second was being swamped quite a lot. We shouted, but they couldn't hear us. We retreated and sat on the top as the final glimmer of light behind the brooding clouds died, and it started to rain again. Once the ropes were coiled and the kit packed, we hung about considering what to do about those below. 'Was that a shout?' Jeremy asked. We both strained to hear above the roar of the sea and the wind. We waited some more and listened some more, wondering how we would extricate them if the need arose. But we were now wet through and shaking with the cold. With one last pause: 'Did you hear that?' we looked at each other and headed off.

We never found out who they were, but they must have climbed out as there was no rescue initiated, and no one was reported to have died. It's now interesting to see that Wodwo is given E2 5b and Umbgozi E3 5b. It's truly amazing what you can climb when you don't know the true grade, and you are gripped.

Eaten Alive by Kevin Howett

A few days later, the storm-force winds abated. The swell, however, did not diminish, and our options still seemed limited as we wandered along the west coast to see what was accessible. Finally, we reached Arch Zawn in the north of the island where we were immediately attracted to the golden wall on its northern headland which shimmered in the weak April morning sun, and two routes at HVS came guidebook recommended. What's more, it seemed our luck was in as the tide was out, the base of the wall was clear, there were no really nasty waves to contend with, and the descent into the cove from the south was clear of any sea. We geared up and dropped in, passing the sea stack and the arch that gave the cove its name and crossed wet, slippery boulders to gain the ledges below the routes. Headline succumbed first. The traverse on the crux on pitch one was steep and exciting but not unduly hard, and we were wired to get back down to do its neighbour, Stop Press, which looked even better. But by now, the sun shone weakly through a veil of grey cloud: a clear sign the weather was changing for the worse again; the promised second storm? If we wanted to climb the route, we had to get back down fast. We raced back round the cove and descended, but now the waves were gathering pace and washing around the arch. It soon became obvious that we had no chance of gaining the ledge below the route without experiencing further miserable soakings, so we backed off and searched for a route further away from the incoming sea.

The back of the cove contained a lot of marginal substrates. Indeed, the routes recorded did not inspire enthusiasm. Footnote (Severe) was described as 'particularly loose'. The Way Out (VS) was described as 'appalling' and not a good way out. And they both looked worse close up. The only thing of any worth and near to hand was a good-looking corner line by a slab. The rock did not look unduly bad; it was an excellent line, and by now our means of access had been cut off as a means of escape. We had no option but to give Purple People Eater a go. We should have guessed what it was like from the name.

In a later guidebook, it is described as 'An extraordinary route with no protection', but there is no reference to bad rock. That came in an even later guide where it is described as 'frightening'. But we were unaware of either moniker. The base of the route sits above some enormous boulders, lying against each other with gaps between that could swallow a climber whole. It was my lead, so Jeremy sorted the ropes balanced on the top of one of the boulders. Everything was wet and slippery, so he threaded a large sling around two kissing boulders and clipped in. As I sorted my rack, he motioned with a nod that I should look seaward. Things were definitely getting lively again.

I pulled onto the steep wall and into a crack where I immediately realised the rock was as bad as anything we had encountered yet, and the crack was not a crack at all. I moved up whilst the crack crumbled down with every pull of my hands. I was there for some time, scraping at it with my nut key in an attempt to manufacture a runner placement, when I became aware of a strange gurgling sound coming from below. I looked down expecting to see Jeremy undergoing some kind of fit, but he was staring between his two boulders from where the sound was emerging as the sea slowly percolated upwards from the briny below to wash his feet. This was a new one on us as the waves were crashing onto rocks some way off, so where was this coming from? We looked at each other with some concern as we clearly had to get a move on. I climbed higher, thinking there had to be protection somewhere, but there was none, and I was simply getting myself into an increasingly serious situation of no safe return.

Then, there came a more urgent gurgling. I looked down to see Jeremy's face straining up at me as he pulled upward against his belay sling, which fixed him to the boulders, trying desperately to gain height as all around him a rising dark grey sea first overtook his legs and then his chest before it subsided into the hollows below. The most inappropriate thought crossed my mind: that would have made a brilliant photo.

Jeremy was clearly shaken, and I had morbid thoughts of him drowning whilst pulling me off the cliff headfirst into the boulders. I had to get down. But how? With crumbling rock and no protection, there was a high chance of falling onto the boulders anyway.

Before I could make a move, there came the gurgling again, but this time much more forceful. I looked down to see the briny wash up Jeremy's body at pace. He took a deep breath, and his head disappeared just as an enormous wave crashed past me at head level and pounded down onto him. How he managed not to pull me off as he was overcome in that maelstrom, I don't know. I remained relatively dry as the water subsided to expose him slumped, dripping amongst the boulders. Fear creates the reaction of fight or flight. I reasoned flight as the better option, so I frantically dug a runner into the 'crack', and Jeremy tentatively lowered me before it pulled out.

We were now in a pickle. The belay was deconstructed, and we slithered off the boulders whilst being chased by an ever more violent sea to the back corner of the cove. The water continued to rise upwards and pound downwards on everything around us. I have never seen the like before or since. There was no choice of escape as all the routes around us were now being heavily washed as we were now pinned into the only dry part of the cove. Without a word, Jeremy took the lead and just set off up steep rubble, rotten rock, and grass. It was by far the most impressive lead of the trip which ended safely when he was able to thread a couple of rabbit holes.

Purple People Eater is now variously graded E1 5a or E2 5a. I know what grade I would give it...

The first ascent of the impressive feature of The Flying Dutchman, E7 6b, was completed after attempts spread out over several days by Nick White and Dave Thomas in 1989. Neil Gresham made a deep water solo version in 2006, taking two plunges into the Atlantic before topping out. The narrow sea channel beneath the arch gives a restricted window of opportunity for the soloist. With spring tides running, Neil was down to the last tide of the holiday before returning to the mainland: 'The Flying Dutchman is one of the most exciting deep water solos I've ever done, and it is a safe and logical undertaking providing the tides are right and the second pitch is avoided. An easy escape can be made rightward from the first belay at HVS. Regarding grades, a few people have suggested that it feels like F7c if you are climbing completely onsight and the route isn't chalked. The difficulties with the Dutchman are more technical than physical, and both my falls were taken as a result of going off-route; it's very unobvious where you're supposed to enter the top crack. However, with the climb fully chalked and with beta, it would probably feel more like F7b/7b+. One of the falls I took was about 40ft, and the other was 50ft.'

The Flying Dutchman by Neil Gresham

Day after day, day after day,
We stuck, nor breath nor motion;
As idle as a painted ship
Upon a painted ocean.

The Rime of the Ancient Mariner by Samuel Taylor Coleridge

With immaculate granite sea cliffs surrounding its perimeter, Lundy Island is one of the finest trad climbing destinations in the UK. In September 2006, I joined a group of top British climbers on a one-week trip to explore the island. Most of the team were keen to use trad methods to pioneer new routes, but I was keen to explore the potential for deep water solo. The line I had in mind was the trad test piece The Flying Dutchman, E7 6b. Tim Emmett, who had tried to climb it with ropes a year ago, had commented that at high tide, it would make an outstanding challenge as a DWS. The Dutchman climbs the edge of a spectacular leaning arch via a series of mean-looking discontinuous cracks; it was not the style of climbing I was used to for DWS. The second pitch was clearly too high to solo, so my plan was to solo pitch one and then create a new variation finish to escape leftwards at about 70ft.

I had waited patiently for calm weather and high tides, and the perfect day came two days before the end of the trip. I had been down to inspect the line at low tide and had figured that there would be just enough water. I returned that evening, but the sea had become very choppy; however, there was an area of calm water on the inside of the arch, which was all I needed to take the bait. With my heart racing, I set off down the abseil, somehow resisting the temptation for a sneaky peek at the holds. The cliff was so steep that I had to swing in and out to prevent from becoming stranded in space above the sea. When I was low enough, I swung in, latched the rock, took a deep breath and disconnected myself from the abseil rope. This was it now; the only option was to climb out or take one hell of a ducking.

As I started the first moves, a wave came crashing in and soaked my right foot. Bad start, but I wiped my shoe on my calf and tried to ignore it. The first part of the route involved a mean overhanging jam crack followed by some technical moves around a small overhang at 30ft. I reached round and pulled over onto the wall above, but the next holds felt too small for me to pull through. I raced to work out a sequence and grimaced as my fingertips bit into the small crystals. At 40ft the exit crack beckoned, but how should I reach it? Surely I'm too far to the right? If only I had more time to think about this. But with my arms fading, I had no choice but to lunge for it. Before I knew it, I was spinning off into the air, and the cold rush of the Atlantic stole my breath. I managed to scramble out after being buffeted around a little, and the realisation that I was OK provided a fair consolation prize.

The next evening, I was back. We would be leaving at midday tomorrow, so it was now or never. The sea was a little calmer, and I felt way more confident. Yesterday's plunge had broken the psychological barrier, and this time I was looking forward to the difficulty rather than dreading it. As I abseiled in, the first half of the climb passed in a flash. It was all going too well as I eyeballed the exit crack. Then, in a cruel twist of fate, I made a wrong move and was forced to lunge leftwards for the last hold. I hit it this time, but both feet slipped from the wall, and all my weight came onto one arm. I gave everything to hold the swing, but it wasn't enough. I went crashing into the sea again.

That night, back at the campsite, I was inconsolable. It had been a great week on the island, but I would be going home without the Dutchman, and I was trying hard not to let it taint the experience. I just had to let it go, and yet a crazy plan was lurking in the back of my head. There was still a chance if I caught the early morning high tide that I could steal the climb in the last hour and dash off to catch the ferry.

Left top: The Flying Buttress. Photo John Cleare. Left middle: Neil Gresham on The Flying Dutchman , F7b+ S2. Left bottom: Neil Gresham flying off The Dutchman. Photos Dave Pickford. Facing page: Neil Gresham on The Flying Dutchman. Photo Simon Cardy.

I would always wonder 'what if' if I didn't at least try. The next morning dawned clear, though with a slight dampness in the air. I downed some coffee and charged off to the cliff top. My body and mind felt completely drained from the previous two days of effort. Surely I was setting myself up for a final downfall?

Somehow I didn't care any more, and I just needed to get whatever was going to happen over with. As I abseiled down, I realised that the rock was damp with condensation. It felt like a sick joke. I thought that I might as well just abseil straight into the sea. But a strange moment of calm followed, and I decided to give it my best shot.

The first crack passed, the overhang, the wall, and now the exit crack. I couldn't allow the significance of the next few moves to register, so I disengaged my brain and just climbed. At the point where I should have moved left the day before, I moved left, reached out... and this time, it was mine. My consciousness came flooding back; I shook my head in disbelief as I pulled out of the shadows into the morning sun. Far down below me, a seal let out a gentle bark as if to say, what was all the fuss about? Who knows or cares now? I had a ferry to catch.

Facing page: Dave Pickford on the first ascent of Outer Dark, F6c+ S0, in Phantom Zawn, Lundy. Photo Pickford collection.
Above: Tom Luddington soloing Dark Power, F6a S3, in Phantom Zawn. Right: Sam Ponsford on The Howling, F6c S1, on the Ghostbusters Wall at Phantom Zawn. Photos Dave Pickford.

But that was nothing to
what things came out

From the sea caves of
Criccieth yonder.

What were they?
Mermaids? Dragons?
Ghosts?

Welsh Incident by Robert Graves

Wales

With not a single beach of sand
To bathe on, and no sight to see
But rugged cliffs, unscalable.
And tides that raced incessantly
Through caverns inaccessible
To all but those who held the spell
That fell on you and fell on me
To let us in to Faerie.

AW Andrews from 'The Coast of Pembrokeshire'. *CCJ*, 1950

The spectacular coastal scenery of Pembrokeshire was little known to the public until after World War Two. St David's, as the most westerly tip of Wales, was always regarded as remote, and this observation was still being made in the 1950s. However, climbing possibilities on St David's Head and on Ramsey Island were noted as long ago as 1900 by AW Andrews, OK Williamson and others. Andrews described the climbing potential of the Pembrokeshire coast in the *CCJ* in 1950. He was particularly impressed by St David's Head and declared that the whole of the coast there was worth traversing:

'Near the sea, the limestone is much harder, as all loose matter has been washed away and all that is easily soluble dissolved. Some climbing and traversing is possible... There are pigeon holes on the cliffs which form a line of recessed pockets in some places. One of these extends for quite a hundred yards and would make a fine hand traverse. I have only examined the ends, but the holds seem good.'

Aladdin's Cave by Grant Farquhar

I've always preferred black and white. Colour is what you remember, but black and white is what you feel.

John Cleare

I'm driving to visit John Cleare in his lair among the hills that fringe Salisbury Plain. I've been there before when I visited him to pick his brains for *The White Cliff,* but I still get lost and have to make a few U-turns before I locate his house. Like the last time I was there, he comes out in a genial mood to greet me before leading me into his AGA-equipped forester's cottage for a coffee before we head out to the pub.

After lunch, John takes me into his studio – an Aladdin's Cave which is packed with all the paraphernalia of a professional climbing photographer with equipment and floor-to-ceiling books everywhere. Meanwhile, throughout, John regales me with entertaining tales of bygone days. John is 86 years old, but there are no problems with his memory.

Quizzing him about his climbing history, he tells me that he had always wanted to climb mountains and was introduced to the 'game' at school by one of his teachers. However, his apprenticeship was not without its harrowing moments, for during his third year climbing, at only just 17 years of age, he found himself handling the difficult recovery of a fatal accident high on Sgurr nan Gillean. In the aftermath, his school rope-mate, Jackie Lewis, was reluctant to climb again, and after leaving school took up the less risky sport of motor racing professionally.

Several years later, after completing National Service at Manorbier, Jackie mentioned that there were interesting cliffs locally; after attending Jackie's wedding in Tenby,

Facing page: Mike Robertson on Anniversary Waltz, F6b S2, at The Castle, Pembroke. Photo Linda Roberts.

John scouted around the area before enlisting John Reece-Jones (a Llanberis native who was working as a school teacher in the Peak District) and together they made the first notable climbs on Lydstep Point in 1962. They also girdled St Catherine's Island off the harbour point at Tenby: 'No one was climbing there in those days, but Colin Mortlock was nosing around in a canoe.'

John remembered the cliffs at Swanage from early family holidays. While studying at Photography School, he drove down with his usual climbing partner, RAF fast-jet pilot Tony Smythe, to investigate. At first, Southampton University students were the only other active Dorset climbers, but John soon introduced several London Alpine and Climbers' Club colleagues, and the Dorset coast became an established climbing venue. John was with Tom Patey when he recce'd Etheldreda's Pinnacle and shows me a photo of the attempt. I'm interested to know how much further Patey got than the photo.

'Not much further.'

What did Tom say about the rock, I wondered?

'Loose chalk.'

As well as his legendary climbing photography, John is a pioneer of sea-level traversing with routes such as Traverse of the Gods, Magical Mystery Tour and traverses elsewhere such as those in Pembroke. His seminal 1973 book *Sea Cliff Climbing in Britain* with Robin Collomb is a meticulously researched and highly accurate presentation of the state of play on UK sea cliffs which was a highly obscure pursuit at the time. He described the ascent of the Eldorado Traverse in the book:

'The outstanding expedition on this versant of the St David's peninsula is the Eldorado Traverse, extending from Newgale Sands to the cove of Porth Mynawyd, a little East of Dinas Fach, and covering a map distance of one and a half miles. It was accomplished after several serious incidents by a large party led by Biven and Cleare in 1970. No less than six Tyroleans had to be made, two pendules and several zawn-swims. The cliffs are made of various igneous rocks, some good, some bad, but consistently steep and forbidding.

Escape routes up or down looked impossible to the party. The clifftop provides no abseil points; you are confronted with a slope of bluebells getting steeper and steeper, merging into vertical or overhanging rock, in some places nearly 300ft above the sea. During the course of the traverse, the party crossed round into one zawn, full of deep shadow and smooth blank walls; then, with a clatter, a flock of white rock doves flew out of a cave at the back and spiralled up into a shaft of sunlight penetrating deep into the zawn. In May, the party suffered from exposure. Constantly taking to the sea, sometimes unintentionally, the cold water was an unpleasant discomfort throughout the traverse.'

Above: John Cleare in his studio in 2022. Photo Grant Farquhar.
Facing page: First ascent of the Elegug Stacks, 2nd November 1970. Ian Howell and Frank Cannings have already descended from the Elegug Spire (left), but the tide is rising and the weather worsening as the rest of the team descend from the rather higher but easier Elegug Tower to return through the surf by Tyrolean traverse. Peter Biven who led the climb is last man on the summit, checking the abseil point composed of clumps of sea-cabbage. Photo John Cleare.
Below: John Cleare's annotated Pembroke exploration map. Photo John Cleare.

Elegug Stacks by John Cleare

The Castlemartin peninsula in Pembrokeshire, Southwest Wales, is an area of limestone downland forming the southern-bounding arm of Milford Haven. It drops into the sea in a long line of excellent south-facing cliffs dotted with skerries and small stacks, the most impressive of which are the two at Elegug: the Tower and the Spire.

Unfortunately, the whole area is in the hands of the military and is given over to the training of tanks; it is here that the German Panzers, whose now welcome visits were at one time the subject of much controversy, undertake their battle training. Thus, access is restricted and usually verboten, although a certain amount of unofficial exploration has been done by inquisitive climbers.

The two Elegug Stacks rise from a shallow cove ringed by 150ft cliffs, and it would seem that only the Tower, the larger and more massive of the pair, is accessible dry-shod, and then only at low spring tides. This necessitates another fairly standard stack-baggers technique. Once reached, and before any attempt is made to ascend the pinnacle, retreat is assured by rigging a Tyrolean rope above the high-tide line.

Stacks are usually best approached at low tide, but retreat at higher water is often hazardous and problematical without adequate safeguards. When we made the first ascents of the Elegug Stacks, it was November, there were spring tides, and a force eight gale was blowing. The sea was cold and wild, and the equipment we needed was reminiscent more of rounding the Horn in a tall ship than climbing, say, the Cima Piccolissima.

The first day we abseiled into the cove in the storm and managed to gain the base of both stacks, where we rigged Tyrolean ropes and made a detailed reconnaissance before, soaked to the skin, we beat a cold and damp retreat back up our abseil ropes and into the nearest warm pub. By the next morning the gale had died, and despite a still heavy sea, the sky was blue and the general atmosphere more conducive to exploratory climbing. The Tyrolean ropes were mostly still intact, so we were able to start work right away.

We had gained the Tower at the base of the landward arête, and it looked as if the best route was above us. An 80ft pitch up a plumb-vertical wall on small holds reached a shoulder where there were sea cabbages and large boulders. There were plenty of good belays, and pegs were unnecessary. A scramble up debris and easy rock led to the base of the final wall which was split by a deep, double crack line. We climbed this until the way was barred by a couple of large, unstable-looking chockstones where a thin traverse left out over the sea and an earthy pull-up on small loose blocks led to the summit slopes, green with sea cabbages blowing in the wind. We reckoned the climb as 175ft and easy VS.

By the time we got to the top, the cliff-top behind us was lined by off-duty Panzer soldiers who raised a loud cheer; there were also a couple of glowering red caps. We were able to look down to where two of our friends were working on the Spire. This is really shaped like a knife, with two vertical and smooth blade-like faces separated by narrow arêtes, and the lads were halfway up the face towards us. They had climbed up to a cracked line of weakness running across the wall, above which a line of overhangs looked very difficult. They traversed the weakness onto the seaward arête, where they managed to make a good stance in a small sentry box. Above them, the arête was split by a wide crack, which was difficult to get into but which led to the narrow summit. They reckoned the climb to be 150ft and HVS.

We were able to abseil off the Tower with little difficulty, apart from fixing the initial anchor, which consisted of linked coins de bois and the stoutest sea cabbages. In those days we had not developed the 'see-saw' abseil, and the boys on the Spire made a rather difficult descent from their own summit cairn and a dubious piton. Apart from someone getting a direct hit by loose rock which put him out of the climbing, it was an excellent expedition. As it turned out, the red caps were not waiting to march us off to the guard house, for we did have official army permission to attempt the stacks, but it would be wise for any future parties to check out with the army before climbing these very exciting pinnacles.

Climbers return from the first ascent of the Elegug sea stacks via a Tyrolean traverse. Facing page: Ian Howell on the summit of Elegug Spire sea stack on its first ascent with Frank Cannings, 1st Nov 1970. Photos John Cleare.

Brother and sister in sea rescue

A BROTHER and sister from Sheffield were among five people airlifted to safety by an RAF helicopter after being cut off by heavy seas on the West Wales coast.

Coastguards say that the three men and two women had been taking part in the new sport of 'coasteering', which involves swimming and scrambling over rocks around a section of coastline.

But the party which included 23-year-old ███████████ and his sister ████ 24, of ████ Road, ██████ got into difficulties at Newgale, Pembrokeshire.

Stranded

They were stranded on rocks and their plight was spotted by a member of the public who alerted coastguards.

The Little Haven in-shore lifeboat was launched but the crew was unable to reach the group due to the heavy swell.

A rescue helicopter from RAF Chivenor, North Devon, was then scrambled and all five were winched on board and flown to safety.

A coastguard spokesman at Milford Haven said: "They were very tired and suffering from cold but were not physically hurt."

Also in the party were 22-year-old ████ ██████ from York, Andrew Walker, 35, of Bamford, Derbyshire; and 32-year-old ████ ██████████ a ██████████ from ████ Pembrokeshire.

Eldorado by Andrew Walker

Gaily bedlight, A gallant knight,
In sunshine and in shadow,
Had journeyed long,
Singing a song,
In search of El Dorado.

Eldorado by Edgar Allan Poe, 1849

The first known aquanists to traverse the 1.5 miles of sea-munched coastline from Newgale to Porth Mynawyd were a large gaggle led by John Cleare and Pete Biven in 1970. The subsequently named El Dorado Traverse was the culminating head of spume after quite a few years of sidling about in North Pembroke, particularly around the St David's Head area. In the style of the era, ropes were used to complicate and lengthen the adventure, with six Tyrolean traverses and two pendulums. Even then, it involved several zawn swims with vague 'serious incidents', including hypothermia and unintended baptismal antics. The latter could not have been much fun with a late-60s rack and polo neck jumpers. Cleare described it briefly in his book. In our case, the lines were unfortunately drawn on a CD case.

We'd been up all night at a party at a nearby house. A lot of drugs had been consumed, not all of them by me. It was fashionable at the time. In the morning, it was felt that we 'must go down to the sea again, to the lonely sea and the sky'. No one was straight enough to dissuade our planned poetry in motion. Our preparation was based on the then topical Twightism of 'go light, go fast'. The sounds of 'racking up' consisted of agricultural snorting rather than jangling equipment. And it was a psychedelic time of life. The only concession to equipment apart from rock boots was stopping at the Newgale surf shop to hire some wetsuits. This was doubly embarrassing as the non-hospitalized survivors were returned there by helicopter several hours later. At least we taught the Welsh a thing or two about 'sheepish'.

From the north end of Newgale Sands, the coast arcs northeast, with funky folded fun features interspersed with three tidal coves, including the broad Pen y Cwm with its waterfall and El Doradoesque colour scheme. The traversing here is reminiscent of the Abraham's Bosum/Penlas Rock area of South Stack and is quite exhilarating if you are speeding your tits off on acid with a gathering onshore wind and incoming tide.

The character and commitment changes after half a mile where the crozzle kicks west and unbroken towards Porth Mynawyd, a mile away as the gannet flies. It consists of at least eight deeply indented boxy zawns, a myriad of smaller cave-like holes and a sea arch. There are no easy escape routes. These features, combined with at least a dozen exposed 'pointy bits' and tidal offshore rocks, belie the 'it's only a mile and a half' assessment, particularly if you've left your brain and sound judgement in the car park – and not noticed the fine surfing conditions. And ignored the three golden words of sea-level traversing: wind, sea and tide.

Six of us set out, but only five were on acid. The only moderately straight person decided to retreat after the initial section, depriving us of what little sanity was left. They then got cut off and trapped by the tide and wild seas. They'd clearly not had enough drugs. They were spotted by a dog walker on the coastal path who phoned the coast guard. I've always loved dogs. This one saved our lives. On winching her into the chopper, she asked, 'Where are the others?' to which the winchman replied,

'What others?'

They had no idea. Like us. One decided to escape the mayhem vertically by climbing out up a disintegrating and heavily tripping rock face (I don't think they ever claimed their new route). The rest of us got rescued by the rescued.

The bit where Cleare describes the 'clatter' of a flock of white rock doves may well have been the location where in our case:

'A clatter of rotor blades flew into the back of the cave and spiralled us up in an amphetamine-crazed rush of adrenalin penetrating deep into our brains.'

It was too rough for the Little Haven inshore lifeboat to get close. Two were winched from a rock in the middle of a torrid zawn, hastily double stropped with their hands held behind their backs, dilated pupils pressed close to each other's babbling white visages. It was at this point that one house-bound post-party goer remembered:

'I was, of course, safe in X's conservatory thinking, that person being winched up is wearing a Buffalo like mine...'

Two more were winched from the mad froth after being repeatedly battered like corks at the back of a drunken zawn somewhere near Pointz Castle. One of them had lost their contact lenses in the sea. I thought they were going to die. They lost extensive skin from their hands in repeated attempts to climb out of the swell and returned from the hospital suitably mummified. Months later, one of the team was climbing on Dow Crag and still had a terrible sense of sea-related dread.

Six Young Friends (sorry Ted)

This one retreated and, cut off
Stood trapped, alone and screaming in a cove,
Then this one, her dear friend
Sought roulette-like escape up a wall of disintegrating
choss;

These two, despite being warned,
Were left clinging together in an aquatic no-man's land,
Trapped and whizz-banged on a wave scrubbed rock.
And these two, a sister and best friend,
Nearly came to the worst, skinned and near-drowned;
But all were saved.

Facing page top: Redacted newspaper report on the El Dorado Expedition. Facing page above: Emerging from a frigid dark zawn in to the light. Facing page below: Walker with fellow aquanists – at this early pre-baptismal stage it looks like only one of us has pissed themselves so far... Above: The search for El Dorado ends with two gallant knights mummified after being shipwrecked at Pointz Castle, North Pembrokeshire. Photos Andrew Walker.

Hunter Killer by James 'Caff' McHaffie

'What are we going to do now?' Jack [Geldard] asked, looking up at a wet wall, with a few bits of gear, and the two ropes hanging from two distant skyhooks.

Caff looked up, then turned to Jack saying, 'You should give it a go on top rope Jack, those hooks are bomber.'

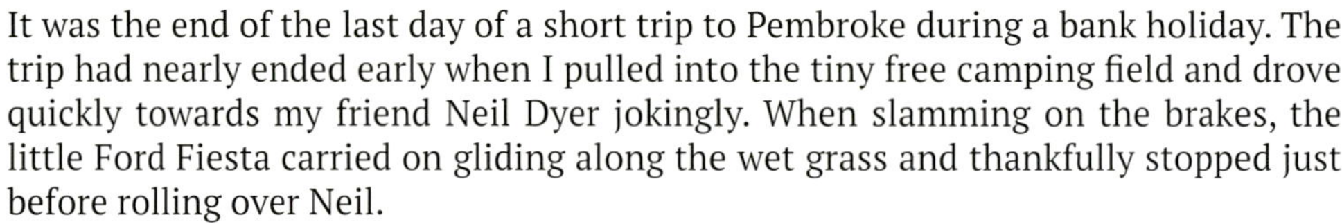

Nick Bullock

It was the end of the last day of a short trip to Pembroke during a bank holiday. The trip had nearly ended early when I pulled into the tiny free camping field and drove quickly towards my friend Neil Dyer jokingly. When slamming on the brakes, the little Ford Fiesta carried on gliding along the wet grass and thankfully stopped just before rolling over Neil.

I'd climbed with Neil at first, but on the last day I ended up partnerless at Huntsman's Leap. Some friends were there from the Lakes, Rob Fielding and others. I'd soloed some easier stuff in the Leap to get going, and then the chalked-up black slab on Darkness at Noon led me to set off up that. I soon found myself past the slab, then spent a bit of time ruminating in the runnel before committing to the steep moves to leave it. On the upper pitch, I was glad I managed to squeeze through a small hole rather than undercutting across on the dodgier rock.

I had a Black Ice rucksack, I think, which you could take the lid off and use as a bumbag. I did this and put my boots and chalk bag in it. Once the tide was in a bit in the Leap I swam out, nervously, as there was a big seal nearby too. I swam rightwards, dried off a bit, then soloed Star Wars (I don't think I had my ab rope handy to set one up to access this route – hence the swim).

I'd pretty much thought to end my mini-break there and went to Ma Weston's for a cup of tea, worrying about the traffic on the way back. There, I met Tim Emmett and Barry Durston who were talking about doing Hunter Killer while the tide was high. Being young, 22, I joined in with their enthusiasm, and we all abseiled into the Leap. They'd both led the route before, but I'd not done it.

Soon, we were huddling on the black ramp beneath the route. Tim went first, having a small fight on the crux and then keeping it steady. Next up was Baz: he said he wasn't in great shape, and this was confirmed when he set off and looked nuclear on every move, shaking, slapping and scraping his way through each section. This made me more than a little nervous; it felt pretty lonely and somewhat dark in the Leap as I eyed up the landing zone and thought about how I'd get back out if I got drenched. I knew the strenuous-looking crux wouldn't suit me too well.

As soon as Baz got to safety, I set off and remember hitting the crux, thankfully, spot on, which meant my weak arms could yard up to the rest in 'the mantrap' (a big hole you can get a leg or two in). I'd been doing a ton of soloing on big routes in the Lakes and Wales for a few years before then, and I remember doing the bit from the mantrap to the top as fast and as confidently as I've ever climbed. It makes me feel nostalgic looking back because I know now I'd be stressed and would climb it slowly and badly if I tried something like that nowadays.

Once out I drove up to Llanberis, and as I was painting Huw and Ursula's pub, Gallt Y Glyn, the day after, I dossed in the woods by the river beneath the Victoria Hotel where Will Perrin used to slackline. It was the first summer I'd spent in Wales, and it was magic: a great bunch of westies to play with.

Left: Gav Symonds off San Simeon, F8a S3, Pembroke. Photos Dave Pickford. Facing page: Tim Emmett on Hunter Killer, F7b S2. Photo Mike Robertson.

Julian Lines watching Tim Emmett falling off from an onsight investigation of what will be a super-hard DWS on the arch section of the through-cave of The Castle. Facing page: Julian Walker on Once Bitten, F7a S1, at Hollow Caves Bay. Photos Dave Pickford.

Mother Scareys by Julian Lines

My first visit to Mother Carey's was with Mikey Robertson. I'd been working nightshift on the rigs and had driven down to Bristol during the night and then driven on to Pembroke in the morning. It was a glorious afternoon as we abseiled down to the base of the crag. The tide was out, and the boulders were crackling away with limpets in the sun. Since I'd had no sleep for 36 hours, my head was spinning, not functioning, like someone who was drunk. I peered around the corner to see the Space Face. 'Crikey!' It was over 100ft high of grotesquely overhanging, primaeval stone. I found it difficult to concentrate just walking over the boulders. I didn't know how I was going to get out. Mikey suggested we soloed out via a superb crack line, which I later found out was Dolorous Gard (HVS). I don't think I'd ever been so out of control in my climbing.

I'd been doing a lot of deep water soloing in the Alien cave, but one sunny morning, I wandered down the ridge adjacent to the Space Face. Maybe I was just curious, but when I got down there and saw the morning spring tide nibble its way up the base of the cliff, I started to get excited at the idea that these routes were marginal solos, where the 110ft cliff is diminished to 80ft on a Pembroke spring tide. Crispin knew of the leftfield possibilities of the face, but Tim Emmett was probably the first to breach the cliff via a solo of Just Klingon. I scoped out a sloping gear-up ledge near the right-hand side. I ran back to the car to get my gear, swam the channel, and got organised on the ledge. It was time to explore and weigh up the cliff. I found the ludicrously steep flake line of Bagpuss on massive holds coming out of the cave. It finished at a big hold at 35ft from where I jumped off.

I was back the following year to try some of the bigger lines, so I went off in search of the start of Hyperspace in a cheeky fit of euphoria and nerves. I think I found it and launched upwards. God it was steep and unrelenting. At one point I felt like bailing, but the guide said 5c, so I carried on finding respite just before terminal lactic acid burn. The big rightwards traverse felt easier, although it gives more time to absorb the surroundings which engulf and scare. I was heading for a big hollow high on the face, but there was smoother rock between me and it, which required technical sequences on smaller holds. I read it well and moved up into the hollow. It looked easier above, but the guide mentioned the finish was loose. I took an age, making sure that everything I was pulling on was solid enough. I didn't fancy an 80ft fall into the sea. A few days later I encouraged Mikey to have a go; he set off on a falling tide. When I heard his screams on the technical traverse above a receding sea from where I could see boulders under the water, I could barely watch.

From my little sloping ledge I could access all the routes; next up was Mother Night. I was quite impatient waiting for the evening tide to come in and set off in a startled rush. I got to a big hold at the top of the groove at about 70ft and almost went to pieces. My brain knew the barriers: I was crossing the line from deep water soloing to soloing. The mood changes, and movement is slower, more calculating; hindering. Above was a technical hanging ramp; I found it harrowing. In fact, I found it the most harrowing of all the solos I would do on this cliff. A few days later, I went on to do Tiger Tiger; this was a different beast altogether. The crux is low, powerful and awkward. The top section of the original route isn't above the water, but I managed to sniff out a lower rising traverse leftwards at 5b to reach the hollow on Hyperspace to the exit I knew I could negotiate. After getting accustomed to the routes, I had this mischievous audacity of trying Fireball XL5, knowing that if I got totally boxed on the lower section I could jump off at the break. It was a morning spring tide; the sea was placid without a ripple and rising slowly like a bath filling up. The scene was set.

At the base, with water lapping at my feet, the threads were dangling out above my head illustrating the brutal steepness of the rock. I pulled onto the starting holds, and in an oblivious minute I was at the break unbelieving of what had just happened. I rested on the ledge, looking at the bulge above with more threads hanging down. With my mind in overdrive, and my stomach in knots, I looked down to the sea and tried to convince myself that it wasn't too far. There wasn't another soul around. It was how I liked it. I charged up and through the bulge which guarded the base of a groove. Then I started stalling. The groove was tricky to access; I was wrong handed. Could I go down? I had to do something quickly, otherwise I was going to blow out. I readjusted and shot up into the groove, climbing in as much controlled

panic as I could muster to finally grasp a massive bucket at the top of the groove. The sea distance was on the margins, a solid 60ft. I wasn't keen on the top finishing wall, so I managed to discover an easy-ish traverse into Star Gate to finish up its groove. I returned two years later to climb Zoony and Just Klingon and found, to my dismay, that my gearing-up ledge had been destroyed by a winter storm. So there wasn't anywhere to crouch and change anymore, which meant I had to resort to plan B and swim in with a bench seat.

Little Big Crag by Martin Crocker

Aptly described as Little Big Crag (or is it a Big Little Crag?) by Pat Littlejohn and Chris Howes in their *Mountain* mag exposé of 1983, Ogmore presents a taunting reminder of the diversity of SE Wales climbing and how extraordinary it can be. As steep as they come, Ogmore's impending walls, roofs, and crisp-like Liassic banding blow minds as well as arms. Such is the physicality of the routes, 'give-it-a-goers' exhaust themselves as they adjust to all the dead-hanging out in space, their payload of cams aiding gravity. All the while, a sense of urgency, if not panic, is unsettled further by the crescendo of manic Bristol Channel tides rumbling and racing in towards them. Today, Ogmore stays relevant, spiritually intact, and accessible. It is the spell-binding monster behind the door you dare not open.

Imagine a world without gravity. Deep water soloing is about as close as you can get to it. By the late 90s, I'd become hooked big-time. Countless days of adrenaline-fuelled excitement were supplied by the cliffs of Ogmore and others, a perfect counterpoint to the routine of deskbound nine-to-five. Even to this day, I'll frequently hang out above the sea, my daughter's 20-year-old Freddie the Frog water wings smiling unflinchingly to either side.

There are two main DWS zones: Tiger Bay and Davey Jones' Locker cave. But soloing here isn't for everyone, and neither is there any expectation that it could ever be. It's super-serious notwithstanding all the science of the tides, depth calculations, and low water walk-pasts – and the sorcery of happy water wings.

Above: Tiger Bay at low tide. Facing page: Martin Crocker on the first ascent of Total Eclipse of the Sun, F7b+ S3+, or E8 6b. Photos Black Planet Photography.

Total Eclipse of the Sun

Up to two or three minutes before totality, it seemed certain that the corona would be completely hidden, and up to one minute thin clouds threatened to cut off portions of it. Fortunately there was time to secure the observations, and about one minute after totality the clouds obscured the Sun.

The Crocker Expedition to Observe the Total Solar Eclipse of May 28, 1900 by WW Campbell and CD Perrine. *Publications of the Astronomical Society of the Pacific*. Vol 12, Number 75, 1900

Almost 100 years later, another Crocker expedition set out to take advantage of the spring tide associated with the total solar eclipse of 11th August 1999. No book about deep water soloing in the UK would be complete without some descriptions of the exploits of Martin Crocker. Born in Bristol, Crocker dominated climbing development in the West Country for decades with cutting-edge new routes in Avon, Cheddar, Devon, Cornwall and Wales while also developing numerous obscure venues along the way.

On ropes he is renowned for wearing out numerous belayers in the same day, and he is one of the most prolific DWS first ascensionists in the UK. His route Total Eclipse of the Sun, E8 6b, takes on the complete traverse of Tiger Bay at Ogmore, and due to the shallow, boulder infested, water was climbed at the peak of the freak, high spring-tide that coincided with the total solar eclipse on 10th August 1999 in the UK. In the photos of him en route, he has his daughter's inflatable swim arm bands attached to his belt which is, presumably, to reduce his water penetration depth in the event of a fall into the shallow water.

This page and facing page: Rob Lamey on Davey Jones'
Locker, F7b+ S2 or E7 6b. Photos Simon Rawlinson.

Davey Jones' Locker by Rob Lamey

The traditional climbing found on the cliffs at Ogmore could briefly be described as adventurous esoterica. Steep routes, rotting fixed protection, extremely tidal, loose top outs and poor belays make it a somewhat unpopular venue when compared to the nearby beachside sport routes at Dunraven Bay. Despite these factors I have developed a strong love for this crag, which was amplified when Simon Rawlinson and I first ventured into the vast cave of Davey Jones' Locker to sample the deep water soloing on offer.

Over the summer of 2014, we repeated many of the existing routes and were lucky enough to add some first ascents of our own, but the main challenge remained elusive. The route, Davey Jones' Locker, by which the cave gains its name, climbs from the darkest depths via a direct line through the steepest section, utilising a crack to cross an eight-foot perfectly horizontal roof. Rounding this roof provides the crux. With a toe hook lodged deep in the crack and my body outstretched, trying to gain every inch of reach possible, I couldn't progress beyond this point. Backsplats, faceplants, kamikaze-esque ejections, and salt water enemas were all endured from that crux over the next five years. FIVE YEARS! Sounds shit, doesn't it?

Well, it wasn't – I loved every second of that process. I'm not one to get all philosophical, but success on that route wasn't paramount because once you abseil down to the gearing up ledge it's a truly magical place. An evening spring tide with the Sun low in the sky creating mottled textures over the sea's surface, a few friends and some post-climb beers is idyllic.

What was the crucial beta to avail me of yet another aggressive splashdown, I hear you ask?

The answer lies deep within the bowels of Davey Jones' Locker awaiting your discovery...

Crispin Waddy on Dynamite,
F6b S1, at Skomar Arch West,
Pembroke. Photo Mike Hutton.

Eternal Damnation by Crispin Waddy

I doubt he was any happier than I was. Or maybe he was. Maybe he enjoyed this sort of thing? Justified anger; retribution. It was the early hours, and I had been awoken from my well-earned slumbers by two large Alsatians and the red-faced landlord of the St Govan's Inn. Still groggy, I was struggling up the hill to the village carrying two large doormats followed by my wheezing judge, jury and, given his mood, potential executioner.

We'd had a superb day. John Vlasto, a regular partner in those days, was a patient belayer who could easily follow anything I could lead and, rather conveniently, preferred it that way. It was a beautiful crisp morning as we strolled over to Stackpole Head through the flowers and butterflies, with the iconic song of skylarks in the cool wind. I'd spotted a steep line right of Swordfish and was keen to try it. It looked worth a go on-sight. I managed to get up to and over the first five-foot roof to a nice, comfy stance.

The next roof, looming directly above, looked more challenging. Maybe 10ft wide, horizontal, and with nothing but a smattering of small holds that would be more typical on a vertical wall. But near the lip was a deep slot, too wide to jam, that looked like it may just provide a solution. Unwarranted optimism can be a useful attribute sometimes.

The chink in the armour was the deep foot jam in the break at the back. This allowed for a full-body span with the other foot pressing lower on the wall, the few holds doing nothing other than to stop you gently flapping to and fro, like a gate in the wind. I could just get a hand up into the slot, groped around at full stretch and found... nothing.

At some point I must have fallen off as I ended up swinging on the gear, clipping in and having a peek inside the slot. My chalk dabs were just tickling the edge of a sinker jug that I'd had no inkling of. I had a few goes from the stance but couldn't make up the tiny extra distance. Eventually, I gave up and finished it with a point of aid, with a view to coming back some other time. The move from that jug to the lip, incidentally, is the only time I've ever actually used a figure of four in anger.

I don't remember what else we did that day, only that by the evening we were too tired to cook and decided to eat in the pub. This was an extremely rare luxury for us, as students living on nothing but meagre grants. At the end of the meal, I must have been leaning back with my eyes closed as the landlady came over and commented, quite aggressively, that: 'It isn't a doss house, you know'. It was certainly true that climbers, especially perhaps the younger ones, in those straightened times were capable of nursing a pint for quite a long time, but on this occasion, we certainly felt that we'd spent a reasonable amount, and this was unnecessarily unpleasant. However it was true that we needed to get some rest, and we used it as a cue to head off.

In the little entrance porch, as we were sorting our rucksacks, she reappeared and, bizarrely, accused us of stealing beer glasses. There was no basis for this at all.

'Go on then, let's see what's in your knapsacks.'

'What are you on about!?'

'Come on, open them up.'

As it was going to be trivial to disprove her weird theory and should resolve the dispute, we complied. She was a large woman, but not tall, and many of you will know how small a space this is. The situation was absurd, comical and infuriating in equal measure. There was no trace of an apology or any contrition of any sort as we watched her head back into the pub, muttering to herself.

John was furious. I watched him look around the porch, wondering what he was doing. Eventually, he realised that the only thing that wasn't fixed down was the doormat, which he grabbed and walked out with. John is a 'collector of experiences'

Above: Crispin Waddy off Release the Bats (E5) in Huntsman's Leap. Photos Dave Pickford.

and may have seized this excuse to make one. This is the man who once crossed the Rio Grande in Juarez, which is the border Between the USA and Mexico, on a rubber ring dragged by a man in waders who was operating a little business.

Uncle Sam surely would've approved of such entrepreneurship – ferrying day workers to and fro for local employers taking advantage of the lower wages. All this was in plain view of the border guards on the bridge and during the First Gulf War. Of course, this is no longer conceivable after Donald Trump's remarkable success getting the wall built, as promised, and even getting the Mexicans to pay. God bless the MAGA movement. Anyway, when we got back to our car there was a note saying: 'Have been arrested by the border patrol, see you in a few years.' He was locked up and deported within a few weeks. But he revelled in the experience of, at least for a while, meeting a variety of hardened criminals and murderers. Sounded quite scary to me…

Back in Bosherston, in those days, the church porch was the sleeping venue of choice. I didn't own any sort of sleeping mat then, but I was perfectly happy sleeping on the rough doormat there. John was properly kitted up, so that night I had the luxury of an extra mat for my lower half. But we were aware that the publican knew that climbers sometimes used this porch and had expressed vehement disapproval of the practice, unlike the vicar who used to wake us up in the mornings.

'Forty minutes to the service,' he'd say, addressing up to four sleeping bags as he tiptoed through the tightly packed porch. 'Can you be sure to have it tidy by then?' Occasionally, he'd ask if we'd like a hot drink and reappear 20 minutes later with mugs of tea and biscuits on a tray.

Clearly, though, this wasn't the best night to be sleeping there, so we went on down to the pumphouse at the lily ponds, the overflow for when the church porch was full. This is how it came to pass that the landlord and his dogs woke me later that night – evidently, he was a bit more knowledgeable about the habits of itinerant climbers than I realised. To be fair, from his point of view (and he was an ex-policeman) it was clear what had happened. I was asleep on his mat and the one from the church; my guilt was incontrovertible. I couldn't and didn't argue the point, and to explain the mitigating circumstances, reflecting so badly as they did on his wife, didn't seem particularly diplomatic.

'Do you know what you can get for stealing from a church?' he asked me as we headed up the hill. Perhaps it was a good thing that I had declined John's offer to accompany me as I sheepishly replaced the mats, as he later suggested that the best answer, if not the most sensible, would have been: 'Eternal damnation?'

(A certain Lancastrian hemiplegic serial adventurer and author, now resident in Tasmania, once ate a single grape from the harvest festival display in this same church and claims that it was struck by lightning immediately afterwards).

Quite near our route there was one of Ben and Marion Wintringham's called Eternal Sin, which had also used a point of aid during the first ascent, I think, hence its name. So, there was no choice really: Eternal Damnation it was. I wish I'd left the question mark in.

'Leave Pembroke and never come back,' he told me, in no uncertain terms. Pembroke, the town, or the county? Presumably, the latter as the former would have been a strange ban. Perhaps it was OK to come to Bosherston but via a bypass? No, I think it was the county he meant. But did his jurisdiction extend that far? Either way, there was no doubt that the pub would be even less welcoming to me in future than previously. Just a few days later, though, I was making arrangements with Andy Lafferty, another Bristol student and occasional climber, to meet up in the St Govans Inn the following weekend. I find it hard to understand, looking back, why I did that. Exam stress compounded by too little revision – as a result of too much climbing – I guess.

It was only when I was approaching the pub that I remembered my ban. I think Andy and his friends were fairly confused to see me trying to attract his attention, but not that of the landlord or his staff, through the windows. Eventually, he came out for an explanation. Andy struck me, in those days, as a slightly otherworldly type. I didn't know him well, but he was agreeable, enthusiastic and happy to get involved in my schemes.

He has a twin brother with whom he claimed to be able to communicate telepathically, which no doubt saved on phone bills. He wasn't very experienced with sea cliff climbing and was open to my suggestions. I had another idea for a new route, this time in Huntsman's Leap. The tidal lower reaches were only partially developed, and I had inspected a line on the far right of the Monster Face which started well below the high tide mark.

The following day we abbed into the Leap, but too late; the tide had already turned. I was studying physics and had a good grasp of how predictable the movement of the moon was and how it affected the tides but had, nevertheless, as usual, failed to get the timings right. You'd have thought that someone could have produced a little book detailing this information?

But there was a narrow ledge a little way up the route which was still accessible. Andy was a trusting soul, and I got him up there and put in a poor nut and an RP 2 for a belay. It was all a little too cramped and unsettling, so I set off up the route with the ropes still uncoiled and my boots undone in the hope of finding somewhere to sort myself out.

There was a tiny crack, maybe 20ft up. It looked perfect for an RP 2. I only had one of them though… It turns out that the RP 3, as I suspected, wasn't well seated. Fortunately, the route had traversed a little over the sea, so it was into its watery embrace that I fell as I lent back to tie my laces.

We escaped up Fitzcorraldo, a good consolation. Afterwards, looking across at it, I noticed a bit of yellow tat under the overlap below the crux.

'Andy, is that my MOAC?' I asked, pointing at it.

'I thought it was jammed. In situ.'

Admittedly, it was well disguised as a piece of old in situ gear, with its tatty discoloured tape, but it was also the last remaining one of the seven that I'd made, along with two sticht plates, as the first rack that got me started climbing with my school friends at Swanage, and as such it held some sentimental value.

The next day, at a more appropriate time, we were in place, and the route was in the bag. I really loved that route – I've always preferred the ones that have something out of the ordinary about them. In this case, there's a section up a shallow runnel, the sombre light from above reflecting off the blank-looking surface. But, almost as if designed, there's a series of deep slots leading upwards. Eventually, they abruptly disappear, leaving no chance of progress up the wall above. But, amazingly, at that point a bulge in the west wall means that it's possible to bridge the zawn and move easily up to the boulder choke in the entrance of the Leap. So, The Ducking Stool came to be.

I was a regular visitor in those days, and a few weeks later I was, for some reason, trying to hitch into Pembroke from Bosherston. A lorry with two blokes in it, coming the other way, stopped for me.

'We're heading back in a minute, just got a few deliveries to make.'

'Great!'

'You can give us a hand if you like,' The driver said.

'No problem. Happy to help; what have you got?'

'Just got a load of beer for the pub.'

Hmm... I couldn't refuse, but it could get interesting. I admit that I do quite like these kinds of situations. So I was in the cellar receiving the barrels when the landlord came down the ladder.

He certainly knew there was something not quite right about me or my presence, but I think it was so out of context that he couldn't really work out what it was. After that, the situation descended into farce. He was also organising a party that evening on a local farm, so we all piled into the three-man cab, me being pretty well flattened by my former adversary as we headed off. I was just thinking about the last time we'd met and grinning to myself.

I'm pretty sure he never quite clocked me for who I was, but a year or so later, months after I had cautiously started using the pub again – it's the only one in the village after all – he suddenly remembered that I'd been banned. But this time, he said that the incident in question had been a week earlier, and my (more respectable) friends vouched for me, that I had been in Bristol at the time, and he relented. A little later they sold up, and that was that.

Above: Ben Heason on Exultation, F6B+ S2, at Stennis Ford, Pembroke. Below: Bob Hickish on Hydrotherapy, F8a+ S2, on the Great Central Hole at Hollow Caves Bay, Pembroke. Photos Mike Hutton.

Martin Crocker on the first DWS ascent of Sorcery, F7a+ S2, Ogmore. Photo Black Planet Photography. Martin: 'For 1979, Sorcery was something else. Pat Littlejohn graded it E5. I upgraded it to E6 when I made the second ascent in the mid-80s after failing on it and taking some pretty hair-raising falls onto a crumbling, twisted Rock 2 placement a few years earlier. Bizarrely, the solo ascent felt easier and much safer.'

Free Masonry by Crispin Waddy

'Well, I'll just have to lower you into the sea.'

Andy Long hadn't seen the route, and I had somehow neglected to inform him that this seemed a quite probable outcome if we were to attempt it. A good partner was a vital requirement for this route, and I wasn't going to risk putting him off, though I doubt it would have phased him anyway. It was clear that it would likely be a committing, possibly impossible, route with limited escape options. The massively convoluted coastline of South Pembroke, with its myriad zawns, holes, caves and arches, is one of the main reasons that the routes here are often so enticing. The routes that fully engage with this surreal rock architecture are guaranteed to be highly characterful and memorable. And this one certainly did that.

I've always been more interested in finding new routes than repeating them. It's not obvious why, really, but somehow, it just seems more satisfying. There is a feeling of creativity about it, but to me, it feels more akin to the creativity that scientists and mathematicians report when they uncover a natural law than to aesthetic creativity. The lines already exist. The forces of nature have sculpted them decades or millennia ago (chipping and quarrying notwithstanding) – all that we are doing by climbing them is discovering and presenting information. I think that's why the greatest satisfaction comes from climbing on sight, as the process of discovery is drawn out and often in doubt.

There are a lot of different factors that go into making good routes – I guess that the different weight people give to them is the reason that opinions vary so widely on the quality of any given route. For me, atmosphere, position and character are pre-eminent. My favourite routes are generally those where it's possible to feel relaxed and comfortable in wild positions – positions that seem unattainable at first glance but, by some magic combination of line, hold layout and rock architecture, are attainable. Ideally, they will be technically interesting but not unduly strenuous, as pulling hard and getting pumped seems a little undignified – misses the point somehow, I feel – and is best left to sports climbers, boulderers and other such 'athletes'.

So when I spot a possibility, these are the lines that excite me. In the same way, a good sequence that solves a difficult crux works because it reduces the load on your fingers or arms to manageable levels, and solving these puzzles is one of the great joys of climbing. But, it seems to me the quality of the most compelling lines is also in the way they require less physical effort than, at first glance, would be expected to gain those otherwise surprising positions. It's the enjoyment of the same type of effect, though on a greater scale, as if climbing is fractal, not just in a physical way, where the shape of holds and features reappear at every scale, but also in the way that the best routes bestow satisfaction is due to the same qualities at different scales.

Exposure is a strange beast, though – height adds to it, of course, but the effect of the layout of the holds and features seems to be multiplicative. So, for instance, there is a perfectly positioned square foothold, maybe an inch above the lip of a jutting block, that forms the crux of The Conger at Swanage – one of my favourite holds on one of my favourite routes anywhere, though soon to be consigned to the sea along with much of the block itself, by the looks of things.[1] This block, hanging drunkenly over the black void of the deep cave beneath, gives the route a sense of exposure out of all proportion to its height, being a mere 25ft above the sea. It feels as if the same effect, that narrow depth of field that fools us when viewing a photo, where only the subject is in focus, sometimes applies on well-positioned routes. It seems to give an illusion of depth and, hence, a greater feeling of exposure. It's as if exposure is a commodity and effort is the currency, and so the best routes – the real bargains – gain wild positions without really forcing you to break a sweat. This route looked like it could be a bargain – but bargains aren't necessarily cheap – and, viewing from the sidelines, there was significant doubt that I (or the rock) had deep enough pockets. The bulky arch that forms the largest of the three sea-level entrances to The Cauldron has a compelling quality. Viewed from the easily accessible landward side, it isn't particularly inviting. All that can be seen from there is its scruffy loose top, but if you make your way down Flimston Ridge and through a surprising wave-carved tunnel

---

1   This fell off sometime in 2020 or so, I think.

Grant Farquhar on Hanging by the Bridge, E5 6b, with Crispin Waddy and Andy Long on Free Masonry, E7 6b, in the background. Photo Steve Mayers.

Murdoch Jamieson on Free
Masonry. Photo Tony Stone.

to the base of its right-hand side (a wrong turn here finds you overlooking an amazing subterranean backlit pool deep within the cliff), it has an impressive aspect. It is cut through one of the bigger cliffs in the area and spans a similar distance to the nearby, and much better known, Green Bridge of Wales. The eye is drawn up the junction between its hideously overhanging underside and the nearer vertical outer walls to a cave that is, amazingly, perched exactly at the apex, like a giant Cyclopean eye socket. It's a destination rather than just a convenient stopping point. And what a place to aim for – an oasis of calm to hide in where you could forget, for a while, your self-imposed predicament.

I had seen this line a few years previously during an ascent of Stone Bridge. As I hung on the first belay, I idly wondered about what would happen if one tried traversing left into the middle of the big arch from there. Stone Bridge is a Mick Fowler and Brian Wyvill route that follows a diagonal fault up the arch, given E3 at the time. Unsurprisingly, given the terrain, it's now settled at E5 6b – impressive for 1980. The line breaking left out of this directly over the arch to its apex had looked improbable then and had long seemed like one of those lines that would always remain a fantasy, though one that gnawed at me. Sometimes, though, the stars align.

There are a few factors that need to come together to even allow an attempt to happen. It's bird-banned until The Glorious First and inside an active firing range, so frequently not accessible midweek. It features a long traverse over the sea and would be exciting for both members of the team. So, for my first attempt, the thing that swung it was cajoling a willing and, as importantly, capable partner. Andy led the crux start of Stone Bridge, and then we were into new ground. The cave was a little above us and some 50ft left. Initially, it was reasonable climbing across various grooves until I was stopped by a blank section. It had been going so well. The cave seemed almost within spitting distance to my left. Had it been nesting season, the fulmars would have been enthusiastically trying to prove this. A little lateral thinking was needed. I looked down. I had good gear, and so I was well protected downclimbing a more difficult groove, feeling a little guilty at only managing to fiddle in RPs for Andy. This groove ended at the lip of the arch, and there I was forced to belay. This rope was sharp at both ends.

After a complicated change over on the belay, I was off, and soon enough, off again. Shadows were lengthening, and I felt justified in taking a rest. The remainder of the pitch went OK, and soon I was in the cave – so long dreamt of – and a more comfortable belay. By now, dusk was fast approaching, and there was no discussion about trying to top out. The grey, choppy sea was uninviting, but, as to an 18th-century drunkard confronted by a press gang, it looked unavoidable. I lowered Andy till he was just above the sea, where he inverted and paddled furiously whilst remaining more or less dry. It's possible to get a massive swing going like this, and this 'dry swimming' technique has often saved me a dunking. But this time, it was a forlorn hope. It was too far, and it was clear that ordinary swimming was the only option.

It all seemed to add a bit of gravitas to the route, and so, somewhat chastened, we squelched through the tunnels and traverses to regain Flimston Ridge. As we trudged, dripping, up this, we could plainly see, as the Sun set, the last kids mucking around in the sea in Flimston Bay; their shouts and laughter rather deflating the impression that what we had had to do was serious in any real way. Many of those same water molecules that had held us up and cooled us down would have, on countless occasions in past aeons, lent their weight to the waves that slowly enlarged the tiny crack that eventually became the arch and previously would have cycled through the bodies of organisms whose skeletal remains made the limestone itself. But so what?

Due to work and other commitments, I couldn't get back before winter rolled around. So that's how it stood for a year: unfinished business. The gulls returned from their holidays and took up residence again in the cave, brought up a new batch of chicks and left. The unclimbed top pitch wasn't forgotten. The following autumn, I managed to persuade George Smith along with the bait of the freeing of the third pitch. Like *Pulp Fiction's* Mr Wolf, his sorting out my indiscretions on various crags around Britain had become something of a regular occurrence. Now, some climbers have strong fingers, excessive stamina, lack of fear, or brilliant technical movement skills, but George's climbing superpower seems to be the spotting and efficient use of knee bars and extreme pain tolerance in his lower thighs. He doesn't look like your classic good climber. Those same scientists who famously demonstrated to their own satisfaction that bumblebees couldn't fly would no doubt have shown that he would always be confined to the lower grades, but here, as usual, and without much fuss, he climbed past my rest point and had made himself comfortable in the cave. I joined him rather less easily.

The eye socket cave has a significant roof immediately above it – a chunky Neanderthal eyebrow – and the obvious way on was once again by an exposed traverse left along an easier-angled wall sandwiched between that roof and the void. This led to a steep corner, which looked like the most notable line of weakness leading to the top. Its inherent weakness was its strongest defence, as it turned out. The left wall of this corner was comprised of appallingly loose, overhanging, collapsing rubble held together by damp mud. After a strenuous grapple with this and the removal of several wheelbarrow loads of assorted blocks, I abandoned it and retired to the cave for a rethink. This foray had at least given me a chance to glimpse a little of the lay of the land directly above the belay, which is hidden from within its relaxing interior.

It turned out that it was necessary to pull blindly straight over the roof with no worthwhile gear but the belay – a cam in my mouth for the illicitly spied hidden break above. There was a brief breathing space before a wildly overhanging groove led to easier ground and the top. 'That's a 7', said George as he topped out. At the time, the hip kids in Llanberis would leave out the 'E' when talking about hard routes, and he was taking the piss.

Neil Gresham on The Wizard.
Photo Mike Robertson.

Pembroke Nirvana by Neil Gresham

I watched that film about Olympiad – it looks nasty getting in to it and nasty getting out of it.

Henry Barber

The first crags to be developed for DWS in Britain were the obvious candidates in Dorset and Devon, which boasted easy access and a broad tidal window. However, as the development of these areas reached saturation around the turn of the century, it occurred to protagonists that some of our finest DWS treasures were yet to be unearthed. In Pembroke, a number of compelling steep walls and caves had remained elusive due to the giant tidal range and the inherent difficulties of gaining access.

The choice was either to leave them to the waves or to develop new tactics to embrace the challenge. As such the use of dry-bags, dinghies, and Tyrolean traverses have brought about a rise in standards and allowed the genre to expand further. Be sure that Pembroke DWS is not for the faint-hearted. With the crashing Atlantic and no obvious means of escape, a sinister and foreboding aura fills the air. The experience can be likened to a fusion of Scottish winter climbing and redpointing – a frustrating battle to get all the variables to balance, yet with nirvana moments lying on the table.

The Wizard F8a S2

Here we go again. I always know what comes next when I spot that tell-tale glint in Mike Robertson's eyes. We're nestling in a dark corner of the St Govan's Arms, sipping the foamy tops of our post-splashdown pints.

'And as for Kato Zawn…' he wipes his lips and leans forward, lowering his voice with theatrical aplomb '…it's a slice of Lulworth Cove, but it's untouched, and there's a king line with your name on it.' I doubt that the lobster-coloured family who are devouring their fish and chips just to our left are likely candidates to commit first-ascent espionage, but I play along in hushed tones: 'Let's go there tomorrow!'

Excitement levels peak as we jog along the Penally coastal path the next morning. Mike gestures towards the cliff edge, and I peer over with bated breath. And there, below, is a cauldron simmering away with all the necessary ingredients for DWS mayhem: a grotesquely leaning wall, 50ft tall, pristine and bullet-hard with the sea gently caressing the base. Mike gives the call to arms: 'Looks like it will get harder the further left you go, and the groove on the far side is the big prize. Surely no change for an eight, and it looks ground-uppable.'

At this point in history, the thought of attacking a route like this from the base, without any knowledge of whether it was possible, seemed pretty out there. I can feel myself being vaporised by nerves on the very spot. Clearly, it would make all the difference if I was to zip down an ab rope to suss out the top moves and chalk the holds, but a strange voice was tapping my brain and telling me to resist. After all, there was nothing to lose if I tried it from the bottom a few times. At moments like this, it's easy to forget that the whole point of adventure is to experience uncertainty.

I ab down the less-steep, righthand side of the wall, disconnect from the rope and begin traversing timidly leftwards. A blind and tricky sequence kicks in as the wall starts to tilt over. I fumble for holds and find something, but it isn't great. A strange fizzing sound. I build my feet and slap, but I'm too focused to notice the jet of colourful sparks which shoots out from a crack and tips me off backwards into the water. I whoop out loud from the release of tension, and my splashes muffle the faint murmur of a menacing chuckle.

Next time I make it to a small roof and attempt a fierce laybacking sequence to enter the hanging groove. My limbs move in staccato as if cursed by a rigor mortis charm, and before I know it, I feel the cold surge of the water again. On the third go, I make it to the top of the groove, arms bulging, confidence swelling. Then some strange ranting from within the walls, and suddenly, everything blanks out. I search desperately for handholds as the pump escalates, and my fingers start to uncurl.

Neil Gresham on Olympiad.
Photo Lukasz Warzecha.

Maybe there's a foothold instead? I look down to see that the water level has dropped, and the seabed is seeping through the ocean's turquoise surface. More mumbo jumbo chanting before I'm ripped clean from the cliff and dumped in the sea. My feet touch the bottom with a tap.

'I think that's it for today,' I shout up to Mike. We are straight back the next morning, caffeine and adrenaline-fuelled, after a sleepless night. But each time I get to the top of the groove, I become lost in the same labyrinth of wrong turns. And each time my gameplan is foiled as if the holds are being moved to different positions. Confused and exasperated, my final slap out left is feeble and futile.

I'm sitting on the cliff top with Mike, resolve draining away as I ring the seawater from my chalk bag. Should I get the ab rope out or just leave it for someone better? I stare vacantly towards the horizon as a fishing boat bobs past. A seagull cries. To create a masterpiece you have to commit yourself to an indefinite timeline, a process which always brings joy and agony in equal measures. A bony finger beckons.

Day number three, attempt number nine, and I'm on autopilot as I reach the top of the groove. First the drop knee, then the pinch. A merry little dance up to the point where I'm left alone to improvise in front of a cruel audience. I set my feet again, pick a target, launch and prepare for the inevitable drop. But this time, I stick something and pause, swaying silently, not daring to move. Then a hollow cracking sound triggers total chaos.

My left foot pops, and I swing out. The wall topples over; the rock starts to melt into the dark ocean, which is raging below. I stick my feet on the return of the swing, and a cyclone erupts, twisting and tubing upwards and sucking me up over the lip of the wall with the scream of a thousand jet engines. Time trips backwards. I collapse, dazed and disorientated on the cliff-top, gasping for breath.

Far below, in the murky depths of the Atlantic, the two broken halves of an ornately carved teak rod come to rest gently on the seabed.

San Simeon F8a S3

From a semi-hanging belay at the base of the wall, I crane my neck backwards as Charlie Woodburn slots in his last wire and launches upwards. Perspective distorted, the wall appears to be toppling into the sea as he grapples over the top, and the sound of his emphatic breathing fades into the lapping tide. Tall and with elegant flow, another Pembroke gem has just been unearthed. Whilst jumaring out, I'm suitably impressed at my pal's ability to place protection on such sustained terrain, and then a devilish idea occurs to me.

So, a year later, I'm back at the base of the wall. Except, this time, I'm armed only with rock shoes, chalk bag and a vague notion of hope. I've built up to this moment by deep water soloing all over the world, but this one will compare to no other. Sixty feet represents the realistic cut-off point for a safe DWS fall, yet the top of San Simeon lies 80ft above me, and the climbing is relentless at F8a. Mike Robertson gives me the nod from the water to let me know he's got my back, and Gav Symonds awaits on the cliff-top, eager to nip at my heels. Silence prevails as I step off the ledge.

The climbing soon draws me in; I'm blissfully unaware of my position until I cross the line at 60ft, and the bubble bursts. I falter on a hard slap for a distant edge, fumble it and take to the air, leaving the door wide open for Gav. Moments later, he reaches the same move and sticks the edge. It's his for the taking if he can just hold it together. All I can do is watch. Yet in a cruel twist of fate for Gav, at 70ft, whilst stretching for a crucial pocket, he changes his mind and grabs a tantalisingly positioned in-situ wire. He hangs there, realising he's voided the ascent. With no choice but to bale or risk climbing onwards, he shouts for advice in desperation. 'Err, don't ask us!' we reply, and moments later he ejects and slams into the water, with the only consolation that his landing was perfect. I'd given up for the day, but the pressure is now planted firmly back on my shoulders.

As the sun slips away, I abseil down for the final showdown, taking the precaution of removing the wire from temptation's way. Every move is now a struggle as I battle with greasy conditions and the stagnant pump, which lingers in my arms. With every metre, the temptation to abandon ship becomes greater until the switch flicks at 70ft, and I realise that I now have to stay on at all costs. The last few drops of reserve drain away as I blunder into the exit moves, fingers uncurling on the final crimps. I lunge for a wobbly block on top of the wall and only just stick it with my fingertips as my feet swing off. Moments later, I'm slumped in a heap on the cliff top, gasping for air. I struggle to locate myself in this lofty position, and it takes a minute or two to banish the images of my body spiralling downwards into a bottomless, watery kingdom.

Hydrotherapy F8a+ S2

In 2011, I launched a siege on one of the area's most overhanging cliffs – the Underworld Cave in Hollow Caves Bay. I soloed the highball and committing Hydrotherapy after a considerable number of attempts to get tides and conditions to tally. If ever there were a route with a swagger that gives the fingers up to all suitors, it's this one. Hydrotherapy presents a rare cocktail of obstacles, requiring high spring tides to coincide with dry afternoon conditions on a day when the range isn't firing, but you're firing on all cylinders. I knew that to climb this route I would need to be in it for the long haul, but by the fourth trip to Pembroke I was starting to question my commitment and sanity. The previous time I was greeted by red flags in the range. The trip before that it was wet, and the first trip was spent swinging around on ropes and failing to get anywhere near the route. What's more, I still hadn't worked out how I was going to get to the base – a dodgy traverse above a reef or maybe a dinghy? I'd vowed that if things didn't come together this trip then I'd throw in the towel, so, of course, it threw me a line. I managed to aid my way in, do all the moves on an abseil rope, and I came up with a crazy scheme for getting into position.

A week later, I'm taking deep breaths at the 'exit point' on the west flank of the cave. Below me, a zip-wire plummets downwards into the darkness. Time to travel in style! 'Yeehah!' As I whizz down the rope, the crazy sci-fi geometry of the cave envelops me. I detach from the rope, gaze outwards and realise that my war charge has dissolved into a David-and-Goliath moment.

The concept of attempting to solo out from my current vantage point back into the known world seems to defy all logic. Nonetheless, I yard outwards in blind faith as the dome, as the trippy acoustics of the dome magnify the breaking waves into a throbbing pulsebeat. I scrape through the first crux but then lose tension on the undercuts and explode backwards horizontally. I thrash in the air like a cat but don't quite manage to right myself in time as I slam into the boiling foam. The taste of blood and salt on my lip was exactly what I needed. This is war now.

Back up at the first crux with my 'send face' on, I wheel-spin carelessly, my foot pops and my body arcs outwards. I reel it back in, praying that the error won't cost me. At the undercuts, I tense every muscle in my body, shouting with the strain, and soon I'm shaking out at a ludicrously exposed perch on the lip. How long to stay here? Don't blow it now! More to the point, dear God, look where I am! I lurch into the final 'heart-break' cross-through, but there's no mistake now. Soloing out, up the upper chimney, my body casts tall, dancing shadows on the rock which mingle with the incandescent reflections from the glistening sea.

Right above and below: Neil Gresham during attempts on Olympiad. Photo Lukasz Warzecha. Below: Tim Emmett and Neil Gresham partaking in the traditional Pembroke pre-climbing ritual: breakfast at Ma Weston's cafe at St Govan's. Photo Mike Robertson.

Olympiad F8b S1

On the day of the opening ceremony of the London 2012 Summer Olympics, I was at Forbidden Head in Pembroke performing a different kind of opening ceremony. This route climbs a very steep, compact wall which had been regarded as one of the last great DWS challenges in the UK.

Six am: The hypnotic hum of the stove as the sun filters through dirty, cracked safety glass into the campsite barn. A splinter pricks the palm of my hand, and the beam creaks as I settle into my third set of hangs. I blow away chalk and cobwebs, pour out the steaming black liquid, close my eyes and sip. The caffeine floods my brain, cocktailing with the adrenaline that has been wreaking havoc ever since my eyes opened.

Half an hour later, I'm cowering in my gloomy cavern, ritually fiddling with gear. Unroll the dry bag, the sickly taste of energy drink, the heady smell of liquid-chalk fumes as they mingle with stale seaweed. I couldn't possibly feel less worthy as I dry my shoes on a soggy beach towel. I may as well get it over with. As I step off the ledge and enter the sequence, an Excocet missile bursts over my head and cleaves the sky in two.

Two minutes later, everything fades as I drill my eyes into the conglomerate ball which signals the end to all of this. All I need to do is stretch out and take it. But, I'm totally unprepared for this moment because the truth is that I never thought it would come. I can tell in an instant that it's too far. It may as well be a solar system away. But somehow, the act of conceding seems to free me from expectations as I explode and launch. I feel it's not enough and prepare for yet another ducking. Hang on a minute… Oh my God! I'm still here. I'm still here.

How can this be? I'm on top of the cliff instead of swimming for the hundredth time. I inhale and scream from the bottom of my lungs. This is way too much to process. In a moment of pure abandon, I strip naked and leap from the cliff edge into the air. Slap! The burning sting between my legs brings me back to my senses as I frolic in the waves, laughing and crying simultaneously.

When I snagged that exit jug and made the first ascent of Olympiad it ended the saga that began a year earlier after Mike Robertson told me that he had made another discovery: 'I've found it. I was paddling at Forbidden Head last week, and there's an insanely steep, blank wall. We're talking upper F8s or harder. It's the one. It's the future.'

The new genre of DWS routes in Pembroke are potentially the most serious and require more respect than any other DWS routes in the world. The tidal range is huge; you can walk around underneath Olympiad and Hydrotherapy at low tide.

They require a little more planning and commitment than going sport climbing or even rocking up at Connor Cove or Lulworth with a towel and a couple of chalk bags. The moves are difficult to work on Olympiad because you need to place runners on abseil in order to get in close to the cliff. Additionally, it's tricky to get to the base of the route. I started off by swimming in with my kit in a dry bag, but it's not ideal splashing around in the Atlantic before tackling a F8b.

In the end, I solved the problem by abseiling into a small inflatable dingy and paddling into the base. Above all else, I was working against a mental barrier because a few people had assessed it as F8b+ or F8c. It took a while to realise that it might not be quite that hard and that I actually had a chance of climbing it. At the time of writing, it is the hardest DWS in the UK. Olympiad was repeated with ropes in 2021 as a trad route at E10 6c by Steve McClure. Edmund Morris made the third (and second deep water solo) ascent in July 2023.

Footnote – the future of British DWS

Whilst Britain's hardest DWS routes might seem relatively modest compared to global testpieces such as Chris Sharma's Es Pontas F9a in Majorca, it should be noted that climate and logistics conspire to create a subtly different challenge. The big question is what the future holds for DWS. It is popular to suggest that we are close to saturation point, and it could well be that the development of British DWS was a two-decade blip in our climbing history.

The main protagonists have explored the key cliffs by boat and have suggested that the obvious plum lines have already been dealt with. But as we have seen so many times in climbing, rises in standards and changes in attitude can open up new possibilities. Who knows what steep, blank walls with V14+ boulder sections are lurking in the backs of zawns in wait for the new breed?

Caveat Emptor by Stephen Reid

Once, while climbing on Holyhead Mountain, sometime in the early 80s, there was a gang of four or five youngsters from Holyhead hanging around asking questions about climbing. These lads were 14 years old or thereabouts.

It was only after they had gone that I realised so had my guidebook. I was just bemoaning this when another climber I didn't know said,

'That's nothing; they've stolen my bloody rope.'

However, this tale had a tragic ending as a month or so later, there was a knock on my door. It was a policeman with my, rather soggy, guidebook. It had been found on the body of one of two youngsters from Holyhead who had drowned while roped together at the foot of the Gogarth Main Cliff.

He wanted to know if I knew them. After I explained, I got the guidebook back. But it had a short life as it slipped out of my tee-shirt not long afterwards while I was leading the hard pitch on Rat Race, I think, and plummeted into the sea.

Neil Gresham on Olympiad.
Photo Lukasz Warzecha.

A Near Miss by Alan 'Richard' McHardy

I had a near miss on Britomartis in 1977 with my wife, Barbara. I had not been to Gogarth for six years, and that visit was with a client who I was guiding. So I was running on a meter and did not spend any time looking at the view or any notice boards. Anyway, back to '77. It was the kind of day romantic poets lust about. The sea really was like a mill pond. The sun was hot without burning, but the ambient air temperature was perfect. Back in the late-60s, I spent a lot of time there, but although Britomartis was on my list of routes to do, it got missed.

Walking over toward North Stack, the place was entirely deserted of people but overcrowded with bees feeding on the gorse which was in full flower. What a place for a wild jump. However, it takes two to tango, and the reader will understand later what getting piss-wet-through did to number one.

Getting to the bottom of the climb involves a longish abseil to a stance above the sea and a little out of line of the route. The tide was very low which exposed a flat platform covered in seaweed. Between the crag and the platform was a channel about a foot wide. Belaying from the platform was comfortable and directly under the line of ascent, so the ropes were running in a straight line. It seemed over-the-top safe, but we found a belay up above on the crag. The sea was so calm there was no noise. The whole thing was just too perfect.

What I did not know about was the ferry to Dublin. A squeak from below made me look down and out. A weird ripple was crossing the green beige of the sea and heading toward us. It was followed not long after by a bigger ripple which washed the ropes away and wet Barbara's legs. The third ripple was bigger still and came up to her chest. Without our belay, she would've been washed away with the ropes and trying to swim.

Meanwhile, I was on the crux and badly needed some slack, but that would have to wait. In my youth, I would often stop on a crux and have to direct my mind back to the job in hand. It was a mental trick I had for coping with fear: my mind would bugger off somewhere else. But, there does come a time when hanging about does lead to fingers uncurling, a feeling of regret followed by: Oh shit!

However, with great speed, our heroine had pulled the ropes out of the sea, spat the salt water out of her mouth and was again belaying me. Granted, she was not too chuffed, but luckily it did not put her off climbing on sea cliffs. She followed the route wet-through without even a pull.

On top I found a pair of glasses that perfectly fitted my eyes. I didn't need glasses, but I thought they might look cool. Her close encounter had led to a dose of morality, so I quickly put the glasses where whoever had lost them might find them. On the way back she was dripping wet and walking some distance behind. The bees sounded like they were laughing at what I was hoping for and certainly not going to get.

The late George Shield's wife, Tina, told me that all climbers were psychopaths. She went on to tell me that the meaning of the word has more to do with not considering the consequence of one's actions – not just on other people but also on yourself.

Left and facing page: Pete Greening soloing Britomartis, HVS. Photo Julian Dalton/Greening collection.

Pimping the Yank by Henry Barber

From 1973 to the mid-80s, I would make regular pilgrimages to the UK to test myself on the classics and the latest test pieces. I developed many friendships along the way, but I was never successful at getting my mates and key objectives to line up quite like I wanted.

I'd read about the famous Joe Brown, Don Whillans, and Hamish MacInnes, and soon a much larger cadre of potential partners emerged. I failed to pick up on one crucial aspect though: nobody wanted to be 'out-machoed', so that usually meant a major piss-up the night before going climbing in order, it seemed, to provide ready excuses for poor performances.

Our weekly meeting place in Llanberis was, of course, the Padarn Lake Hotel. As the goals for the weekend got loftier, the crowd of suitors got thinner, and more and more people were at risk of extreme hangovers bordering on alcohol poisoning. The modus operandi was to see if everyone could get enough takeaway alcohol to entice Al Harris into another night of debauchery at his place above the famous car luge run: Fachwen.

He did not need much persuasion. The morning after was dominated by bouts of tea and greasy food while the lads tried to figure out another excuse not to climb on a blue-bird day.

All of a sudden, someone had the brilliant idea to do a sea-level traverse. Of course, this seemed like the perfect solution. They could get the dead-keen yank out on something that wasn't too hard or serious. Climbing on the test pieces was deadly serious in some cases, but everyone seemed to think climbing should be anything but serious – after all, it was meant to be enjoyed.

Now, I've always liked traversing, and I've always liked sea cliffs, so a sea-level traverse seemed like the perfect solution. We decided to head for Gogarth. The routes there are all big adventures at whatever grade you are climbing. That is what makes sea cliff climbing so special.

The good news was that Joe Brown never went up to Big'el (sic), so he was ready for most non-serious outings. I think he called them 'a lark', and everybody ended up eating humble pie trying to keep up with him as he, as usual, was miles ahead. Joe told me that if you touch barnacles with your hand, you can use them to stand on because they are used to 'tensing' for a wave. I tried it, and it worked on the bigger ones that stick out at least a quarter of an inch. I've since used that technique elsewhere on sea cliffs in Maine. Joe was incredible to climb with, but he never slows down.

On this particular day, which happened to be the first of April 1976, I was climbing in my jeans. I watched Joe move across a sloping ledge, but when I stepped on it I slipped and fell into the Irish Sea. I was treading water while holding my Pentax above my head when Joe popped back to check on me.

'What are you doing pratting about in the ocean?' he asked.

'Get me the fuck out of here,' was my reply.

On another similar occasion, we set out on the Range traverse to Trearddur Bay with an all-star Llanberis crew including Joe Brown, Mo Anthoine, Al Harris, Ken Thoms, Troyd, Nick Escourt and Jim Curran. Something that should have been light-hearted inevitably turned into something more serious, and it was an epic crossing. Curran, in particular, was a nervous wreck.

Right: Henry Barber climbing in Bermuda in 2018. Photos Grant Farquhar.
Facing page: 'My last day climbing with Nick Escourt before he went to K2. Notice the foot dragging in the water.' Photo Henry Barber.

This was around the time we filmed my solo of The Strand. Nick rolled his car three-and-a-half times up and over a bridge coming home from the Black Boy Inn in Caernarfon. It was funny as hell, except that my Aussie girlfriend was in the car. Troyd comforted her, if you can call that comfort.

Maybe tales like this sound hard to believe, but no one can refute my story because everyone in the photograph of that day is dead. This was the last time I saw Nick as he was about to leave for K2. Troyd tragically committed suicide. Al Harris sort-of committed suicide in a car crash, and poor young Ken Thoms died of cancer. Jim Curran died in 2016, and good old Joe passed in 2020 at the age of 89 years. I'm the last man standing.

Also during those years I climbed a long traverse called Zodiac at St Govans with Pat Littlejohn. He said he would lead the crux pitch, and I thought, initially, that was generous of him. The thing is, it might have been protected for the leader, placing protection before the crux, but it certainly wasn't protected for the second (me), and after I took the runner out, I was facing a massive fall when I did the crux. Once again, I got pimped by the Brits which was not an unusual occurrence in those days.

By the way, I never got to climb with Whillans, but we did dance twice, and that was a serious affair.

The Sea-level traverse of South Stack Island
by Barbara James

Nothing more than V Diff. All the seriousness of the Alps but escapable. An ebb tide, warm sun and an easy walk down to the promised first tyrolean. Surprise. A calm sea means that it is climbable around the back of the zawn. A squeeze through a narrowing and out into the sun again. More scrambling. This is great. Oh, here we go. Climb down slings to a greasy ledge, then another committing move around a corner, but I must take out the protecting chock before launching out. Hmm... easier than it looked. This is the first escape place. Continue across friendly, sun-warmed slabs, flatteringly easy until a hand-traverse ensures pause for thought.

Round a headland to a zawn. Tyrolean unavoidable. Cast. Miss. Cast. Miss. Eventually, a chock lodges and stands the tests. Ron descends down the rope from the belay (best for the heaviest to do the testing!) and begins to pull across the zawn. Chock shoots out, and Ron narrowly misses a soaking. Cast. Miss. Cast. Miss. At last. Ron crosses, and soon I follow. More delectable scrambling to arrive below a big slab rising 150ft from the sea. Continue around a corner and into the shade. The rock is greasy, and suddenly, it's not fun. Let's leave the rest for another day.

Facing page: Graeme Desroy on the Main Cliff traverse. Photo Jethro Kiernan. Right top: Joe Brown and Mo Anthoine. Right middle upper: Nick Estcourt. Right middle bottom: Joe is miles ahead, and we have thrown the rope so everyone doesn't swim. Left to right: Al Harris, (not sure), Nick Escourt, Pete Minks, Mo Anthoine, Troyd, Me and Jim Curren. Right bottom: The same crew – check out Harris with the skateboard. Photos Henry Barber.

The Jellyfish Wrestlers by Martin Crook

Even though I know the walking brain that is Crispin Waddy, it could not be said that this in any way reflects an association with deep water soloing other than having admired from an armchair impressive images showing single contenders suspended between overhangs unfettered by rope or rack. The price of pump out or unexpected hold break being uninterrupted AirDrop into certain submarination. Caught as it were twixt devil and deep blue sea. Places like Portland, Pembroke or Rhoscolyn's Electric Blue sea cave highlight the essence of the discipline that some years ago entered climbing's lexicon and is nowadays reverentially referred to by its acronym: DWS.

Alas, personal experience of this undoubtedly exciting adjunct remains sadly lacking yet does include the lesser-known exalted minority cult of metamorphic high water soloing, or MHWS. Which, as we shall see, does not simply mean turning up at any given sea cliff premeditatively psyched on soloing targeted routes, either on-sight or with prior knowledge. Admirable as this may be, the true MHWS sets off initially with a kind of blank mind only achieved by long years of study under Zen Buddhist Masters or, more commonly in the Western World, shortcut methods where during a single evening conversation descends so far into uncharted bull depths that early morning hours creep in unnoticed.

Such states arrived at involving tangential meditation, liquid libation and, or medication prescribed by chemical warfare specialists, sometimes labelled AK-47, and wheelchairs are perhaps unsuitable for affluenza types on a mighty path of goal attainment yet, of course, remain an option. Whatever the case, MHW soloists adapt to the day with a degree of spontaneity so that sports plans, if they could be called such, hinge primarily on somewhat optimistic spur-of-the-moment decisions in which fate's hidden spirit, as it may be gleaned from the following account, often deals a less than perfect hand.

Gearing up atop the crumbling descent gully which leads towards Gogarth Main Cliff was, by July 1992, a ritual familiar to Noel Craine and me. We were, however, unfamiliar with leaving this haven unbound by dominatrix gear-skirts essential for big-route roped-ascents. The reason being: 'Half a league and half a league onward,' a sea-level traverse beyond the shining cliffs to Parliament House Cave. This idea, foretold in the Book of Revelations better known as the Gogarth guide, was proposed and agreed on. It would require lightweight human-crab scuttlers to don only rock shoes and chalk bags for a crossing, as it were, naked before the beasts. So, after climbing down ledges then sidewinding leftwards on sea-washed rock we were soon under them.

Dinosaur, Mammoth, Hunger, Extinction, The Big Sleep: climbers folklore woven into the great overhanging Bayeux Tapestry of quartzite that reared above us sparkling white in the sun for over 100m. Squinting aloft, one hand visoring as if in salute. I saw in the mind's eye beatnik cool Alan Rouse – all Michael Caine glasses and bell bottom flares – going up determined, a MOAC between his teeth during Positron's first ascent in March 1971. How must it have felt I wondered, uncertainty, nagging exposure, skull-pumped climbing into headwall legend.

Today, it would seem a fear of frying had caused temporary abandonment, leaving the cliff now bathed in an eerie silence. Passing belay cracks below Citadel's starting wall recalled an unhappy hour weeks earlier involving Noel's floating trainers' partial immersion, rescue attempts whilst crimping on barnacles, lasso fishing, wet ropes and shivering. Now, however, no real difficulties other than moderate accompanied the scene, yet after gaining the block marking Pentothal's departure a much steeper section deposited us first into niches where looms Heroin and Hypodermic then respite platforms before Hustler's open book corner goes vertical. From here claims the 1990 Gogarth guide: 'The traverse peters out, and the crag merges with Easter Island area'.

Not in your white horses' dreams. In fact, far from petering out, there now comes 20m horizontal clinging across leaning rock: a kind of bouldery crux which, alas, even in hindsight Noel and I cannot truly grade since halfway along, rising tide levels forced us to abandon the coveted limpet footholds which Lesley Shadbolt had

suggested could be used as portable aids during his sea-level girdle of Sark in the Channel Islands. This did not work, and waves lapped slicked rock shoes so much into uselessness that we skated in vain against the clock of the inevitable. Those wet-suited and prepared would have little problem. Yet Noel and I asserted that we belonged to the cadre of punters calling themselves 'rock climbers' so that any bypassed moves would amount to points of aid. Swiftly rose Poseidon's power and with it the realisation based on misread tide tables that, as King Canute found out, oblivious to foolhardy human effort, gravitational moon-pull cannot be stopped.

Even so we clung on to the bitter end, climbing at times with water at midriff whilst spume wash covered hand holds, and then we were in, slumped onto the aid point, swimming, shocked cold against weird undertows. In truth, little distance remained before the happy sanctuary of the Easter Island Crags. But as soon as we flotsam and jetsamed into Wonderwall's box zawn, we faced another Hellespont, a second aid point, getting succour only after rejoining terra firma beyond Ormzud's weird slot-world.

Who can remember Ken Wilson's original *Hard Rock?* – nowadays a classic book noir. Peruse page 90, and there's Ed Grindley 'hastening to a resting place' on North Crag Eliminate in the English Lake District. On the face of it nothing to do with sea-level traversing in Wales, right? Yet it struck me that we, him, him, us – like many climbers in dramatic circumstances, whether on dry mountain classic or wave-washed seacliff ad infinitum – for as long as climbers climb, there will be 'hastening to resting places'. Even in the age of Tick Tock, Twitter and Instagram, training regimes and climbing 'studios', what else would grimpers do when pumped or drowning? That said, how long this bout of jellyfish wrestling might be sustained was anyone's guess, but a bailout plan based on gaining shadowlands underneath Spider's Web and then escape via careful loose rock behaviour beyond Genuflex and Blowout would, I hoped, hasten us to nirvana offered by strollable clifftop paths. Alas, it was not to be since as soon as we turned a corner into Wen Zawn, well, unwelcome was the sight that greeted us.

Noel, with resigned calm, simply uttered, 'Oh my word,' since the entire curvature of cliffs, slab, back wall, and opposite promontory what's now submerged by wave after wave of angry white pyjamas sinking without trace any further notion surrounding our plan B get-out clause. It was as if great booming voices from the Treorchy Male Voice Choir hidden inside the sea arch were singing: 'Hear us when we pray to thee for those in peril on the sea,' which sucked forth out of the caverns of wrath resonated across the rolling deep in mocking harmony.

Two climbers, one stance hung, leaning out so as to see the leader grappling with Conan the Librarian's Janitor Finish, were no doubt engrossed with their own gripping situation. Yet we could not help but envy the fact that they were roped together, which was more than could be said for our own nicht, nein, Untermenschen dripping drama. What happened next elicited claims, however frivolous, that the new sport was unlikely to have many addicts other than those confronted by similar scenarios. Metamorphic high water soloing, we called it. In an eye blink, a day out sea level traversing Bosch-morphed into one requiring gear-shift psychology, a psychology that said soldering heads on please, forget the sun-drenched low level because from now on, it's 80m straight up. At any rate seals were gathering as if looking forward to the afternoon's entertainment, and as it was not in our nature to disappoint them, we set off.

Few routes could be more aptly named than the one now offering salvation. Described as an alternative approach to the upper slabs of Ddu, Hydrophobia is a 45m 5a pitch luckily endowed with reasonable holds and steady, slightly off-vertical terrain perfect for escaping human sponges paranoid about engaging with anything likely to cause a glimmer of pump out. Noel went first. We seemed to move at an evolutionary pace as if retracing the steps of the first amorphous jelly blobs heading into an undiscovered country, crawling with purpose towards two-legged futures.

Eventually, we progressed to such an extent that we joined Ddu's upper half where the full exposure given to those marooned without comfort from umbilical-like ropes was soon sharply embraced. Nevertheless, with the great white slab acting as a surrogate radiator, dripping clothes which had initially handicapped ascent had, as if there had been a passing shower, dried into the realms of the merely damp. What transpired next was hard to fathom. One minute Noel was there; the next he wasn't. Disappeared, or so it seemed. Not fallen but suddenly vanished. That's when his head appeared intact looking down from the cliff top. My ashen face, rubbernecking over a small overhang on Ddu's final furlong, realised too late that he'd managed to traverse rightwards onto easier ground near the Dream of White Horses abseil and thus effect a better escape than the one now causing me some concern.

By a quirk of fate Noel and I had, in recent memory, made a drizzle-laden tandem solo of Spiral Stairs, the classic Dinas Cromlech V Diff in Llanberis Pass, for reasons barely based within mountaineering practice. For that episode we were, however, prepared with premeditated psyche. For the one in hand, we clearly were not. Anyway, cautiously I took moves over Ddu's 5a overlap. Astonished at the situation, once established in grooveness beyond, I was able to claim, perhaps aided by an adrenaline-fuelled release, that Spiral was harder. How my old friend, the soloist, Jimmy Jewel would have laughed when told about these notions if he'd not so tragically left us. An Autumn afternoon in 1987. Tremadog, Poor Man's Peutery, a slip on wet rock, miles within his grade. Gone.

Now, there remained only a steady plod of a type well known by the cadre of climbers. A slow advance, pausing to chuck off any throwaway holds, with gulls hanging in the updraft, kite-like and watching until seagrass finally merges with the last vestiges of broken rocks into horizontal standing places and you're out, welcome to the club. 'Oh my lord,' Noel said. So much for the jellyfish wrestlers.

Fear of Water is Fear of Sex by Andrew Walker

Did you rush into the sea or just paddle? The amount you get involved with the sea represents how in touch you are with your emotions and your sexuality.

Jane Alexander

The Great Eskimo Vocabulary Hoax identifies 50 or 100 different Inuit words for the word 'snow', and like the Great Moon Landing Marketing Campaign, it's not true. But there should be that many words for the multifarious patinas, sensory textures and colours of the seething silvery sea that are found only at the briny White Cliff sea-roots: the gloopy jelly green stillness at the back of some quiet sunlit zawn, like Flytrap; the glutinous irrational sucking of an invisible tidal pull, like off South Stack; that hard black opacity of a cold sunless swim through a dark sloshing tunnel, like from Parliament House Cave; the aerated saltiness as the sun sparkles the airborne froth before the swell rips your epidermis off, like under Annie's Arch; and it's never blue when you're in it. Freud classified the connection between sexual deviance and water as a foetal one, and Jung saw the relationship with water as representing our subconsciousness. I would have loved to have seen them both rubberised up and pontificating about this while traversing Smurf Zawn on a choppy day with an incoming tide.

There are thin lines at this seaside. Like the invisible, ever-changing thin line of sideways sliding crozzle between the leering walls and the briny Irish Sea that underscores these folded multi-coloured miles. Like the line between aquaphilia and aquaphobia. Aquaphobia is an irrational or disproportionate fear of water, especially anxiety in deep water or when submerging one's face in water. And unlike the topo-fied vertical lines, these cruxes, starts, ends, and grades for this 'crabbing about' sometimes slide around or may not even exist at all. Or not even matter.

Abrahams Bosom to South Stack: I went back to this mono-dextrously with a reliance on the side stroke and a strong French guide. Not for the first time, I'd failed to relate the loss of an arm with any potential decrease in, well, having two arms. As I was flushed backwards and westerly out of a narrow channel by a typically Waddy-esque current under Penlas Rock, I screamed for help. Joel's shouted reply Franglais'd over the whooshing undertow was: 'But I am not a strong swimmer!' Hydrophobia is an extreme or irrational fear of water (although, as a symptom of rabies it is optional). We finished hours late at South Stack, in the dark, guano-covered and bleeding. Our dyspeptic lighthouse-lit families were minutes from phoning the coastguard. Again.

Yellow Wall to South Stack: The extinguishing of that Altrincham All-Star under the leering Red Wall's colours has been documented elsewhere[2]. The limpet-crimping sea-level action here is so much more fun than the roped ego-gnarl above. But only if you're not afraid of water. There is a bobbing bright blonde-redhead cocktail in the burnished ochre of Mousetrap Zawn. Noel was up with the guano and damp on the Hysteresis wall with a very young Leo. It wasn't clear how many days he'd been up there – possibly several, as he was already halfway up the first pitch. As we swam past, he wasted further time by warning his partner in his censorial social worker Oxbridge: 'Leo, never go partying with that man.' From anyone else, this would have been a compliment. And heckling should always go upwards. At Castell Helen, I realised that the soloing abilities, unlike the aquaphilia, were not equally dispersed in the team. Aquaphilia, by the way, is a form of sexual fetishism that involves images of people swimming or posing in water and sexual activity in or under the water. We gate-crashed the random end of a rope to find George belaying us at the top, and far more happily, his client had a hip flask of rum.

South Stack channel, a shipwright's grave, the saviour drowned. He'd put the 'Creag Dhu' in Creag Dhu Wall, survived a Himalayan hunger march 'expedition', smeared virgin Etive granite in shite footwear, recovered another drowned Scot's gear from above yet another unforgiving sea, and forever changed the climbing of

---

2   *The White Cliff.*
           Right top: Andrew Walker and Huw Perkins on the Main Cliff traverse.
           Right middle: Perkins and Walker on the South Stack to Main Cliff traverse.
           Right bottom: Tim Fryer traversing from South Stack to Main Cliff. Photos Walker collection.

Above: Andrew Walker soloing the final pitch of Dream to finish the South Stack to Main Cliff traverse. Below: Walker and Huw Perkins entering Wen Zawn. Photos Walker collection

frozen water. In the 1950s, long before we invented the Round Anglesey TT Races under the pretext of going to Gogarth to climb, Cunningham was a regular and poor competitor in the Creagh Dhu TT races along Loch Lomondside. He couldn't swim, either.

South Stack to Main Cliff: We found a wrecked boat on this coast, and it temporarily stopped the communal shivering. How it caught fire by the seaside is still unclear. A place for seamen and salty, slippery slithering. Medical News Today: 'Sex in open water is the type of water sex with the highest risk. Strong currents can make it difficult for pleasurable sex, and people are at a much higher risk of drowning.'

Main Cliff to North Stack: It is unclear if the soloing options here are 'deep water' or just 'deep shit', considering the potential distances and some of the 'landings'. Ending one long traverse from South Stack and weeping wetsuit-water on the holds of Dream, it was perhaps typical of that era to hear the heartfelt cry from the Wen promontory: 'Fall off and die, you fucker!' And perhaps typical of a more recent era to hear whining from the roped and bumbling about: 'You're getting the holds wet'. Wen Zawn was beautiful. Pablo's warning, and we both had hearing issues.

Two flies caught in the Spiders Web, a High Tor epitaph for the drowned. They both fell 100ft into the sea when their three-peg belay failed. Robert Brown disappeared into the sea. Arnis Strapcans died on the Brenva Face years later. We also had a fall on Spiders Web, and my partner that day died on the Badile years later. So does my selfish sunshine playground succumb to the bumbling commercialised be-helmeted 'death by numbers' that parts of Pembroke have become? The jet skis and tourist boats are already in Gogarth Bay. A thrusting throbbing rib penetrates the flickering green stillness of Flytrap Zawn, with attendant shouts: 'Get out of the way!' The seals had no choice. The diesel stained the air and the jelly-green water.

Primordial in the Parliament House back passage, swimming through to meet a wild north-coast chop, dead-legged by irate breeding seals, reptilian thalassophobic stirrings. Interesting word that. Thalassophobia is a specific phobia that involves a persistent and intense fear of deep bodies of water such as the ocean or sea. North Stack Promontory with an incoming storm is a good place to experience this: smell the cries of gulls, taste the cracking whip of the wind, and view the sounds of hysterical synaesthesia (The 'North Stack on LSD Family Adventure Experience' is currently unreviewed on TripAdvisor).

North Stack to Holyhead: The North Face. Buy the jacket, it's easier. Get a GoPro (waterproof). Use Social Media. Spoil it. There's some 'otherness' about these sea-level shenanigans that reminds me of old-school alpine broil. The dry bag was for spliffs, just like the haul bag was for the beer; insurance was unethical, the first aid kit was two explicit white powder or pink pill options, and, no, we're not taking a fucking radio (Generation X) or phone (Generation Y).

Tickling the Dinosaur by Grant Farquhar

You come out of medical school knowing bugger all. No wonder August is the killing season.

Dr Claire Maitland in *Cardiac Arrest* (TV series)

In the early 90s, I was a junior doctor in North Wales. I didn't choose Ysbyty Gwynedd as the site for my junior doctor training due to its academic credentials or for the niceness of the weather. However, that particular day, in May 1992, as I looked out of the kitchen window of the doctor's accommodation in Ysbyty Gwynedd, it was a glorious blue-sky evening beckoning me outside to sample what I was living there for: the quality of the climbing.

That weekend, I'd been on call. On the Friday before the 56-hour shift, I went for a pint in The Heights pub in Llanberis and picked up a hitchhiker on the way back to Bangor. We got chatting, and I told him that I was starting work at 9am the next day. The hitchhiker wanted to know what time my shift finished and was incredulous to discover that it would not be until 5pm on Monday. I guess the bereft expression on my face as I relayed this information convinced him that I wasn't bullshitting.

Thirty years later, I still occasionally have dreams in which I'm a junior doctor performing Kafkaesque labours involving blood, guts and death during endless sleep-deprived shifts in an amalgam of those wards I worked on in Ysbyty Gwynedd: Aran, Glyder, Tryfan, Conwy, Dulas and – yes – Gogarth Ward. So, after yet another brutal on-call weekend, I was chomping at the bit for some real Gogarth action rather than more Gogarth Ward action. Quite a few of my hospital colleagues were regular climbing partners, and Calum Muskett's dad, Chas, was the charge nurse on Alaw Ward, but I didn't have a climbing partner on that particular day.

Sea-level traversing, in particular the traverse of the Main Cliff, had always appealed to me, but I'd never done any at that point. I decided that day was going to be the day. I wasn't going to slay any Welsh dragons, but I could, perhaps, tickle the Welsh dinosaur. I had no idea about what gear to take, apart from rock shoes. I also had no idea that the water temperature in Wales in May would be several degrees below fucking freezing.

My sleep-deprived self behind the wheel of my white boy-racer Vauxhall Cavalier SRi probably constituted a major road hazard as I weaved and overtook along the old A5 to Holyhead. But I managed not to fall asleep behind the wheel and soon pulled up at South Stack. As I ran across to the Main Cliff, the wind was dying on the deserted crag, and the last rays of the setting sun were casting a golden glow. It was going to be dark soon, I realised, and hurried down the descent gully.

The initial traverse along the base of the Main Cliff was familiar, and I was still dry. I wasn't committed yet and could always reverse, I reasoned. Towards the end of the Main Cliff, I started traversing into new territory for me. I'd heard that the section between Main Cliff and Easter Island Gully was supposed to be English 6b. I couldn't immediately see what height the traverse went at, and I didn't have time to faff about. So, this was going to be the first swim. Not having a clue about such things as wetsuits at that point, I donned my windproof. The water felt so cold that I thought diving in might kill me due to cold shock. As I gingerly lowered myself into the water, I found myself making an involuntary sharp intake of breath.

Driven to swim by the water temperature like a schoolboy being pursued by Jimmy Savile, I arrived at Easter Island Gully where I'd already done some of the routes. Next stop was the Flytrap Area, where I'd done very little, but I didn't hang around to check it out because I was getting uncomfortably colder. Wen Zawn was definitely the highlight of the traverse due to its magnificent rock architecture. I would've liked to linger there and explore the obscure depths, but I couldn't because the day was rapidly turning to night, and I was getting COLD!

Right: Graeme Desroy on the Main Cliff traverse. Photo Jethro Kiernan.

On the next swim, I was startled by a seal popping up to check me out. Fuck off you whiskery bastard, I thought. Getting out of the water was now becoming a grim struggle with chilled muscles. Shit, was this a good idea? But it was too late for regrets; I was committed.

It was difficult to fully appreciate the awesomeness of Parliament House Cave due to chattering teeth and full body shivers. The sea was quite rough in the through-cave behind North Stack Wall, but this was the final swim between me and the finish.

I breaststroked frantically through the tunnel and picked my egress. It was just as well that this was the last time that I would have to climb out of the water as I felt – almost – too hypothermic to make it.

I ran up the slope and back along the path to South Stack and the car. The heater was on full blast all the way home. I was still cold and shivering when I got there, and I immediately ran a hot bath.

I wanted an adventure, and I got it. Tomorrow, it would be back to Gogarth Ward to deal with the cases that I'd admitted during the on-call weekend: heart attacks, diabetic comas, asthma attacks, pancreatitis, meningitis, and so on.[3]

Electric Blue by Grant Farquhar

Electric Blue at Rhoscolyn is a good example of a route that is a lot easier to deep water solo than it is to lead, traditionally, with ropes. First climbed by Stevie Haston (with ropes) in 1983, it was graded E4 5c in the guidebook. In the hundreds of ascents since recorded in the UKC logbooks, it is telling that prior to the year 2000, nearly all the ascents were lead with ropes, and after 2000, the vast majority of ascents were DWS which is a good illustration of the change in mindset that occurred around that time in the UK in relation to deep water soloing. For the vast majority of climbers, it simply wasn't on the radar prior to then.

On Electric Blue, which gets a DWS grade of F6b+ S2, once you step onto the route you are almost immediately above deep water. The low-down crux feels a lot harder when you have to hang around and fiddle in gear rather than just climb through to easier ground. The S2 is, presumably, for the height which is getting a tad high for DWS by the time you top out although the climbing is easy at that point, and if you got through the crux you would have to be doing something stupid to fall off high up. It's a quality route and probably the best DWS in North Wales.

3 If you want to know what it was like being a junior doctor in the NHS in the 90s then watch the first six episodes of the *Cardiac Arrest* TV series. Pretty much everything that happened to the housemen in those episodes happened to me: being completely unprepared for being a junior doctor from medical school; insane hours with total mental and physical exhaustion; the attitude of management, the consultants and the nurses; the endless blood-taking and IVs; chemotherapy disasters; excessive guilt at the deaths of patients which happened every day, and, finally, the suicides of colleagues.

Left and facing page: Charlie Woodburn on Electric Blue, F6b+ S2, at Sea Cave Zawn, Rhoscolyn, Anglesey, North Wales. Photos Grant Farquhar.

Things To Do In Lleyn When You Are Daft by Martin Crook

An account of The First Ascent of Headlander 100 metres XS, April 1992, Martin Crook and Alistair Hughes.

'I do believe that areas like the Lleyn could represent a facet that has all but disappeared from British climbing,' said Steve 'the general' Mayers commenting on the on-sight ethic in the early 90s. Al had said that the crack 'looked about Severe'. The tide was out, and we stood beached between seaweeded rocks, cannonball-shined, wet from the outgoing tide. Having descended via a grass ridge into what was later described as Three Caves Zawn, such were the obvious topographical features undercutting the cliff that towered for 50m above us.

Here then, beyond Rhiw's lonely village and its queer crocodilian edge, where Bardsey Sound's tide race conducts unknown depths with awesome power round the mouth of hell, cutting off the main landmass from a small mysterious island bearing its name, where in 1188, Gerald of Wales had noted, lived an order of devout monks (pilgrims, I knew, still journeyed there). We set out to worship at the vertical shrine of our own cult, watching our companion figures morph over far headland as we did so. Pure in line, our route lay in devil-disguised disfigurement which, austere in its fractured neutrality, soon proved a tortuous path where better judgement on a different day might have countenanced retreat.

The rock type hereabouts is granodiorite and typified by a compact lichenous nature. It makes runners and belays difficult to arrange without pegs, although it generally provides solid holds. Thus, the true heart of multi-pitch adventure in its most testing form requires travelling further out down the peninsula's long arching arm. Only then can one become familiar with the great orange slopes within the vicinity of the head itself: Cilan.

In fact, if a mention of Cilan Main did not cause a momentary shudder, chances are you hadn't yet experienced the bizarreness it represented or were trying to forget it, perhaps in general withdrawal from the everyday world. Big and serious, Cilan is where horizontal grit and shale bands take the eye before craning neck muscles allow one's gaze to take in the massive capping black roofs 200ft above a tiny beach at the crag's undercut base.

By the late-80s and into the early-90s, the roll call of climbers making exploratory routes or repeats began to increase, and whilst the peninsula did not experience a trendy 'place to be' scenario, a few Llanberis-based teams initially spearheaded by Ray Kay and Dave 'Skinny' Jones in the company of John Toombs and Leigh McGinley did much to arouse a curiosity in others that would increase the Lleyn resume. Pat Littlejohn, with various partners, had also been quietly operating in the area for some time, and thus, with a Culm Coast seal of approval, it was clear that those entering this lonely realm would be ill-advised to do so without a certain apprehension. Sparsely documented, there seemed a magnetic charm purveyed by the only guidebook, a slim off-yellow paperback, published by the Climbers' Club in 1979 and compiled by Trevor Jones, who after editing the known information into 48 pages reminded would-be acolytes that 'the descriptions and in particular the grades should therefore be considered as provisional and treated with some respect'. It was good advice.

After using this tome on a number of bittersweet occasions, we found a number of zawns apparently untouched by previous explorers. To reach these, there were difficult sea-level traverses and unfeasibly treacherous fishermen's paths. It was in one of these arenas that I now confronted the so-called 'Severe-looking' crack in the zawn of the three caves. It was marked 'Bytilith' on OS maps, yet a most defining feature signing our approach centred on a defunct pipeline emanating from a short red-brick wall atop and left of the cliffs when looking out to sea. This forgotten edifice plunged in decay down a disturbing couloir for over a hundred feet, ending its fall on beach boulders, from where at low tide it was also possible to get round into the semi-cauldron zawn in which rested our point of interest, a crack line whose apparent ascetic charm was difficult to ignore.

Above: Steve Mayers on the first pitch of Terrorhawk on Cilan Main Cliff. Facing page: Cilan Main Cliff. Photos Grant Farquhar.

Anyone venturing out on the short climbs offered by the nearby gritstone edge at Rhiw might be forgiven for thinking that the sea cliffs in the vicinity might display a similar, generally solid, nature. Yet, above all hope, this form of false consciousness would swiftly be annulled unless concentrating solely on the sea-rumbled boulders. Thus, the crack, once engaged, rendered a gear shift on my part so that my mind, hit with fresh information, returned to Cilan mode after seeing designated footholds explode or delaminate when nominally weighted in the architecture each side of the alarming fissure.

I think therefore I jam, or, in a less than grand philosophical sense, which might nevertheless have great repercussions on a personal level, I thought about how jams should be best placed. It wasn't that baffling technical difficulties suddenly caused a long pause in proceedings, but that typically such ascents rely on a slow probing up and down after securing, at least psychologically, any available protection. Of this, deep inside the crack's mud butter, better-crystallised rock gave home to a couple of Friend placements, and a big sideways Hex biting the fracture's doubtful outer edges backed them up.

When leaders hardly move for half an hour, it might be no surprise to hear the second shout in encouragement: 'Go for it.' While possibly galvanising action in sound-rock settings with bomber gear, the matter in hand was more likely to succumb after a long mental war of attrition. Knowing this, Al shuffled atop the highest boulders as the turning tide began cutting off our escape and maintained a silent vigil. Cruel were the impediments barring way to the imagined haven promised by a beckoning ledge, where steepness was temporarily postponed. This was the belay, proportioned

with a horizontal crumble line taking a Friend and some nuts on which I braced to bring up Al. A shipwrecked monk lamenting the stone boat's sinking, no longer paying out.

On the lunatic fringe of the next pitch, a crack/groove took a Rock 4, biting in solidly first go. Bridge out, lean in, left shouldering and a reach with the right gains faulty finger-locks. Moving in reptile shapes I make a position under an overlap, where things – as Glenn Robbins was fond of saying, and George Smith would later name his Shale City test piece – were 'getting ugly'. Shadows tell the sun is losing us, and I can't pull over – a man hanging from the gallows of his own making. This surplomb was half mud, half biscuit, half past dead Whymper... earth. Stuck below it, excavating fragile layers with one hand sapping, exposure bites as an invisible pig suckles strength away. But here now is the animal in its most savage form with survival its only goal. Half standing amidst the left arête's museum porcelain I commit, chest first, to a gator-roll mantle and twist body parts over drip-fed Hammer House horror, emerging a white-faced phantom on the uber steps as uncloaked rock grass becomes airborne, famous amongst the shearwaters.

We have no head torches, and the overlord on the incubus steps is waiting. It feels like I am treadmilling a smashed escalator, and with the sky empty of birds I find a grim oasis, but there is no warning and suddenly you're dead. Double vision? No – blood. Only a little, but the hard-baked pudding stone had struck my skull a direct hit. Its bass rhythm sent a shock wave through the jaw, attempting to dwarf me. Stunned, my knees buckled, and I stepped for a moment off the round world's edge... deaf.

But I am OK apart from rope drag on non-extended runners, causing a final crawling technique which gains relief only after clipping in under the rim on the cold cliff as Rutger Hauer says 'not yet, not yet', clenching his Roy Batty fist. Alive. In a sort of muck-lined crevasse formed by rabbit warren honeycombed banks, I face out towards the Irish Sea ready to bring Al up from the cirque of the unclingables. He comes on like a medieval abbot surveying monastery ruins. Schwarzenegger-big in his coat. 'An insensitive oaf,' a girlfriend once called him, but he picks a way through the Herculean Jenga-pillars without causing collapse when the merest indelicate touch would have caused regret. With wind blast, hoods go tight on the drawcord, and below a crusader zeal fuels the oxygen of escape as he passes over the fairy-tale roof with hands hard-grasping.

'Extremely Severe', he would later say. But then we are there avoiding the head landers, where there are none except those captured in memory. Burdened only by the hillside's incline, we must set a zig-zag course away from the pipe wall and, taking a breather, become conscious of the slopers at dusk in the RS Thomas necroscope night.

Down in the dark zawn... at the end of Wales.

It has often been suggested that climbing centres must exist in Ireland equal to those of Scotland, Wastdale, and Snowdon. The possibility of this may now, we think, be rejected finally since recent exploration of the most out-of-the-way parts has shown plenty of good walking country, but rock suitable for serious climbing is almost entirely lacking.

Conor O'Brien, *CCJ*, 1912

Ireland

Might I make so bold as to ask what these things like lobster pots are?

Oh! The soul cages is it?

The what? Sir!

These things that I keep the souls in.

Arrah! What souls, sir? Sure the fish have got no souls in them?

Oh! No, that they have not; these are the souls of drowned sailors.

The Soul Cages by Thomas Crofton Croker

The Sturall Headland, an outstanding sea cliff feature in Donegal, was the scene of one of the first rock climbs in Ireland. The first recorded climb to the summit was made by Walter Parry Haskett Smith in about 1890. His route ascended the skyline ridge from the landward side and provides a very exposed 400m mountain ridge scramble with some very loose rock. Similar to developments in the rest of the British Isles, climbers tended to focus on the mountain crags at the expense of the sea cliffs. However, Doug Scott's ascent of Main Mast on Sail Rock in 1967 marked the beginning of the movement away from the granite highlands and so began the development of the immense potential of the Irish coastline from Fairhead in the northeast to The Burren and the Aran Islands in the southwest.

The Duke's Head by Iain Miller

'I may have found you an unclimbed sea stack if you're interested?' came a text message from Marble Hill resident and all-round good guy, noble brother Ian Parke.

'You have my complete and undivided attention. Do tell, sir,' was my instantaneous reply.

The Duke's Head is a 20m-high sea stack off Dundonnell Head, which is just north of Marble Hill beach in Sheephaven Bay, County Donegal. The sea stack is set in a perfect little amphitheatre of surrounding sea cliffs. Directly inland from The Duke is the mouth of a huge, hidden sea arch, which runs under Dundonnell Head for over 100m.

At the back of the cave is a perfect 'pirate's treasure' raised shingle-storm beach. The day after receiving Ian's text, I launched my kayak from Marble Hill beach and went for a look. The plan was to do a recce to see what would be involved in getting to the top of The Duke.

Paddling with an easterly wind and tetchy north-easterly seas, I rounded Dundonnell Head and followed the coast to the lair of The Duke. A final, narrow, steep-sided and slightly bouncy channel took me into the amphitheatre. Alas, the stack's close proximity to the surrounding cliffs was creating a whitewater rage around its base.

Facing page: Sam Hamer on The Crozzle Monster, F7c+ S1, at Ailladie, The Burren. Photo Mike Hutton.

Above: Fiona Nic Fhionnlaoich on The Duke's Head. Photo Iain Miller. Below: Taming The Dragon (Scealpan Bui) on Owey Island. Photo Alan Tees.

A semi-submerged skerry was helping to create an almost continuous wall of white water in the channel between the stack and mainland Donegal. Landing on the stack was not an option at this stage, and so a visit to the sea cave with a good look at the possible landward access became the next objective.

A couple of days later I made a return trip, this time in the company of Ian Parke and a local climber, Fiona Nic Fhionnlaoich. Ian was our local access guide as the clifftop approach is a veritable minefield of fenced fields and other potential access problems. Our cunning plan that day was a clifftop recce.

We crossed the fields to an outstanding viewpoint above the monster cave overlooking The Duke. Alas, it became immediately obvious that Neptune was still in residence in the amphitheatre. A swift decision was made to abandon any attempt that day and have another look the following day.

As the tide was around mid-ebb, we decided to pay a visit to the storm beach at the back of the cave. Fiona and I descended the near-vertical grass to the non-tidal ledges at the landward base of the stack. From here, the channel between us and the stack was suitably atmospheric with a constant churning of white water; attempting a crossing that day would most definitely have ended badly. A visit to the back of the cave looked a much more promising option.

We tied in and racked up in the mouth of this huge cave. To get to the storm beach in the semi-darkness at the back of the cave involved a greasy, wet, Irish traverse above the semi-submerged boulders about 20ft below. This took us to a short abseil and downclimb to the less submerged boulders in the last 50m of the cave. A further subterranean boulder-hop took us to the distant shores of the pirate's cove. From there, the outside world looked very far away.

The following day, with much calmer seas, Fiona and I once again found ourselves on the sea-level non-tidal platforms facing The Duke. We had opted to repeat the previous day's vertical grass descent, and this confirmed our opinion that this was an awful way to descend. The short sea-passage across the narrow channel was flat calm, so we inflated the dinghy and set sail to the larger tidal-platforms below the east face of the stack.

On the seaward face, we climbed the large stepped-corner, which was the most obvious full-stack-height feature. This stepped corner provided an excellent climb to a very square and flat-topped summit. In the strong easterly winds, the summit of the stack was not in the lee of the surrounding sea cliffs, so simply standing on the summit was a bit of a wobble-fest.

After the obligatory summit photo session, we rigged the abseil and returned to sea level and our mighty vessel. We opted for a slightly longer passage across the mouth of the cave and climbed a series of ledges back to terra firma.

Colm Shannon on Skin Deep,
F7c/7c+ S1. Photo John McCune.

Ireland's Magic Isle by Alan Tees

I nearly walked off the wall many years ago, not seeing it until the last minute; thus Holy Jaysus! The name stuck, but It remained unclimbed, and I thought I would never see it climbed in my lifetime.

Alan Tees

I first went to Owey in June 1989 on a Northwest MC midsummer island camping trip, its first to that particular island. In the event there were only two of us, plus my two small sons, who braved the waves and canvas with the rest of the club turning up the next day. The weather was glorious, the sun sparkled on turquoise seas, and there was an island to be explored. The last permanent residents left between 1975–1977, and the population had steadily decreased from its maximum of 152 in 1911, but the houses were still left open, with many of the contents still inside. The schoolhouse was still there, unroofed. Daniel O'Donnell's mother was born on the island, but don't hold that against it.

Now, many of the houses have been repaired (mercifully in the manner of their original state) and are used as holiday homes, mainly by families during the summer months. If Gola is a rock climbers' island, Owey is one for adventure and exploration, and we were stunned by the array of high cliffs, sea stacks, pinnacles, arches, small coves, narrow channels, and, of course, caves, after which the island gets its name. Owey is, literally, riddled with holes below the waterline, many of which link or open out into huge caverns. Then, of course, there is the 'lake below the lake'. My climbing buddy Bill and I were about to catch the boat back to the mainland (well, Cruit Island is near-enough the mainland, being connected by a bridge) when we met an elderly couple of former residents who mentioned this with a bit of a twinkle in their eye. We weren't sure whether it was a wind-up or not but couldn't wait to get back to investigate. Perusal of a map showed a small lake at the far side of the island, and further exploration revealed a fissure in the cliff beyond it, which led down, and down, to a large cavern with a lake in it. We have been down since with an inflatable dinghy, paddles, and a powerful flashlight and even crossed to the far side. Remarkable, but the waters have receded recently, leaving an increasing shoreline of silt.

The cliffs on Owey are very impressive, but many of them suffer from Feldspar leaching in that the granite is sugary and loose, and many the mouthwatering line turns out to be more of an exercise in survival. There is, however, 'Gola quality' granite in a number of areas. David Walsh visited in 1991, putting up the excellent Nordkapp amongst other routes, returning in 1993 with reinforcements, and by the end of that year there were almost 30 climbs on the island.

The Colmcille Climbing Club landed in 1999 and 2000, climbing in a big bay on the southwest, which they called Dragon's Bay (the correct name is Scealpan Bui). The climbing was adventurous and spectacular rather than technical due to the nature of the rock, but great for photography. They caught the Owey bug and have been coming back ever since.

Probably the most striking feature on the island is Stackamillion at Cladaghroan, a teetering pinnacle in a bay partly surrounded by two sea arches, which amazingly was climbed in 2003 by two passing Poles. Even more amazingly, they left the route description of their new route, The Blade, E2, in the golf club at Cruit Island. Well you would, wouldn't you?

At Torglass, there is another stiletto of clean granite also climbed by the Poles and, more recently, by Dave Millar and Martin Bonar in 2009 (after several suitors, including myself, had been repelled by lack of protection and advancing tide). Just west of this is the aptly named Holy Jaysus Wall (Na Farragain), whose plunging vertical ramparts are come upon suddenly and usually given a very wide berth indeed. The original publication of this article in the 2013 *Irish Mountain Log* initiated a lot of interest in the island crags, and amazingly, Holy Jaysus wall has since been climbed,

Left: Kevin Kilroy and Michelle O'Loughlin abbing the Holy Jaysus Wall during the filming of *Feather in the West* in Sep 2019. Photo courtesy Iain Miller. Facing page: Darren Ryan on the summit of Stackamillion AKA Tor an Mhadadh Uisce (Otter's Rock). Photo Iain Miller.

giving two hard routes: The Second Coming and Immaculata, both E6, as well as extensive development of the big west facing cliffs. A comprehensive and up-to-date list of routes on the island is available on www.uniqueascent.ie, who also frequently use the island for their commercial activities. In 2010, we came into possession of a beautiful hand-drawn map by John McGinley (one of several of the west coast of Donegal). From this, we have been able to learn the correct local names for many, but not all, of the geographical features.

For walkers, a circumnavigation of the island will provide a fascinating and photogenic short day's walk, but a day trip really cannot do Owey justice. The proper experience involves immersion in the timelessness of a deserted island, paddling, exploration, swimming, fishing (maybe even some climbing?), barbecuing, then watching the sun set on the Atlantic and the light fading on the screes of Errigal Mountain to the east. Sea pink on the grey and yellow granite and the cry of the gull – absolute magic. If you have a kayak, bring it, as the circumnavigation of the island is one of the best trips in the country.

Owey is a place of contradictions. Seaward, you can feel an amazing sense of remoteness and adventure amongst the spectacular island rock architecture, but as you walk over the hill, back towards the camp by the harbour, suddenly, there are golfers just across the channel going about their business or loading their clubs into the car to go home, whilst you look on from what seems like another planet.

Right: Colm Shannon making the first ascent of The Jelly Situation, F7c+/8a S1/2. Photo Joshua Willett. Below: Jules Lines on The Power of Hobo, F7c S1. Photo Lines collection.

The Devil's Castle by Chris Harle

Mick Fowler and I were about to cancel our climbing trip to the west coast of Ireland when Andrew Barnwell and Jonathan Edwards came to the rescue. Their accommodating and flexible nature and their enthusiasm for an adventure reassuringly ticked all the boxes. It also helped that Andrew's VW California was easily big enough for four people and the considerable amount of equipment needed to attempt the main objective – a sea stack called the Devil's Castle.

We were in Ireland for four full days, and despite the variable weather, we achieved most of what we hoped for, including an ascent of Ireland's highest mountain, Carrantuohill in the MacGillycuddy's Reeks, via Howling Ridge – 300m of spectacular mixed scrambling with V Diff climbing sections.

But, our main holiday objective was The Devil's Castle, a sea stack near Ballybunion. It had been on Mick's radar for some time with its potential for a first ascent. He is an avid researcher of climbing possibilities, and he suggested that the Devil's Castle might be one of the last unclimbed sea stacks off the British Isles. Mick's definition of a sea stack is quite precise – it must have a pillar-like appearance, be near vertical on all sides, and not easily accessible. However, aligning people and their availability, plus the right sea and weather conditions, was no easy task.

On Ballybunion beach, Mick's unbridled enthusiasm infused the team with energy, and soon the boat was assembled and inflated, the engine installed, climbing gear loaded, and four over-60s were ready for action. We chugged along the dramatic coast for two miles to the Devil's Castle, which clearly fitted Mick's sea stack criteria. Two circuits confirmed that there was no straightforward route to the top, but as there was only one small landing platform, the route was chosen for us. I was first out. The minor swell made it awkward to get established on the platform, and the barnacle-covered rock took its bloody toll on my legs in an effort to stay dry. The climbing dry bags followed, but not before one was dropped into the sea. Fortunately, it floated long enough to retrieve, and once Mick landed, Andrew and Jonathan retreated to a safe viewing distance.

Mick took the first pitch. Steep rock with Jenga-looking blocks of looseness raining down at the slightest touch. Mick seemed unperturbed even with sparse, unreliable friend protection. Rope drag and an obvious cave stance gave Mick the chance to bring me up. We swapped gear, and I pulled up into a short crack that led to a grassy ramp. Here I hammered in a warthog (a long ice peg) to calm the rising panic. I avoided the final tottering tower with an unprotected detour. Mick followed on a more direct line, and we were soon celebrating success and turning our attention to getting down safely. However, we accidentally dropped the large rope-sling that we had intended to use for the abseil, but no matter, a length of the climbing rope was duly sacrificed. Back at sea level, our crucial support team were waiting.

Obviously, without Jonathan and Andrew's boat skills, and patience, our attempt would have been impossible. I made the leap back into the boat without mishap while Mick decided to check out the sea temperature when his leap came up short. It was a happy team that returned safely back to the beach landing place. The route was approximately 30m with a tentative grading of E1 5a to reflect the overall seriousness of the climb.

Facing page and right: Mick Fowler and Chris Harle on The Devil's Castle. Below left: Mick on the summit. Below right: Jonathan Edwards and Andrew Barnwell crewing the boat. Photos courtesy Chris Harle.

The Burren by Julian Lines

Fine deep water solo venues are a rare commodity, so when I heard through the grapevine that Andy Long had done some E7s on The Burren, which he said would make good deep water solos, my ears pricked up. However, The Burren was far away on the west coast of Ireland, and it was a long way to travel on a whim with barely any information. Then The Burren guide came out in 2008, and the pictures of Ricky Bell soloing Daka Daka and The Littlest Hobo fired my intrigue. So, I got in touch with Ricky. Now, there is someone who could inject inspiration and psyche into the laziest couch potato. He sent me descriptions of further solo shenanigans that he and Si Moore had been doing up to F8a and notes of potential solos; they even went as far as checking out the water: 'We also snorkelled under most of the wall (An Falla Uaignech) up to Sea Bird and it is f***ing deep!'

I arrived during a rare good spell of weather and didn't really know where to start; I was spoilt for choice. So I started soloing existing routes on the high tide and then went scoping out other walls on the rope during the low tides. What piqued my attention the most was the blank wall that carved straight down into the ocean beneath Ricky's excellent rising traverse of The Crozzly Show. On the left side of this wall was Si's Identity Crisis, but there was room for more lines here on the most perfect crozzly rock. The water depth wasn't a problem; the wall is perhaps 20m high, of which the last 8m is easy, so that also worked. The only problem was that if you fell in, you would have to swim over 100m to escape the Atlantic. I had become prepared, so out came the toys. I abseiled down to the tide line with my chunky bench seat, a rubber ring and a dry bag full of gear. I hung the seat on a jumar and knotted the bottom of the abseil rope. On its end, I tied the rubber ring and let it float around in the water. Maybe it was overkill as the sea wasn't that choppy. I changed into my climbing gear and launched off the seat onto the pristine wall. I had seen the holds, chalked them and thought it would go, but due to a mixture of uncertainty and not being warmed up, I fell off. I felt far more refreshed, and next time round I experienced the most beautiful set of sequences on pristine rock to join and finish up The Crozzly Show. I was utterly blessed, and King Crozzle is full of superlatives.

One morning, I met a local climber, Colm. He was young and keen and intrigued about all my toys and how they helped me to get into and out of places that would take massive swims if you didn't have them. He turned up one day, and on peering along the cliff, there I was, just finishing off Skin Deep, another superb line just to the right of King Crozzle. He was probably a little aghast of how I just soloed alone. If something went wrong, it could seriously go wrong, and there was no plan B or rescue. I understood this, calculated it, and accepted the risks. It was all part of living or the feeling of being alive. The first week at the Burren was absolutely stunning weather, and it gave me time to work out all the solo possibilities on the crag. But then things changed, and it resorted to west coast of Ireland weather. There were two further routes I wanted to do before I left, so I hung out in the van and waited and waited. The days were passing one by one...

One of the lines was the groove to the left of Black Widow, which you traverse into via a finger rail from the Black Widow belay ledge, climb the groove and finish out right into Black Widow or go direct via a very high and very hard sequence. I practised the moves in the groove on a rope. I laughed as I just couldn't believe how the climbing was so exquisitely good. The other line was the wall to the right of Skin Deep, which led into Bing Crozzly. I'd worked this on a rope also and even went down to start it, but a squall came in and soaked me. One morning I was lying in the van pondering over another damp day when the van door opened. I was in shock, as was the old man who opened the door. He was very apologetic and thought the van had been abandoned and was investigating. He was a local and frequently went down to the cliffs to fish, so he said. The next morning he turned up with a freshly caught and de-boned mackerel; it was the best mackerel I'd ever had.

The Irish weather wasn't abating, and sadly, I was ousted before getting my routes done. I knew I would come back at a later date to finish them. Year after year passed, and then I heard that Colm had done the routes. The groove was called Wonder Dragon, and the wall Crozzle Monster. Part of me felt wounded, but the other part of me felt elated as I'd managed to inspire Colm, just as Ricky inspired me to go there in the first place.

Ricky Bell on The Crozzly Show, F7b S1. Facing page:
Ricky Bell on Bing Crozzly, F7b S1. Photos Craig Hiller.

The Jelly Situation by Colm Shannon

When I returned home from my first year of college, I realised that I lived five minutes from the perfect limestone of Ailladie and got stuck into the trad routes there. At the time, I was vaguely aware of Ricky Bell and friends having put up a few DWS, but I didn't know exactly where they were or the logistics of how to go about doing them, so I never gave them much thought at the time. However, my eyes were opened to the DWS potential at Ailladie in 2010 when Julian Lines visited. Jules appeared for a month, and it blew my mind seeing him working lines like The Power of the Hobo or Skin Deep. He showed me how he set up a rope to get out of the water, as the nearest exit is a fair swim away. After he left, I made my own rope ladder, bought a dinghy from Smith's and went to work. I started with Jules's route King Crozzle, which I managed to do ground up, and then climbed The Hobo shortly after. King Crozzle is brilliant, with some long reaches off good edges, and it was a perfect introduction to the climbing on that wall.

After returning from Ceuse with some fitness in 2012, I finally succeeded on The Crozzle Monster having first looked at it two summers previous. It's an amazing complex line, probably the best line on the wall, and I was delighted to have added a route of my own to the cliff. It's since had at least three repeats, and it's been cool seeing how others have solved the crux differently, from beastly lock-offs to full-on dynos. At this point, DWS at Ailladie was still relatively under the radar, and I'd been working away in my own little bubble, but I shared a few videos and pictures and that got some other climbers interested. The following summer, during a heat wave, we had an impromptu DWS fest. It was great fun, with a rope on every line and plenty of splashes. I brought a gas BBQ from the house to the top of the crag, and after some burgers and beer, I was able to make the third ascent of Identity Crisis – another line I'd worked on and off during previous years.

I was back in the summer of 2015 with two routes on my to-do list: The Vein, originally E7 6c, but apparently F7a as a DWS, and Skin Deep, another route of Jules's that's graded F7c/7c+. Home for a weekend, I popped out on a Friday evening to check out Skin Deep. The tide was perfect, and the sea was dead calm, but the wall was too hot. After an hour swinging about on a rope with sweaty fingers, struggling with the small, sharp holds, I packed it in and went home. But, the next morning, the wall was lovely and cool, the tide was perfect, and the sea was calm. Unfortunately, my skin was trashed. I dialled in my beta and retreated back to the house to grow some skin. I optimistically returned that evening, but this time the sea was too rough. Sunday morning dawned with perfect weather. I threw a rope down The Vein to check it out. It's another amazing route with bouldery moves between good breaks. I was psyched, and after getting everything set up for a double send, I sat back and waited impatiently for the tide to come in. Six o'clock eventually rolled around, bringing the high tide, but the sea was now too rough. I left that weekend empty-handed – part of the process when trying to climb DWS on the wild Irish coast. I was back a few weekends later and managed to make the second ascent of Skin Deep, with John McCune on ab getting some amazing pictures. This one might be the hardest line on the wall. By 2016, having climbed most of the established DWS lines, I'd started looking for more. So in June, with magically dry weather out west, I knew where I wanted to be.

On a Friday evening, with a high tide, calm-ish seas, and a glorious setting sun, The Adventures of the Wonderwagon, F7b+/7c was born. It's a great line with some fantastic, unique climbing, but it's not the safest of the DWS lines. The sea churns a lot under that section of the wall unless it's properly calm, and the ledge at the bottom of the groove is a bit disconcerting, as are the boulders in the water, so I doubt it will be a popular one, but I was really excited to have gotten it done. Then, with just a few days left in the country before I was due to head off to Europe for a road trip, I found a final ridiculous line right up the middle of An Falla Uaigneach. That route became The Jelly Situation, F7c/7c+. At a height when the other routes start to ease off, this one punches through a roof, followed by a pumpy section of headwall. It's best done at a very high tide, and I'd also recommend having someone stationed on the Seabird ledge ready to act as a lifeguard. Looking back, I'm not sure how safe I really was on this one...

Facing page: Colm Shannon making the first ascent of The Jelly Situation, F7c+/8a S1/2.
Photo Joshua Willett. Right: Sam Hamer on King Crozzle, F7b+ S1. Photo Mike Hutton.

Or where the northern ocean, in vast whirls,
Boils round the naked melancholy isles
Of farthest Thule, and the Atlantic surge
Pours in among the stormy Hebrides.

Autumn by James Thomson

Scotland

Farquhar knew by the serpent's wisdom that he had, when he laid his finger under his teeth; the king was cured and had all his doctors hung. Then the king said that he would give Farquhar lands or gold or whatever he asked. Then Farquhar asked to have the king's daughter and all the isles that the sea runs around, from point of Stoer to Stromness in the Orkneys, so the king gave him a grant of all the isles. But Farquhar the physician never came to be Farquhar the king, for he had an ill-wisher that poisoned him, and he died.

Fearachur Leigh from Sutherland in *Popular Tales of the West Highlands Vol 2* by John Francis Campbell, 1890

Scotland has an abundance of sea cliffs, especially on the islands, which provide excellent climbing but will never be a mecca for deep water soloing due to the temperature of the water. The Scottish sea stacks have long been famous for aquatic exploits. The earliest sea cliff climbers, however, were the fowlers, especially of St Kilda who were climbing routes of, at least, a modern grade of Severe in the 17th century.

As the summer wind goes droning o'er the sun-bright seas,
And the Minch is all a-dazzle to the Hebrides,
They will skim along like salmon – you can see their shoulders gleam,
And the flashing of their fingers in the Blue Men's Stream.

Hebridean Boatman's Song (trad)

Legend has it that the Blue Men of the Minch were Storm Kelpies that were mainly found in the waters between Lewis and the Shiant Isles. This 'Blue Men's Stream' was also named 'The Current of Destruction' because the Blue Men were blamed for the loss of so many ships in the rough waters. Sailors were always on the lookout for the Blue Men, who dwelt in caves, they said, at the bottom of the sea, and many sailed around the Shiant Isles instead of taking the shortcut between them and Lewis. According to the myth, when the Blue Men were ready to attack a ship, their chief shouted, Sphinx-like, two lines of poetry to the skipper of the boat, and if the skipper did not add two lines to complete the verse, the Blue Men seized the ship.

A Letter from the Sea

The St Kildans were surely the pioneers of rock climbing, certainly in Britain… It is interesting that their rope techniques developed in parallel to that in Britain and in the Alps, where the party would be led by the most able man, moving together on easier rock and one at a time when it became difficult.

Hamish MacInnes

St Kilda is an isolated and heavily-weathered granite and gabbro archipelago situated 40 miles west of Harris in the Outer Hebrides in the North Atlantic Ocean. The islands have the highest sea cliff (Conachair,1,410ft, on Hirta) and the highest sea stacks (Stac an Armin, 643ft, and Stac Lee, 564ft) in the British Isles. It has long been thought that St Kilda was continuously inhabited for over 2,000 years, but in 2015, evidence of earlier Neolithic settlement were discovered – shards of pottery and stone tools – suggesting that the islands were settled around 4,000 BC. Despite 6,000 years of settlement, the population probably never exceeded around 200, with

Facing page: Julian Lines on Shere Khan, F7c+ S1. Photo Dave Cuthbertson.

the peak in the late 17th century. There is no longer a resident population since the islands were evacuated in 1930. Like the inhabitants of the similarly remote Easter Island, the St Kildans were, by necessity, a 'bird culture'. St Kilda has a greater number and more species of sea birds than any other island group in Europe, and Stac Lee is possibly the biggest gannetry in the world. Their economy was primarily based on the produce of sea birds: solan geese, gannets, fulmars and puffins, with the St Kildans deriving their source of food, oil, fertiliser and revenue from the colonies of sea birds which they shared the islands with. The islands were historically part of the domain of the MacLeods of Harris, whose steward was responsible for the collection of rent, which was paid for in feathers.

The first written record dates from 1202 when an Icelandic cleric wrote of taking shelter on 'the islands that are called Hirtir'. In a paper communicated to the Royal Society of London by Sir Robert Moray and published in their *Philosophical Transactions* in 1678, Moray described the dangers connected with the capture of sea fowl by the 'men of Hirta' on the apparently inaccessible 'Stac Donna' (although he is most likely describing an ascent of the highly impressive 240ft Stac Biorrach rather than the nearby and much less impressive 87ft Stac Dona): 'After they have landed, with much difficulty, a man having room but for one of his feet, he must climb up 12 or 16 fathoms high. Then he comes to a place where, having but room for his left foot and left hand, he must leap from thence to such another place before him, which, if he hit right, the rest of the ascent is easy; and with a small cord, which he carries with him, he hauls up a rope, whereby all the rest come up. But if he misseth that footstep (as oftentimes they do), he falls into the sea, and the company takes him in by the small cord, and he sits still until he be a little refreshed, and then he tries it again; for everyone there is not able for that sport.'

Martin Martin was a Gaelic-speaking Scottish writer from the Isle of Skye best known for his 1703 work *A Description of the Western Islands of Scotland*. Martin witnessed locals climbing in the Barra Isles but was most impressed by those from St Kilda:

'The Inhabitants of St Kilda excel all over those I ever saw in climbing rocks. They told me that some years ago, their boat was split to pieces upon the west side of Borrera isle, and they were forced to lay hold on a bare rock, which was steep and above 20 fathoms high. Notwithstanding this difficulty, some of them climbed up to the top and, from thence, let down a rope and plaids and so drew up all the boat's crew, though climbing this rock would seem impossible to any other except themselves. This little commonwealth has two ropes of about 24 fathoms length for climbing the rocks, which they do by turns. The ropes are secured all around with cow's hides salted for the life, which preserves them from being cut by the edge of the rocks. By the assistance of these ropes, they purchase a great number of eggs and fowl. These poor people do sometimes fall down as they climb the rocks and perish. Their wives, on such occasions, make doleful songs which they call

lamentations. The chief topics are their courage, their dexterity in climbing, and their great affection which they showed to their wives and children.'

Writing in his *Late Voyage to St. Kilda,* published in 1698, Martin also described an ascent of 'Stac Donna', although, once again, this is probably Stac Biorrach:

'A mischievous rock... for it hath proved so to some of their number, who perished in attempting to climb it. It is much of the form and height of a steeple. There is a very great dexterity, and it is reckoned no small gallantry to climb this rock, especially that part of it called "the thumb", which is so little that of all the parts of a man's body, the thumb only can lay hold on it, and that must be only for the space of one minute; during which time his feet have no support, nor any part of his body can touch the stone, except the thumb, at which minute he must jump by the help of his thumb, and the agility of his body, concurring to raise him higher at the same time, to a sharp point of the rock, which when he has got hold of, puts him above danger, and having a rope about his middle, that he casts down to the boat, by the help of which he carries up as many persons as are designed for fowling. At this time, the foreman, or principal climber, has the reward of four fowls bestowed upon him above his proportion, and perhaps, one might think 4,000 too little to compensate so great a danger as this man incurs. He has this advantage by it that he is recorded among their greatest heroes; as are all the foremen who lead the van in getting up this mischievous rock.'

As well as praising the wonderful dexterity of the fowlers, which he witnessed first-hand, Martin also observed the early initiation of their offspring in climbing: 'The young boys of three years old begin to climb the walls of their houses; their frequent discourses of climbing, together with the fatal end of several in the exercise of it, is the same to them as that of fighting and killing is with soldiers, and so is become as familiar and less formidable to them than otherwise certainly it would be.'

Lachlan MacLean, writing in 1838 in his *Sketches of the Island of St Kilda*, described the birds as 'a way of life, and often enough, when they broke their necks trying to catch them, a cause of death'.

James Wilson visited St Kilda during *A Voyage Round the Coasts of Scotland and the Isles,* published in 1842, and observed the St Kildans on their crags:

'Suddenly, we could hear in the air above us a faint huzza-ing sound, and at the same instant, three or four men from different parts of the cliff threw themselves into the air and darted some distance downwards, just as spiders drop from the top of a wall. They then swung and capered along the face of the precipice, bounding off at intervals by striking their feet against it and springing from side to side with as much fearless ease and agility as if they were so many schoolboys exercising in a swing a few feet over a soft and balmy clover field... In this manner, shouting and dancing, they descended a long way towards us, though still suspended at a vast height in the air, for it would probably have taken all their cordage joined together to have reached the sea. A great mass of the central portion of the precipice was smoother than the wall of a well-built house, and it was this portion especially which was not only perpendicular but had its basement arched inwards into an enormous wave-worn grotto so that anyone falling from the summit would drop at once sheer into the sea. It was on this smoother portion of the perpendicular mountain that one or two of the cragsmen chiefly displayed their extraordinary powers because as there was nothing to interrupt either the rapid descent of the rope, or its lateral movement, or their own outward bounds, we could see them sometimes swinging to and fro after the manner of a pendulum, or dancing in the air with a convulsive motion of the legs and arms (presenting a painful resemblance to men hanging in the agonies of death), or tripping a more light fantastic toe by means of a rapid and vigorous action of the feet against the perpendicular surface of the rock. These men merely capered for our amusement but caught no birds, for such was, in fact, the adamantine smoothness of the surface that not even a winged inhabitant of the air could have found rest for the sole of its foot.'

In *St Kilda Past and Present*, 1878, George Seton described fowling as 'the principal avocation of the St Kildans, and the great ambition of every male on the island is to excel as a cragsman':

'The exploits of the cragsmen on the cliffs of Borrera and Soa and adjacent stacks are, if possible, even more astonishing than their performances on the main island. Stack Briorach, the pointed rock between Soa and St Kilda, is regarded as the crucial test of a fowler's pluck. Here, the rope is of no avail, and the rock can only be climbed after the fashion of the celebrated steeplejack, lately gone to his rest. The man who fails to accomplish the ascent never gets a wife in St Kilda.'

Seton also documented that the St Kildans supposedly had an unusual physiological characteristic: 'The great toes of the cragsmen are widely separated from the others, from the circumstance of their frequently resting their entire weight on that part of the foot in climbing.' Richard Kearton, in his 1897 book *A Camera on St Kilda,* described that the ankles of the natives were 'tremendously developed' and illustrated this observation with a photograph of his own ankle in comparison with that of a native. From centuries of scaling cliffs barefooted after their prey, the St Kildans were said to have adapted to their environment and evolved prehensile feet. Observations of the St Kildans' prowess in climbing and their purported selected adaptations to their natural environment bring up some intriguing questions: Can human climbers evolve on isolated islands like Charles Darwin's Galapagos finches? Is that scientifically possible or a myth?

Right top: The St Kilda parliament. Right middle: Stac Lee and Hirta. Right bottom: A group of St Kildans. Photos George Washington Wilson (1823–93) courtesy of The University of Aberdeen photographic collection and licenced under the Creative Commons Attribution 4.0 International Licence.

Above left: Grant Farquhar bouldering on Easter Island. Like St Kilda, this remote Pacific Ocean island also had a 'bird culture'. Above right: Birdman carving at Orongo on Easter Island – site of the bird cult. The islands visible are where the annual birdman ceremony harvested eggs from. Photos Eloïse Pitts Crick. Facing page: Getting the Fulmars on St Kilda. Photo George Washington Wilson (1823–93) courtesy of The University of Aberdeen photographic collection and licensed under the Creative Commons Attribution 4.0 International Licence.

Ornithologist Charles Dixon visited St Kilda and, in contrast to other commentators, was not particularly impressed with the climbing there stating that 'beyond the celebrated stacks there are comparatively few cliffs that a tolerable climber could not explore unaided by a rope'. Dixon was also, apparently, unimpressed by Darwin's theory of natural selection and expounded on this in his 1885 book *Evolution Without Natural Selection,* in which he emphasised the additional important evolutionary mechanisms of isolation, climatic influences, the law of use and disuse of organs (Lamarckism) and sexual selection, and from what we now know about epigenetics this may be highly relevant.

The historical descriptions and observations of life on St Kilda above arguably provide numerous examples of such evolutionary forces at work, including natural selection 'the struggle for life', and sexual selection 'the struggle for wife'. Regarding the latter, Lachlan Maclean observed in his *Sketches of the Island of St. Kilda* in 1838: 'The man who cannot climb [Stac Biorrach] never gets a wife in St Kilda.' Richard Barrington wrote that 'it is not unreasonable to suppose that a girl of St Kilda, having in view her future welfare, should establish some sort of test whereby to judge her lover's ability as a climber' and therefore provider. He further elaborated that: 'Climbing was looked upon as a great feat amongst the islanders, where for hundreds of years the chief food of the inhabitants was obtained from the lofty precipices and stacks. In fact, there is no part of the world, as far as I am aware, where the practical advantage of being a skilled cragsman was so well recognised.'

An adolescent rite of passage for St Kildan males occurred at the Mistress Stone, an exposed rock on the cliff-top with a 400ft drop to the ocean below. Martin Martin was challenged to perform this traditional feat:

'Upon the lintel of this door, every bachelor-wooer is by an ancient custom obliged in honour to give a specimen of his affection for the love of his mistress, and it is thus; he is to stand on his left foot, having the one half of

his sole over the rock, and then he draws the right foot further out to the left, and in this posture bowing, he puts both his fists further out to the right foot; and then after he has performed this, he has acquired no small reputation, being always after it accounted worthy of the finest mistress in the world.'

Evolutionary adaptations to specific environments are well documented in human populations around the world. The most well-known example is that of the high-altitude adaptation of the Tibetans, Andeans and Ethiopians. More recently, research into the Sama-Bajau 'sea gypsies' of Southeast Asia found that their spleens are 50 percent larger than normal. This is due to a genetic mutation and confers an advantage for freediving. This is only one published study,[1] so we can't put much weight on it, but if it is replicated then it will add evidence to support the possibility of similar adaptations in the St Kildans.

With only the anecdotal observations described above to go on, we can only speculate as to whether the isolated St Kildans had, indeed, micro-evolved to become more specialised for – and therefore better at – climbing than other humans. But it seems entirely possible. St Kilda has its own unique wren, as well as a sub-species of mouse. Humans inhabited remote St Kilda for around 6,000 years and would have been subject to the same evolutionary forces. We can even take it a step further and speculate that had their isolation not been ended by the modern age then, perhaps, they would have developed into a new species of human: *Homo ascendo.*

Author and photographer Norman Heathcote (1863–1946) visited St Kilda in 1898 and 1899 with his sister, Evelyn. At that time, St Kilda was owned by his uncle, Reginald MacLeod of MacLeod. Heathcote went on to write a book about St Kilda and, owing to descriptions of climbing in his book, was asked by the Scottish Mountaineering Club to contribute an article to their journal. He also published A Map of St Kilda in the *Geographical Journal* of 1900.

1   Ilardo et al, *Cell* Vol 173 Issue 3 p569-580, 2018.

Climbing in St Kilda by Norman Heathcote

There are few places in the United Kingdom where more attractive climbs are provided for the enterprising mountaineer or where the venturesome rock climber can find better opportunities of risking his neck. Those who have exploited the Cuillin, Ben Nevis, and other climbing centres of Scotland may find a fresh field for their labours. I feel sure that anyone who does not mind roughing it, and has time to spare, may spend a most enjoyable fortnight in St Kilda. If he approaches them in the right way, he will find the people easy to get on with, gentlemanlike in manners, pleasant companions, and first-rate climbers.

However, I do not wish to encourage too many people to go there, as there is not room for more than a few people at a time, so I will begin by pointing out some of the disadvantages of St Kilda from a mountaineering point of view. In the first place, it is a long way off, and when you have got there, it is by no means certain that you will be able to get away again on any given date. The steamers *Dunara Castle* and *Hebrides* sailing from Glasgow include St Kilda among their places of call about once a fortnight during the summer months, but if it happens to be blowing at all hard from the southeast, it is quite possible that passengers may not be able to land.

Then, there is no accommodation to speak of. The native houses are very good – for native houses – but it would be preferable to occupy one for a week than a month. Personally, I stayed in the factor's house, which at least has the merit of being empty, but I gather from the accounts of those who have tried them that the St Kildans are not the only inhabitants of the other houses. They also have a pleasing habit of cooking all kinds of food in the same vessel, so until you have learned to like the flavour of fulmar oil, the food is apt to be unpalatable.

But after all, these are minor matters. The most serious objection to St Kilda as a climbing centre is that all the most interesting expeditions have to be done by boat. Seeing that St Kilda is a small island right out in the Atlantic, and that the sea is never calm, even in the calmest weather, it is obvious that this is a great drawback. Even in June, which is probably the finest month, one might spend a fortnight there without ever getting an opportunity of landing on Stac na Biorrach or Stac Lee. You cannot even be sure of being able to take advantage of a favourable day, as you must be dependent on the natives for a boat, and if they want to look after their sheep, cut turf, or catch birds, they would be quite capable of refusing to go for love or money, though the latter is a fairly good persuader even in St Kilda.

It is not possible to see St Kilda and its subordinate islands or to form any idea of the prowess of the natives as cragsmen without climbing. A good many of those who have written about the island have obviously not been mountaineers and have often contented themselves with sitting in a boat and watching the natives ascend the rocks or have given thrilling accounts of the difficulties at second hand. I don't profess to be an expert or venturesome mountaineer, but having done a lot of climbing in the Cuillin Mountains of Skye, and a certain amount in Switzerland, I may fairly claim to be a tolerable climber; but I should say there are comparatively few cliffs that I should care to explore without the assistance of a rope.

The St Kildans never wear boots when climbing, and I very soon found that I must imitate them in this respect. They walked with the utmost ease and rapidity along smooth sloping ledges where I had to squirm along with the greatest care, looking out for every crack or excrescence that would serve as a resting place for my nailed soles. When I tried the same ledges in stockings, I found that I could get along something after their fashion, though until I had learnt what thoroughly firm foothold the rough texture of the stockings gave, there was still an inclination to squirm. On dry rocks, probably, bare feet are even better than stockings, also, if possible, more painful, but for landing and climbing up the slippery places near the sea, the natives generally put on a pair of socks.

Facing page left: A Map of St Kilda, surveyed and drawn by Norman Heathcote in 1900. Originally published in *Geographical Journal* 15 (2): 142–144, 204. Also published in 'Climbing in St Kilda'. *Scottish Mountaineering Club Journal* 6 (5): 147–152 and *St Kilda* by Norman Heathcote.

Facing page right: A photogravure illustration from Heathcote's 1900 St Kilda book entitled 'Boating in St Kilda'. Source: Wikimedia Commons. Public Domain.

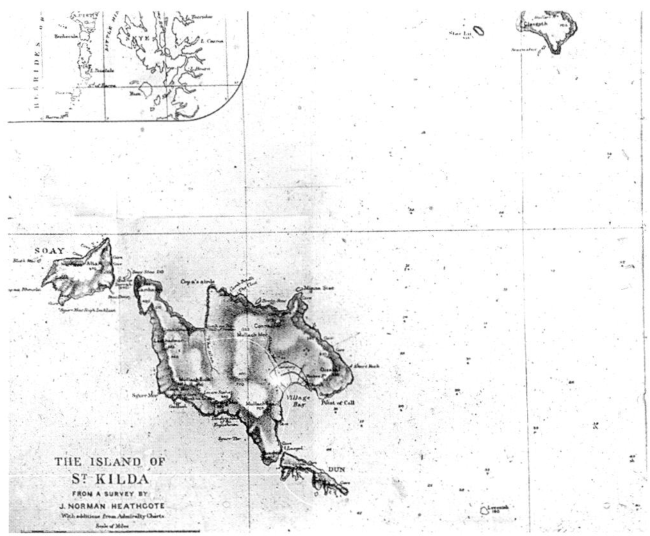

The fact that there has been no serious accidents on the cliffs within the memory of the present inhabitants is pretty good testimony to their skill as cragsmen and to their prudence as men. They often climb alone in places where most people would not care to follow them, but whenever it's a case of negotiating a really difficult and dangerous cliff, they always go with a companion and make sure that the rope is sound. Tales are told of hairbreadth escapes and heartrending accidents in times gone by, and if, as one writer asserts, they were in the habit of entrusting themselves to ropes made of straw or heather, it is not surprising. Not that I believe that for a moment, but probably they were not so well found in the matter of ropes as they are now and used them longer than was consistent with safety. In Martin's time, there were only three ropes on the island, made of cowhide salted and wrapped around with hempen rope to prevent fraying on the rocks. Horsehair ropes came into use later on; in fact, there are still one or two on the island. I have had one on occasionally for purposes of landing or embarking, but, though they are very durable, I would rather trust good manilla hemp on difficult climbs, as these particular ropes must be getting fairly venerable.

The natives are very careful of strangers. They generally use two ropes at any difficult place and take plenty of time, but they have not grasped the principle that in coming down, the strongest climber should be last on the rope. I have seen them leading down places where a slip on the part of their protégé might have been very awkward, and in the case of inexperienced climbers, a slip is always possible. It is said that, on one occasion, a long time ago, two men were let down on the same rope. Presently, the higher of the two noticed that several strands had given way above him and seeing that it was a case of one life or two, cut the rope below him. His companion was dashed to pieces, but the slender support on which his own life depended just sufficed to bring him to the top of the cliff and he was saved. It is an awful thing to do. It may be justifiable, and I would not blame anyone for doing it, but I trust that if I were in a similar situation, I would rather die than do it.

A still more tragic fate befell another young St Kildan. The rope broke, and he fell down the cliff, but, unfortunately for him, he did not reach the bottom. He managed to cling to a grassy slope just above the final precipice, and there he remained, expecting every moment to be rescued. Alas! There was no rope on the island long enough to reach him; there was no possibility of helping him from above. The cliff below was inaccessible; there was nothing to be done. At the end of three days, he began singing Gaelic songs, and then he died.

Most of the climbs in St Kilda are easier than they look because the rocks are always firm. Anywhere within 200ft of the sea, the winter waves sweep away all loose stones, and it is almost superfluous to test a handhold, but even above this level there are not many places where the rocks are rotten. On the main island there are, of course, precipices which can be climbed – also some that cannot; there are grass slopes steep enough to make your hair stand on end and ledges narrow enough to try the steadiest head, but it is impossible to get along the shore anywhere, and so all the climbing has to be done from above, which to my mind is not satisfactory. There is great satisfaction in getting to the top of anything, and the more difficult the ascent, the greater the joy, but to go down a cliff in order to come up again is not at all the same thing – I would do it to get a photograph of a bird, or to find a good point of view from which to watch the waves, but not purely for the sake of climbing.

Stac Lee is perhaps the most interesting spot in the St Kilda group, but its chief interest is for the naturalist, not for the mountaineer. It is the principal breeding place of the gannets, the whole of the top, which is of considerable extent, and every available ledge being packed with their nests. It is about 530ft high, the ground sloping gently from the top on three sides and then falling away. Anyone looking at Stac Lee would say that it is practically perpendicular on all sides and obviously inaccessible, and yet it is not a difficult climb. The route is up the left-hand skyline and, looking at the steepness of the slope, the statement that it is not really difficult seems sufficiently remarkable, but the explanation is that there is a friendly ledge caused by a basaltic dyke which provides an excellent staircase the greater part of the way, and the only qualification necessary to negotiate this is good head, while the principal difficulty is to avoid treading on eggs. A stiff climb up steep rocks for 150ft or so brings you to the dyke which forms a ledge leading diagonally up the face of the cliff until it emerges on the easy slope near the top. The landing is the worst part, and I must say that if I had the option of leading the way, or letting someone else be the first to land, I should willingly resign the honour. It is a most appalling undertaking, at any rate under the conditions that prevailed when my sister and I landed there.

A stanchion has been fixed on a ledge some 20ft above sea level, and when a rope has been thrown over this, the boat is brought close to the rocks. The pioneer has to jump onto the face of the cliff – which at low tide is overhanging and covered with slippery seaweed – and haul himself up to the ledge. For such acrobatic feats, the St Kildans always take off their boots; in fact, for all climbing operations they either go barefoot or wear a pair of coarse socks. I tried climbing in boots at first, but very soon came to the conclusion that their plan was the best. It is bad for the stockings and painful to the feet, but there are so many sloping ledges to be negotiated on which nailed soles can get no foothold, but where one can walk with ease on stockinged feet, that anyone wishing to climb in St Kilda must be prepared to sacrifice their stockings and their feet.

There are one or two steep bits of climbing before you reach the ledge, but nothing so difficult as the so-called Inaccessible Pinnacle in the Cuillin, and as my sister and I have both been up that, Stac Lee presented no terrors to us. She shares the honour of having attained the top of the pinnacle with, l believe, two other ladies.

Stac Lee is more than four miles from St Kilda, and it is advisable to select a fine day for the expedition and to keep an eye on the weather. On one occasion, when we went there, it came on to blow from the west, and we found it was impossible to get home again, the result being that we had to spend the night in the boat on the lee-side of Boreray. Though an interesting experience to look back upon, this was not altogether pleasant at the time and might very easily have been more serious than it was.

There are several other stacks worthy of the attention of the climber. The route up Mian a Stac was pointed out to me, and I must say it looked very difficult, but climbs of this sort where the rocks are absolutely firm are generally easier than they look. There is an easy way up Stac Levenish, and we only attempted the difficult side because the swell had increased to such an extent during our stay there that we found it quite impossible to get into the boat at the usual landing-place and so had to climb to the top again and get down to the lee-side of the rock. It is not really a difficult climb, though it might not be easy to find the way without a guide. There is a steep and narrow chimney which requires some care, but the rocks are very firm, and there are no loose stones as during the winter gales the sea washes right over the stack and clears away everything except the solid rock. It is about 200ft high.

There should be some interesting climbs on Boreray. It is of the same geological formation as the Black Cuillin and is not unlike them in form, except that the slopes are covered with lovely, soft turf instead of the abominable screes that are such an unpleasant feature in those otherwise delightful mountains. I followed the northern arête for some little way down from the top, but being alone, and not having much time to spare, did not go far enough to see whether there would be any chance of getting down to the sea. The average slope is not too steep, but from what I have seen of the arête from other points of view, I should think there would be some awkward bits.

There are several ways by which the top of Soay may be reached from the sea. The easiest route entails a certain amount of climbing, and some of the others look as if they would satisfy the most exacting seeker after difficulties. The first time I visited Soay, it was impossible to land at the usual place owing to the swell, and we rowed around to the north side of the island where the cliffs are about 1,000ft high and look most formidable. The natives undertook to pilot us to the top, but the climb looked so much more suitable for a wild goat than for a lady (my sister accompanied me on all my expeditions) that we did not venture and watched the St Kildans scrambling up the rocks and crawling along the ledges, till they looked like flies on the face of the cliff. We eventually landed on a shelf of rock on the east side and were told that it was possible to reach the top from there. This route looked much more difficult than the one we had just refused, but I have no doubt it would afford an interesting climb.

The most difficult climb ever undertaken by the natives is Stac na Biorrach, an isolated rock 240ft high, situated in the passage between Soay and the main island. It is only possible to land here when the swell is less big than usual; indeed, anyone but a St Kildan would say that it is quite impossible to land under any circumstances. And yet, the landing is not the most difficult part. Martin Martin described how, at one point, the climber has to hold on by one thumb while he swings his legs from one ledge to another. He was speaking from hearsay, and I must say I did not put much faith in his statement until I saw an account of the ascent written by one who actually accomplished it. This shows that there is little or no exaggeration in Martin's description. It is not often attempted now. Two men were up it last year, but it is seven or eight years since the previous ascent. In former days, apparently, they used to climb it every year but always rather by way of a tour de force than in the ordinary course of business. It was looked upon as the test of a St Kildan hero.

At the time of writing, Richard Barrington is the only man not born in St Kilda who has been to the top of Stac na Biorrach. He says that he considers it the most dangerous climb he ever undertook, though he has been to the top of most of the big Alpine peaks. Of course, it is short and is free from the special dangers incidental to a long mountain excursion, but as a test of nerve and agility, it is not easy to find its equal.

The Ascent of Stack na Biorrach by Richard Manliffe Barrington

The chief topics of conversation on this out-of-the-way island are climbing and birds. A visit to Switzerland in 1882, during which the Schreckhorn, Finsteraarhorn, Jungfrau and Matterhorn, and an equal number of high passes were negotiated within 10 days, and the fact that a certain rivalry existed between myself and an elder brother who first ascended the Eiger, induced me to visit St Kilda in 1883, as I wished to test the ability of the natives as cragsmen, to compare them with Swiss guides, and to study the fauna and flora of this remote island, of which little was then known. It is 30 years since I ascended Stack na Biorrach, and therefore, I trust I shall not be accused of hasty self-advertisement; indeed, my chief object in writing is to give the members of the Alpine Club an account of a climb which the older writers have attempted to describe on second-hand information, and, moreover, I fear that even the St Kildans themselves will soon cease to ascend the rock, as they no longer subsist to the same extent on sea birds and there is not the same necessity for dangerous rock climbing.

Formerly, communication with the mainland was of rare occurrence. Since David McBrain's steamers began running there, from 30 to 40 years ago, intercourse with the outer world in summertime has been frequent, if uncertain. Fearing I might be left on the island all the winter, I arranged with McBrain to send a special steamer to take me on in September for the sum of £30. There was no necessity to take advantage of this arrangement, as his ordinary steamer took me off in good time. Not knowing Gaelic, I brought an interpreter with me from Glasgow, but, as he was afraid to go within 10 yards of any cliff and did not understand the St Kilda dialect, he was useless, save as caretaker of an old Crimean tent which we pitched on the only level patch (about 10 yards square) near the landing-place.

The natives could not speak a word of English, and it was nearly a fortnight before they permitted me to accompany them in catching fulmar petrels on the ledges along the face of the great Conachair; they wanted to test my ability. I remember one day walking along its edge and seeing a stout stick firmly embedded in the earth about three yards from the face, with a rope around it. I was sure someone was below catching birds, so descending about 100ft, I came upon another rope, also fastened around a stick, embedded in the next ledge. This I also descended and came to a second ledge on which two men roped together were busy catching birds with long fishing rods, to the end of which horse-hair nooses were attached. Having obtained permission to try my hand and being rewarded with success, the natives became very friendly. Of course, I had my boots off. If you don't take them off, it is done for you compulsorily. For, on another occasion, after landing on the island of Boreray and proceeding to climb without removing them, I felt myself pulled down from behind, one of the islanders grasping my arms and waist together while the other proceeded to unlace my boots. The ropes by which the men descended the Conachair cliff were of hemp and rather heavy, but the line between the two men on the ledge was made of horse hair and was light.

In 1883, there were no horses on St Kilda, but many cows. Macaulay said in 1764 that 'a rope is the most valuable implement that a man of substance can be possessed of in St Kilda. In his will, he makes it the very first article in favour of his eldest son… it was reckoned equal in value to the two best cows on the island'. The rope alluded to by Macaulay appears to have been made entirely of cowhide. I brought two Alpine Club ropes with me, the red central thread being regarded as a great curiosity by the natives. They would use neither of them. But they were very useful when attached to the top of the tent, preventing it from being blown into the sea on two very stormy days. After repeated entreaties, and when the natives had tested my ability in various ways, they consented to bring me to Stack na Biorrach. The wind was light, and the entire able-bodied population assisted in pushing the boat over the rocks into the sea. During this operation, a crowd of about 20 dogs barked furiously.

There were eight rowers, my nephew and myself in the boat. The natives are very religious, and a prayer was said before starting. We rowed around the Doon and under the tremendous cliffs of the western face of St Kilda, the great Atlantic swell making a white fringe along the rocks and booming in the great caves. In about an hour's time, we came to a narrow sound between the island of Soay and the large island, and the

Above: Richard Manliffe Barrington (1849–1915) was an Irish naturalist and ornithologist. A member of the Alpine Club, he ascended many peaks in the Alps, including the Eiger in 1876. He also had a penchant for travelling to remote islands off Ireland and Scotland, including an expedition to Rockall in 1886. Photo from *The Irish Naturalist* Vol 1, 1892. Public Domain. Below: Stac na Biorrach. Photo Marc Calhoun.

"

boat unexpectedly stopped before a perpendicular and, in some places, overhanging stack which looked to me absolutely inaccessible. The men talked in Gaelic, not a word of which I understood. One of them put a horse-hair rope around his waist. I could not imagine what they intended to do. For to ascend the rock immediately opposite appeared an utter impossibility, and my heart sank within me when they shouted: 'Stack na Biorrach, Stack na Biorrach!'

Donald McDonald, the man with the horse-hair rope around his waist, stood in the bow of the boat. Another man held the rope slack, and, watching his opportunity as the boat rose on the top of a swell, McDonald jumped on a small ledge of slimy seaweed below the high water mark. There was a momentary stagger, but he kept his balance and fastened himself to the rock by holding on apparently to the barnacles with which it was covered. He then proceeded upwards by sticking his fingers and toes into small wind-worn cavities on the western face. The rope was gradually slackened, and at a height of about 30ft, he turned to the east, getting on a small narrow ledge, unseen from below, which could not have been more than two or three inches wide.

The whole of this performance was remarkable, especially with regard to its surroundings: the steeple-like rock rising from the ocean off the very wildest part of this remote island; the boatmen shouting in Gaelic to the climber; the great surge of the Atlantic threatening every moment to drive us against the cliff, and the horse-hair rope alternately slack and tightened as the boat rose and fell. At a height of about 30 or 40ft, McDonald stood on a projecting knob, about two feet square, right over the boat. He hauled up another rope and fastened it around the knob. There were now two ropes to the boat. Donald McQueen tapped me on the shoulder and explained by signs that I was to ascend with boots, of course, being first taken off.

At that time, I could ascend a rope easily hand over hand; the swaying of it between the boat and the cliff made it less perpendicular at intervals and, therefore, easier. I recollect every incident as if it only happened yesterday; I soon stood on the knob beside McDonald. He pressed me against the face of the cliff, and, to my horror, Donald McQueen now proceeded to ascend the rope. For the life of me, I did not know where he was going to stand, and to this day, I am puzzled to know how we three men contrived to stand on this projection. Fearing every moment that I would fall, I shouted to pull the boat from the rock so that in case of accident I would drop into the sea and not into the boat from a distance of about 40ft.

McQueen now put the rope around his waist and took the lead up a ledge two feet wide, wet with spray, which sloped at a very steep angle upwards. Having ascended this, he grasped a narrow horizontal ledge about four inches wide and sloping outwards so that the fingers slipped readily, and, with his feet dangling in the air, proceeded to jerk himself along this ledge by getting a fresh hold every time with each hand alternately. It was about 15ft long. McDonald held the horse-hair rope, which was around McQueen's waist, in his hand. This, no doubt, gave him a false sense of security, but otherwise was absolutely useless, for had McQueen fallen they would have both tumbled into the sea.

McQueen now stood on another projection of a more satisfactory character than the first, about 70ft over the sea, and beckoned me to follow him. The horse-hair rope was placed around my waist. With McQueen on one side and McDonald on the other, holding the rope, I proceeded along the ledge, dangling without any foothold. Had it not been slippery with the droppings of guillemots, I might have succeeded. But midway, I slipped and, unable to recover my grip, would have fallen had not the two men simultaneously tightened the horse-hair rope with a powerful jerk, raising me a foot, during which I caught sight of a small lump sticking up and, grasping this anxiously with one hand, was soon safely landed by McQueen at his end of the ledge. Whether this slight projection, which really makes this traverse possible, was 'the thumb' referred to by Martin more than 215 years ago, I cannot say.

McDonald now came along with apparent ease, and we all stood together for the second time. There was more room here, but the cliff above was overhanging, and I was curious to see what would happen next. The rope was unloosed from everybody, and one of the men made a lasso of it and proceeded to throw it around a projection about 14ft overhead. After five or six failures, it was successfully lassoed, and the rope tested by vigorous pulls to see whether it would give way. Having satisfied themselves that it was secure, McQueen ascended. I followed, and then McDonald, all hand over hand.

We were now about 80ft above the water, and as the stack was no longer perpendicular or overhanging, I shall not give minute details of the remainder of the climb, which was not more difficult than many first-class clubmen could contend with. It was interesting, however, and the view from the top was very fine. Hundreds, almost thousands, of guillemots scurried and fluttered or flew into the ocean below. The top was not flat, like the pinnacles on the Farne Islands, but weather-worn and uneven. My thoughts were not, I fear, ornithological but rather concentrated on the problem: How shall I ever get down? However, the descent was accomplished with less assistance than the ascent, and I caught 'the thumb' this time. The boatmen exclaimed: 'Sauna!' which, being interpreted to me, signified that I was a great climber, like a famous St Kildan of that name.

My last visit to St Kilda, in 1896, was very brief and was made when returning from an expedition to the still more remote island of Rockall, 170 miles further west in the Atlantic; nobody has, I believe, been able to land on this island for over half a century. I saw Donald McDonald, then looking very poorly, and I believe Donald McQueen was dead, for I could not find him.

Above: The northeast face of Conachair. Photographed soon after dawn in winter by John Cleare in 1970.

The Edge of the World

The highest sea cliff in the British Isles, Conachair on Hirta, was reconnoitred for climbing possibilities by Tom Patey, Joe Brown, Pete Crew and John Cleare in January 1970. Their intention was to stay for three days, but because of the weather and mountainous seas, they were marooned on the island for nearly three weeks. They believed the rock to be gabbro, and the party had therefore been hopeful of finding solid rock. On closer inspection, they found it to be composed of loose gneiss and basalt and concluded that a massive effort involving artificial climbing would be necessary, compounded by the logistics of the approach.

John Cleare: 'On St Kilda the army was very good to us – but that's the BBC, of course. We sailed out on an army minesweeper, landed at midnight on an open beach, and halfway through unloading (much of it food rations), the skipper had to up-anchor and run for shelter to Mallaig; our marooning commenced. The resident CO was a Gunner Lt – as was I once – and Tom [Patey] was, of course, an ex-Commando Surgeon-Commander RN, so we were persona grata. Meanwhile, John Peacock was CO of the Hebrides Rocket Range at Benbecula, and he'd been an actor in the BBC's *Last Blue Mountain* film that I made. Thus, we got on very well with friends in court. The total height of Conachair is a shade under 1,400ft (the BBC figure was 1,397ft), and the top 100ft or so is vertical grass – at Tom's instigation we took crampons and used them on the grass. St John Head on Hoy is 1,250ft, and Conachair is 100ft higher, but the swell is something else!'

Conachair was finally climbed by Pete Whillace and Ian McMullen in September 1987, naming their, probably unrepeated, E6 route The Edge of the World.

Other modern rock climbs have been climbed, but the St Kilda stacks have seldom been climbed by modern climbers, so it's difficult to get an idea of how hard the passages climbed by the St Kildan pioneers were. According to UKClimbing.co.uk, Stac Lee's Original Route is Severe and Stac an Armin is Moderate.

Hamish MacInnes wrote in his 1997 'High Stacks' article in *The Herald Scotland* about Stac na Biorrach: 'This route is graded Severe in modern standards.' This grade was confirmed in 2023 by Robbie Phillips and friends who sailed to St Kilda with their sights set on climbing a new route on an unclimbed 200m-high overhanging crag on Soay. They also climbed the legendary 'thumb' route on Stac na Biorrach.

The Thumb by Robbie Phillips

In 2023 we are armed with such technological advancements that the St Kildans could only dream of: lightweight dynamic ropes, a full rack of specialised protection, and a powerful engine strapped to a thick-skinned rubber dinghy. But, still, this was one of the boldest moments of my life, leaping onto that rock and committing wholeheartedly to this remote pillar of volcanic rock.

I followed the line of least resistance, aiming for the currently empty birds' nests, as this is what the St Kildans would have done. It all came together like the final pieces of a jigsaw puzzle or more like solving a century-old conundrum. The climbing got steeper as we traversed around the pillar, and I found myself imagining being in the shoes of a St Kildan from centuries before (metaphorically, of course, as they climbed barefoot).

It wasn't easy – they'd have to have honed their skills before attempting this, but I could certainly imagine it. Then came a pretty bizarre move involving a tricky rockover onto a mushroom-like platform from which you lost all handholds and had to thumb your way to the next hold. This was it: the thumb!

If you did not possess climbing skills, to know how to move on rock, you would fall off. The moves were thin and technically interesting – not the kind of thing you could do as a beginner. The modern grade felt about Severe.

15m above the waves, I reached out precariously onto another ledge and scrambled up the thickly covered guano of countless millennia to our first belay. The final couple of pitches eased up, and we found ourselves standing on top of Stac na Biorrach.

Sailing to St Kilda, searching for, finding and subsequently climbing 'the thumb' with an amazing group of friends is a memory I will cherish for the rest of my life. Jumping from a boat in a strong Atlantic swell onto a remote Scottish sea stack, sharp wet barnacles biting into my skin, then repeating those hallowed movements so revered they named the climb for it; for a brief moment in time, I felt that I, too, was St Kildan.

The Great Stack of Handa

On Ederachillis's shore

The grey wolf lies in wait.

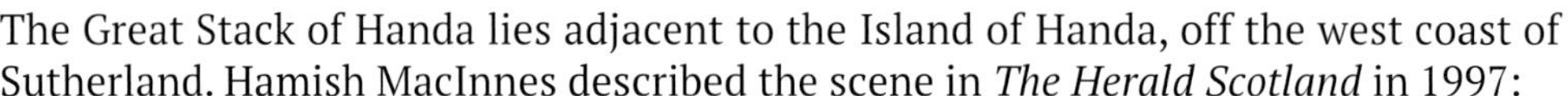

The Book of Highland Minstrelsy, 1846

The Great Stack of Handa lies adjacent to the Island of Handa, off the west coast of Sutherland. Hamish MacInnes described the scene in *The Herald Scotland* in 1997:

'As well as being host to an overcrowded bird colony, it is situated in one of the most scenic corners of the world. The stack has been hewn by the ocean out of the island's bedrock of rose-red sandstone leaving a narrow passageway, an alleyway, 115m deep which in the breeding season every available ledge is occupied, and it resembles a formally dressed capacity audience in La Scala. The stack is a fascinating lump of rock sitting in its alcove like a fat lady on a chamber pot. It actually rests on five stunted columns which can be seen at low tide. In fact, one Alistair Munro, a fisherman from Tarbet, passed beneath the stack at low tide in his boat. It was Alistair who took us to the Great Stack in August 1969 when, with my friends Graeme Hunter and Douglas Lang, we made the first ascent.

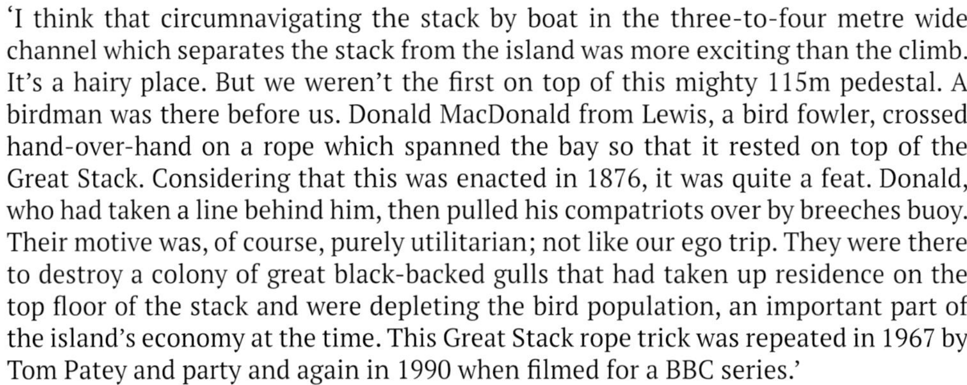

'I think that circumnavigating the stack by boat in the three-to-four metre wide channel which separates the stack from the island was more exciting than the climb. It's a hairy place. But we weren't the first on top of this mighty 115m pedestal. A birdman was there before us. Donald MacDonald from Lewis, a bird fowler, crossed hand-over-hand on a rope which spanned the bay so that it rested on top of the Great Stack. Considering that this was enacted in 1876, it was quite a feat. Donald, who had taken a line behind him, then pulled his compatriots over by breeches buoy. Their motive was, of course, purely utilitarian; not like our ego trip. They were there to destroy a colony of great black-backed gulls that had taken up residence on the top floor of the stack and were depleting the bird population, an important part of the island's economy at the time. This Great Stack rope trick was repeated in 1967 by Tom Patey and party and again in 1990 when filmed for a BBC series.'

First Crossing by WH Murray

In July 1967, Tom Patey (aided by Chris Bonington and Ian McNaught-Davies) made the second crossing by rope to the top of the Great Stack of Handa. He wrote to me early in 1968, saying that he'd heard that Donald MacDonald of Lewis, whose father had made the first crossing in the 1870s, was alive, had recently visited Handa to view the scene of his father's famous exploit, and now lived at Dunoon (having been headmaster of the grammar school). Would I please visit Donald and ask if he remembered any details of the first crossing? Tom still felt shaken by his own 120ft traverse across the 350ft-deep gulf. The big sag in the rope had given him a tough fight to get any footing on the stack. He could scarcely believe that the first crossing had been made without aids, for he could not have done that himself. So what aids had been used? My visit to MacDonald was fruitful, and my resulting letter to Tom went as follows:

Loch Goit, 26th March 1968

Dear Tom,

I have at last seen Donald MacDonald. I called on him yesterday at Dunoon. His wife died three years ago, just after his visit to Handa, and he now lives alone. One son is a doctor in Fife, and another I forget where. Donald, in his mid-80s, is fresh

Left top: Hamish MacInnes, Graeme Hunter and Doug Lang on top of the Great Stack of Handa. Photo Graeme Hunter. Left middle: Alistair Munro dispensing a wee dram to celebrate the ascent and his birthday. Photo Hamish MacInnes. Left bottom: Graeme Hunter on the first pitch. Photo Hamish MacInnes. Facing page: The Great Stack of Handa by Graeme Hunter.

complexioned and fit, with all his faculties except for a slight deafness. He's not frail-looking like Tom Longstaff at the end and is equally alert in mind. He remembers the whole story of his father's crossing to the Great Stack, which had been told to him many times by his father and friends. He can't remember the month, which would be prior to September. The year was 1876, when his father was 26. His father's name was Donald MacDonald, fisherman and crofter at Ness near the Butt of Lewis, of mixed Norse and Gaelic stock – like all the other families around, fair-haired and blue-eyed.

Donald, the son, still has strong impressions of how hard his father and all these men worked, out on the sea in all weather, line fishing, yet still managing to wring a good product from the croft; enough to put sons through universities before the welfare state had been thought of. He remarked that the mainlanders' idea of the islesmen as idlers is quite wrong, the reverse of the truth, and that few mainland farmers or crofters could begin to cope with the work these men did. But they might appear idle if seen after a voyage when they'd stand around for a brief spell with hands in their pockets relaxing. Their one outdoor recreation was rock climbing.

His father, in his 20s and 30s, was in constant practice on rock, very strong in the arm, and supremely confident in his physical fitness. He and the other men of Ness learned their rock climbing initially in hunting the birds on the sea cliffs of Ness and nearly 50 miles north on Sula Sgeir and North Rona, which islands they never referred to by name but always (in Gaelic) as 'The Lands Out Yonder'. This kind of naming seems to be a typical Norse idiom, just as they originally referred to the Scottish Western Isles as 'Havbredey', which means in Norse 'The Isles on the Edge of the Sea'. Hence, 'Ebudae' picked up in Lewis by the Roman navigators of 129 AD and rendered thus in Latin by Ptolemy.

Donald, the son, says that the men of Ness did not merely cull seabirds for the pot. They liked birds in the same way as modern bird watchers. They observed and studied them and could tell you as much about the life of seabirds in detail as any amateur ornithologist today. Hunting the birds for food was another matter which, of course, led to their rock-work, but they did also have a genuine love of birds for their own sake as they did, too, of rock. His father would often be working out on the croft – and when the main work was done, he would suddenly disappear – he'd be off to the sea cliffs just to enjoy the climbing. All his spare time went to the rocks.

The best cliffs near Ness were those down the east coast to Tolsta. On the west coast, from the Butt to Europie, the cliffs were lower. The men spent much more time than they do today in culling the birds from The Lands Out Yonder. Now, they spend only a day or two. Last century, they'd spend at least a fortnight every September taking the gugas (young gannets) as they 'ripened', salting them in the barrel as they took them, so that they were nearly all cured by the time the boats returned to Ness. They tasted, said Donald, not unlike good kippers. The birds were exported worldwide.

The young men rejoiced in these expeditions, hard living but a wonderful change from the croft-work and deep-sea fishing. They went wild with delight and would naturally throw off some feats of daring that would make your hair curl. One of these was the Handa epic.

This was not organised by Donald MacDonald but by his neighbour, Malcolm McDonald, (no relation). Malcolm, then in his 50s, was a natural leader, full of resource and bright ideas. He planned the visit to the Great Stack of Handa and decided how it would be done. Donald was the man chosen (or maybe the volunteer) to make the first crossing; he had the needed nerve and strength.

The stack had already been reconnoitred with a full intention of climbing it from the sea upward. But no line had seemed practicable. The idea of nailing rocks had been thought of and been not acceptable. Their method of crossing was in-general-outline the same as yours. A long rope – 5 to 600ft – was carried end-outwards to the farther points of Handa until its centre crossed the top of the stack. The ends were then secured to stakes. It could be they used the same boulder as you at the northeast end. Donald then crossed from the west side hand over fist, bare feet curled around the rope. He used no waist-loop, footloop, or safety line at the waist. He carried nothing at all. The fixed rope was a thick fishing rope normally used for securing the deep-sea fishing lines to buoys – he gave it a Gaelic name that I can't remember.

There was a tremendous sag in the middle and worse under the stack. Donald had a hard job at that last bit. This was the only time when he thought he might fail. It was made especially difficult because he was unable to make his landing where he had hoped. There was a point where the stack sloped abruptly down toward the gulf, and the rope had slipped while he was crossing until it hung over this sloping ground, where the steep, loose rock gave no firm hold for the foot when he tried to make lodgement. He was now fighting for his life and tapped the reserve of strength needed to pull himself up. When he was rested, Malcolm threw him a line, by which he pulled across two stakes, a block-and-tackle, a breeches buoy, and baskets. Donald fixed the stakes, attached the tackle, and his two companions crossed by breeches buoy. They culled the sea cliff birds and filled the baskets. All were then able to return to Handa by breeches buoy, leaving the stakes behind on the stack, where they could be seen during the next 70 years.

This ploy became one of the wonders of Ness for many a year, but no one thought that the islemens' high-spirited play would interest mainlanders. So that's the story. There are lots of details one would like to know more about. For example, how did Donald hammer in his stakes? And what was the thickness of the fixed rope? Given a very thick and taut rope, crossing hand over fist is not in itself difficult for a skilled man, as we both know. In western Nepal, hillmen habitually use this method for crossing rivers much wider than the Handa gulf. But, their grass ropes are one-and-a-half-inches in diameter, stretched so tight that little sag comes in the middle and lead to easy landings. The rivers may be killers if one falls in, but with nothing like that fearsome drop at Handa.

Eight years later, Malcolm McDonald, the prime mover, quarrelled with his Presbyterian minister at Ness. Rather than submit to his rule of the parish, Malcolm chose self-exile to North Rona. A fellow crofter, McKay, went with him. The island had lain uninhabited for 44 years. They arrived early in the summer of 1884 and occupied a ruined house. Their friends at Ness, feeling uneasy about them after a stormy winter, sent out a boat in April 1885, when Malcolm and his friend were found indoors, dead of exposure. I hope to see you at Easter.

Yours, Bill Murray.

My report to Tom is of interest to climbers on two counts. First, it discloses a double error in our club's guidebook, *The Islands of Scotland* (1989), which ascribes the crossing to 'inhabitants of Handa' thus ignoring that Handa had been cleared in 1848, and also quoting John Alexander Harvie Brown's *Fauna of the North-West Highlands and Skye* (1904), which, itself erroneous throughout on the Handa passage, gave the credit to 'two men and a boy from Uist, at the request of the late Mr Evander McIver, because the great black-backed gulls were becoming destructive and were increasing in numbers'. In fact, it was done by three men of Ness in Lewis on their own initiative. Second, and more importantly, it revealed a more advanced state of climbing ability in the Outer Hebrides than our club's members had realised. Even Patey, comparing his own hair-raising traverse with that of last century, wrote in the *SMCJ*: 'It is even more certain that no mountaineering amateur of that era would have committed himself to such an undertaking.'

He might be right. But, his words draw notice to a moral ripe for plucking – and for slow digestion. Climbers of every period share a common frailty (I, too, being guilty in past days): we imagine that our fellows of an earlier age could not have been as bold and skilful as we. From one century to the next, our self-flattering hearts offer up that much-beloved toast: 'Here's tae us! Wha's like us? Nary a one and they're a' deid.'

Second Crossing

The second recorded crossing to the stack was made by Tom Patey, Chris Bonnington, and Ian McNaught-Davis on 1st July 1967. They used a similar method to the original crossing, however their attempt to climb the stack from the sea up was unsuccessful. Patey commented in the *SMCJ* (1970):

'The Stack is 350 ft high, cut vertically on every side, and lies on an inlet on the precipitous north side of the island of Handa. The enclosing walls are equally vertical or overhanging, and the gap between the cliff top and the flat summit of the Stac is 50ft at the narrowest point. We stretched 600ft of linked nylon ropes from side to side of the inlet so that the centre of our rope lay across the top of the stack. To use the best anchorage points, we had to cross the gap obliquely, lengthening the Tyrolean traverse to 120ft. Without the aid of jumars, it would have been difficult to overcome the considerable sag in the rope. Additional excitement was provided by numerous sea birds cannoning into the taut nylon and by two guillemots on the stack who started pecking the rope which had invaded their territory. The appalling 350ft chasm underneath made this a most impressive occasion.

'I found no traces of the earlier visit although local reports suggest that the posts which were fixed on top of the stack by the original visitors, to provide a permanent rope anchorage, could still be seen up to 20 years ago. The pioneer traverse was certainly a remarkable 19th-century exploit, perhaps more so than the feats of the St Kilda cragsmen. It is even more certain that no mountaineering amateur of that era would have committed himself to such an undertaking and possibly, despite modern equipment, many climbers will still find the traverse a thought-provoking experience. The son of the youthful pioneer now lives at Dunoon, and he recently made a pilgrimage to Handa in order to view the scene of his father's triumph.'

***

On 1st August 1969, the stack was climbed for the first time in a conventional manner from the sea by Graeme Hunter, Hamish MacInnes, and Doug Lang directly out of a boat supported by Alistair Munro of Tarbet.

Facing page: Dr Sue Richards making the first full Tyrolean traverse in 1990. Photo Graeme Hunter.

The Great Stack of Handa by Graeme Hunter

More people have walked on the Moon than on the summit of the Great Stack of Handa.

Handa warden

1969 was an eventful year: the first manned Moon landing, Woodstock, Robin Knox-Johnston sailing non-stop around the world. Harold Wilson was in Number 10 and – on the first of August – the Great Stack of Handa was climbed for the first time. Planning this adventure were Hamish MacInnes, Doug Lang and myself. We needed a boat and an expert boatman to land us on the stack, which is about three miles from the small port of Tarbet, near Scourie, and is subject to dangerous currents and swells. With this in mind, we stopped off at Rhiconich to have a word with our friend, the local policeman, to see if he could recommend someone. The two fishermen who had taken Tom Patey and Chris Bonington on their earlier attempt on the stack had drowned the previous winter.

'Alistair Munro's your man, without doubt,' he said, 'what he doesn't know about the sea and Handa is not worth knowing.' So we phoned him at Tarbet (no smartphones in those days), and he willingly agreed to help us. As well as being the warden of Handa, he operated (with his sons) boat trips for tourists and ornithologists. We met Alistair coming up the jetty from his boat: a white-haired, mahogany-faced man with a cheerful grin. Nothing was too much trouble for him. Of course, he would take us to see the stack.

'Would you like to see the Great Stack tonight?' It was then 9:30pm. We hurriedly agreed, and within a few minutes, we were speeding into the sunset, aiming for the north end of Handa, skirting the great sea cliffs. The Great Stack loomed into sight: a great wall of sandstone echoing with the cries of thousands of seabirds. The prospects looked grim for an ascent on the east face, which did have a good landing platform. Alistair took us slowly round through the narrow channel between the stack and the island where the swell tossed us up and down. In such conditions this passage is more frightening than any climbing.

The only possibility that we could see was a route up the northeast face, which meant actually climbing directly from the boat onto steep and quite difficult rock covered in green slime and barnacles. This looked as if it led to a ledge about 50ft up, below the overhangs. The next morning, the first of August, dawned fine and bright for our attempt. We roped up on the jetty – a somewhat unusual proceeding which provided much amazement for the gathered tourists.

Having circumnavigated the stack the previous day with Alistair Munro, we decided not to start the climb on the most obvious landing stage on the northeast face. This had been tried by a strong party previously and was unsuccessful. Instead, we chose a less obvious vertical wall on the entrance to the cave leading under the stack. Both Hamish and Doug quickly made it clear that they were non-swimmers, and therefore handed the sharp end of the rope to me, plus all the heavy ironmongery ballast, etc.

It was high tide with only a gentle three-foot swell as Alistair, manoeuvered the stern of the boat next to the wall. As we bobbed up and down, I worked out the sequence of holds to break loose from the security of the boat. When the boat reached the high point of the swell, I stepped off onto the stack and climbed up to drier rock where I quickly placed a piton for assurance realising that if I slipped at this stage, I'd go straight to the bottom.

Afterwards, Hamish recalled that he had a vision of his old friend John Cunningham, then probably Scotland's best rock climber, who met his end on a sea cliff. He, too, had a full complement of ironmongery, which ensured he drowned when he went into the water; John couldn't swim.[2]

2   This story is recounted in *Creagh Dhu Climber* by Jeff Connor and *The White Cliff*.

Left top: Graeme Hunter on the diving board. Left middle: Doug Lang abseiling from the diving board directly into the boat. Left bottom: Angus Munro hooking Graeme Hunter with the boat hook. Photos Hamish MacInnes.

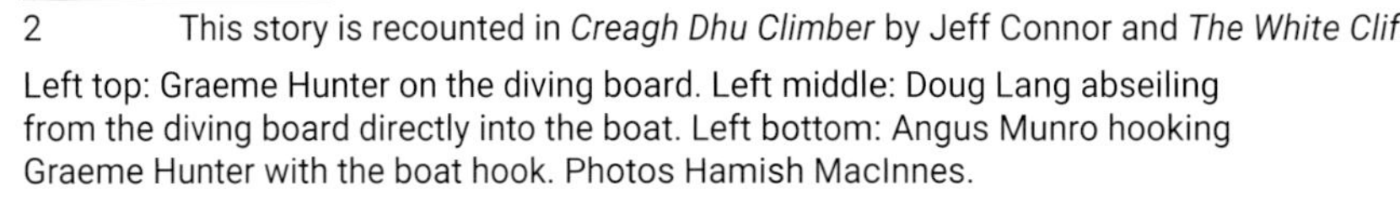

I reached a belay stance on a shoulder about 40ft above and assembled Hamish and Doug to tackle the real crux of the climb. Making room for Doug, Hamish climbed a short pitch to secure himself below the great overhangs above. I took the lead and climbed up to the overhangs, traversed left past a confused fulmar resting on a ledge above, which thankfully didn't spray me with its fishy stomach oil as they are known to do when threatened. Maybe he felt sorry for my predicament.

Anyway, the situation was now rather exposed, quite challenging and with little prospect of effective protection on the compact sandstone. I made progress up and left, through an overhang, to reach a small protruding ledge which was the most exhilaratingly exposed diving board protruding out 150ft above the waves. I secured a belay and brought Hamish up to enjoy the view. By this time, a small flotilla of boats and yachts had gathered to watch this spectacular event or just to witness our predicament.

Again, Hamish continued above on easier ground to allow Doug to join me on my minuscule ledge. Hamish secured himself 150ft above and shouted that it was only a short, easier pitch to the summit. Finally, we all stood on the summit on the island side to the total amazement and consternation of the people on the main island who had been admiring the spectacular view, until we appeared.

After a short time on the grassy summit, which was bird-free, we started to plan our exit from our comfortable situation, entertaining the crowds. Due to some loose, unstable rock we were unable to abseil back down to the ledge where we had left our rucksacks and gear. We had to return to the Diving Board with the prospect of a rather spectacular free abseil directly into the boat. By this time, the tide had gone out, and the swell was more like six feet. Hamish shouted down to Alistair, asking if he could manoeuver his boat to catch us on our descent.

At our end, I had a large off-cut piece of Hamish's mountain rescue stretcher, which I'd made into a wide nut. I jammed it into a suitably wide crack and prepared the abseil rope. Hamish then gently lowered himself into thin air to land on a wet, barnacle-encrusted ledge just above the high water mark. The next trick required Hamish to swing out and quickly slide down the rope when the boat was at the bottom of the six-foot swell, hoping to meet up and let go of the rope at the top of the swell and safely into the boat. This worked fine for Hamish and Doug, but I didn't quite synchronise and had to be winched in with a boat hook, just a little damp and missing skin on my hands.

Alistair had demonstrated exceptional boat handling, and now we could all relax on our short trip back to Tarbet. Alistair, who was fond of a wee tipple, produced a bottle of whisky and announced: 'That was grand, lads, a fine thing on my birthday.' So we all drank to Alistair, to the stack, and a wonderful day. We decided to call the route Munro's Birthday Route as he was such a vital member of the team.

Tragically, three years after our ascent, Alistair drowned at sea. His five sons also died at sea in various accidents, and they are all buried in Scourie Cemetery. This was a devastating tragedy for the whole family to die in such terrible circumstances, emphasising the dangers of this beautiful, rugged coastline.

Fifty years on, the Great Stack looked much the same when we made a family visit with the help of Roger Tebay, who operates the ferry to Handa. I was able to show him exactly where Munro's Birthday Route reached the summit. To his knowledge, it hasn't been climbed since. One of the visiting wardens told me: 'More people have walked on the Moon than on the summit of the Great Stack of Handa.'

Climbing the Great Stack is more of an adventure than simple rock gymnastics. The objective dangers of the sea are obvious, and the scant protection provided by the rounded blank cracks on the crux, with the basic equipment of the day, highlighted the adventure. Sadly, my two companions didn't make it for the 50th anniversary. Doug Lang was killed in an avalanche in Corrie Fee above Glen Doll in 2011, and Hamish, who would have loved to have been there in his 90th year, was unable to travel from his lair in Glencoe to the North West.

The Great Stack of Handa is, understandably, overshadowed in popularity by the Old Man of Hoy, which is some 69ft higher and a little more accessible. At 380ft, the Great Stack, unlike Hoy, is surrounded by turbulent seas, which only adds to the drama. An expert and willing boatman is an essential member of the team.

In 1990, a Tyrolean traverse across the entire bay from Handa Island via the Great Stack, around 140m, was completed in September 1990 by Dr Sue Richards from New Zealand who was killed shortly afterwards on Mount Aspiring in the Southern Alps of New Zealand. The rope was set up by Hamish MacInnes, Graeme Hunter and party. A televised re-enactment of the original traverse was performed by Dave MacLeod and Dave Cuthbertson for the show 'The First Great Climb' for BBC2 in October 2011. Their traverse, like the original, was hand-over-hand but with the (sensible) addition of a safety rope.

The First Great Climb by Dave MacLeod

I was to play the part of Donald MacDonald. My comrades in the recreation were Cubby and Donald King. We had a blast dangling about on fishing ropes above the big drop into the sea. Or should that be, we got blasted by the full wrath of the Atlantic squalls every 10 minutes for the best part of a week. I started off the week thinking that my costume of a tweed suit, bunnet and old leather boots was quite good outdoor attire. But by the end, I must say I couldn't wait to see the back of soggy tweed and soaking boots and get back into modern kit and ropes that weighed less than 40kg. The week was, for me, an old lesson re-learned: the generations gone by had less to work with but had no less boldness, courage, or ingenuity.

Sea Stacks by John Cleare

Go in and take your chance like any man or bird or fish.

Ernest Hemingway. *The Old Man and the Sea*

The last virgin summits in Britain are sea stacks, and for that matter they are probably Europe's last virgin summits, too. Our island has thousands of miles of unique and varied coastline, and around many stretches of it, sea stacks are scattered in profusion. Some, like the famous Old Man of Hoy, are high and well known; others, like Pyatt's Steeple, off the South Hams of Devon – so called by southwest sea cliff climbers after the editor of the *Alpine Journal* who first drew attention to it, thus differentiating it from another Steeple a mile further east are known only to a handful of specialist stack-explorers, and at the time of writing are still virgin. The coasts of Devon and Cornwall are well charted and populated, and yet if the frequency of discovery of new stacks on this coast is anything to go by, then there must be hundreds more awaiting discovery hidden off the more remote coasts of Scotland and the Northern Isles.

For several years I have been working on a guidebook to British sea stacks, incidentally a project which seems never-ending thanks to the new material which it continuously generates, and I was forced early on to consider a suitable definition of the feature. One progresses easily into a deep and philosophical argument along the lines of: when is a sea stack not a sea stack? When it's a skerry, say, or a clett? But eventually, I came up with four points to be satisfied.

Firstly: shape. A stack should be higher than it is wide in at least one of its planes. This covers shark-fin type stacks such as the highest stack in the British Isles, the 627ft high Stack an Armin in the St Kilda archipelago, which in one direction is pyramidical but seen end-on is a slender needle, and of course, further afield, Ball's Pyramid, 1,750ft high, of the Lord Howe group in the Pacific Ocean.

Secondly: height. Within reason, there is no minimum height, except that if it is high enough to be worth climbing, then it's a stack. This might preclude a shapeless, easy-angled skerry, perhaps 100ft high, but include a steep and interesting 40ft needle.

Thirdly: a stack should have been formed by the action of the sea. This rules out certain land pinnacles that happen to be near the coast but embraces all manner of interesting back-shore pinnacles which may – at the present geological time – be sea-washed only at spring-tide gales. The Old Man of large, slabby Hoy, after all, is not an island but stands on a plinth linked to the base of the mainland cliffs by a huge causeway of massive blocks.

The last qualification is really an amplification of the third: that a climber falling from a sea stack in at least one direction should land on sea territory, either the water or the beach if he is unlucky. This effectively rules out the products of ancient seas miles inland but ensures that features of great interest, such as Aleister Crowley's Etheldreda's Pinnacle, high on the chalky face of Beachy Head in Sussex, are included. It is possible that Etheldreda's has not been climbed since Crowley's original ascent in 1894; at least Tom Patey and I couldn't climb it in 1969. Perhaps it was climbed using occult aid? If so, then that alone makes it worthy of inclusion in any list of British stacks.

Facing page and left top: The Old Man of Hoy. Photos Grant Farquhar. Left middle: John Cleare and Tom Patey. Photo John Cleare. Left bottom: Max Dutson and Will Wykes with Max's girlfriend at the time. Photo Grant Farquhar.

The famous Duncansby stacks are supposed to have been ascended by drunken Norwegian sailors, but the accounts of climbers working at the Dounreay Atomic Energy plant are more reliable. There are vast unexplored cliffs on Whiten Head, involving several miles of walking from the nearest road. In a group of offshore stacks which must be reached by boat stands The Maiden. Here in 1970, Tom Patey fell to his death from a mishap when abseiling after making the first ascent. Patey was the most original explorer and climber on the Scottish scene for 20 years after the Second World War. When he started to explore the wild coasts of Sutherland, he was drawn at once to the virgin summits of these pinnacles which proliferate in the region. Individual enterprise is a necessary driving force in progressive mountaineering, and Tom Patey showed more individualism than most. His style was also coupled with an ability to mix and harmonise with the extremes of other movements and personalities in mountaineering. None more so than those flourishing south of the border. His companions were often English, and parties frequently travelled several hundred miles to join him in his exploits.

By focusing attention on the sea stack, Patey not only introduced a new kind of objective to mountaineering in Britain, he also revitalised the romantic space of the pastime which had been going stale for decades. It is true that those who followed in his steps tended to use brutal tactics to achieve their goal, but such techniques that may be considered proper for the ascent of sea stacks need not detract from the special pleasures felt by those drawn to sea cliff climbing. In their nature, sea stacks are inclined to pose unusual maritime approach problems and give hard climbing on suspect rock. Because they are different, so the scope of sea cliff climbing is increased, and this becomes the great attraction. Nothing is further removed from conventional mountaineering.

To the South of Cape Wrath, down the west coast of Sutherland, lies the classical domain of sea stacks where Patey pioneered the way. At the southern edge of the remote Sandwood Bay, Am Buachaille stands 220ft high and is one of his finest discoveries. Two alloy ladders lashed together, and carried over the hills for four miles, were used to bridge the swirling channel between the rock beach and the platform on which the stack rises. Patey used ladders for approaching several sea stacks because he was not a swimmer.

Although, without doubt, the Old Man of Hoy is one of the classic climbs of Britain, there are many stacks which, though much smaller, have a similar ambience and offer as exciting a climbing experience. I find that stacks provide something that mere cliffs do not – a feeling that is only achieved on a true summit – a feeling that is only shared by stack-climbers and mountaineers. Not for nothing have stacks sometimes been dubbed 'sea-mounts'.

After a certain amount of preparation on the lower pitches for the final climb, the first ascent of the Old Man of Hoy took place on 18th July 1966. The party was Tom Patey, Rusty Baillie and Chris Bonington. The route went more or less up the east face – the landward side – and the climbers gave it an artificial grading of A3. In his research, Patey uncovered the fact that 300 years ago, the Old Man did not exist. It was merely a narrow, gigantic promontory jutting from the straight line of cliffs along the coast. About 150 years ago, large portions of this had collapsed in rockfalls leaving a great arch linking the outer end with the main cliff. Finally, about a century ago, the remains of the arch crumbled and left a monolith intact on the beach. The poor quality of the rock in the first two-thirds of its height on the landward side is attributed to this recent formation. From this, it might be concluded that further modification to its shape could occur before the end of the present century.

Following the first ascent, Tom Patey was solely responsible for persuading the BBC to film an outside television broadcast of another ascent. This programme was mounted on a grand scale, and the whole effort seemed to come from Tom's enthusiasm. Apart from climbing teams that would be featured in action in front of the cameras and drawn from the cream of British rock climbing talent, even technical assistants for the assault, load carriers, called 'sherpas', and most of the persons helping but unseen on the television screen were mountaineers of repute. So it was that the most powerful publicity medium known to man brought the news of climbing on Hoy directly into the homes of the public. It is perhaps no accident that very little information had been circulated about the first ascent before the TV programme was put out in 1967.

North of the Old Man, the coastal cliff on Hoy reaches its greatest height at St John's Head, about 1,140ft above the sea, and just about the biggest in Britain. Tom Patey had picked out a line on it. His idea was to sail down the coast in a rubber dinghy and land on the rocky boulders below the cliff. Unfortunately, he did not live to carry out his plan.

The west and north coasts of the mainland, immediately north of Hoy, present a similar but lower rugged coastline. Here, in 1967, Joe Brown and others climbed the Castle of Yesnaby, aided by members of the BBC television team from the Hoy saga. In 1970, another party took North Gaulton Castle in hand, using elaborate rope techniques similar to Patey's method for reaching the top of the Great Stack of Handa. A month later, Stack o' Roo yielded to a swimming approach in the best sea cliff tradition. In August 1970, the important stack known as Standard Rock off Costa Head fell to a pair of young men called Roberts (no relation to each other) from Caernarfon. It is worth recording the achievements of their visit to the Orkneys. They first climbed Stackabank on South Ronaldsay; a loose, serious but not very tall stack linked to the mainland at low tide. The following day, they crossed this small island to climb the easier Clett of Crura stack. Two days later, the party made the second ascent of Stack o' Roo. The next day, it was the turn of the fine shark-fin-shaped stack called Standard Rock, mentioned above. Reaching it involved a difficult abseil followed by a swim of nearly 75 yards to the pinnacle.

The Old Man of Stoer by Tom Patey

Faith, quo Johnnie, I had sic fegs,

Wi' their claymores and their philabegs,

If I face them again Deil brak ma legs,

So I wish you a good morning.

'Hey Johnnie Cope'. Traditional

'Can we borrow your ladders again, George?'

My friend, the hotel proprietor, agreed without batting an eyelid. Years in the catering trade had taught him the customer is always right, however irrational the request. 'I hope they bring you better luck this time, lads,' he remarked affably. 'Useful piece of equipment, a ladder. I've often wondered why they never used them on Mount Everest.'

'It would have been too much of a novelty,' I suggested, glossing over the horizontal ladder in the ice fall on Mt Blanc and, of course, étriers. Ladders stick up, étriers hang down, the principle's the same. However, we lashed the ladders to the car without prolonging the discussion. George was not alone in misinterpreting the role of the ladder in the plan of attack. Our sporadic visits had become a source of speculation to the crofters of the Stoer peninsula. It's not every day that peat-cutting is agreeably interrupted by the passage of four wild-eyed men carrying ladders over the moors towards Stoer Point. They formed their own conclusions.

'Climbers, that's what they are, with climbing tackle. And them thinking to climb the Old Man of Stoer with a wee bit of a ladder that won't reach a quarter of the way up the side of it. They will be asking for a 200ft ladder next. And they will not be finding it here; no, nor in Lochinver either.'

The peat diggers were not in the least surprised when the first two attempts failed, and we returned like Sir John Cope, bearing the news of our own defeat, sagging under the yoke of two ladders, twin symbols of our incompetence. Tales of intractable tides and gale-force winds carried no conviction; they had not even crossed the hill to watch. Once a West Highlander forms an opinion, he stays beside it. Today, there was only a solitary digger at the peats. 'Back again,' I shouted enthusiastically, 'Third time lucky.' He nodded wisely and continued his digging.

Brian Robertson muttered into his embryo beard: 'There's one thing I've been noticing about you,' he grumbled, 'everything we do, it turns out to be an epic. Even the simplest things. Take yesterday, for instance…' We had gone for an hour's run in the car and had not returned until long after dark. This was because we had encountered a gate that was usually unlatched but, on this occasion, was firmly padlocked. To save time and energy, I had tried to nudge it open and shattered the car's headlights…

'That was sheer misfortune,' I claimed, 'today we've left nothing to chance.'

'You can say that again,' interposed Paul Nunn, half-throttled by the rungs of his ladder. 'How often do you do this sort of thing? I've been on some daft games in my time, but this is the daftest. We carried less gear in the Dolomites.' Paul Nunn and Brian 'Killer' Henderson from Buxton and Tyneside, respectively, were two leading English tigers whom Robertson had ensnared on the Etive slabs and persuaded to come with us: 'Wait till you meet Killer,' he had promised me. 'The man's a born craftsman. He makes all his pitons out of scrap metal.'

Killer lived up to his reputation. 'How about this, then?' he said, producing from voluminous pockets yet another sample of his genius. 'I reckon this little job'll come in handy today. You've probably never seen one before, have you?'

I had to confess I had not. I could not even guess its function. 'Nice little job,' he repeated, returning it to his pocket. 'A nice clean finish, too.'

'How much?' I asked.

'Some things are beyond price,' said Killer.

The Stoer peninsula is a huge golf course with short, cropped grass; not a tree for miles. From the little hill above the Old Man you can see all the Northwest peaks spread out like inverted flowerpots. Amid this pastoral setting, the Old Man of Stoer rears up with startling suddenness.

'My God!' said Killer (or words to that effect). His reactions were understandable. Well-thumbed Northumbrian breastworks pale before a 200-foot virgin pinnacle. 'It's like a kind of a symbol,' he mused, 'what a size…' He enlarged on the idea at some length, with apt recourse to Greek mythology.

'How do we cross the water below it?' interrupted Nunn, who had been regarding the problem from a less aesthetic standpoint.

'That's the purpose of the ladders,' I explained. 'Lash them together, and you've got a 30ft span.'

Manhandling the ladders down the mainland cliff was quite difficult. To save time, we scrambled down a 200ft fixed rope with the ladders balanced awkwardly round our necks. The penalty for a slip was instant decapitation. It was going to be a race with the tide. Despite consulting the West Coast Pilot and other official sources, it seemed I'd made yet another miscalculation. In less than an hour, the reef on the far side of the channel would be submerged, leaving us without a landing place for the ladder.

I was secretly worried about the ladders. I had a shrewd suspicion that as soon as the first man reached the rope lashing, that secured the junction, the two ends would fold up like a clasp knife, bearing the unfortunate pioneer straight down to the sea bed. Then, I had a better idea. 'Do you see that rock in the middle? If I can reach that with a single ladder, I can pull the other after me and use it to cross the next gap.' Once committed, I wasted few minutes. By the time I had clambered onto the reef, both launching pads were awash. The 15ft wall on the edge of the reef presented a tricky problem for bare feet until I remembered I happened to have a 15ft ladder with me. Elementary, indeed.

Earlier on, there had been talk of swimming across. Warm sun and placid sea beckoned, but the three landlocked climbers resisted the temptation. A seal moved in for a closer inspection. This mammal is harmless, but of playful disposition; his attentions can be quite embarrassing. We seemed to have reached an impasse. Robertson was in an uncooperative mood. He was not prepared to get wet. He was not even prepared to get his rope wet. Killer was reluctant to surrender one of his artistic specialities to anchor the Tyrolean

rope, for he had guessed (correctly) that it would be left behind. I was obviously in a hopeless position to bargain, so my rope and pitons were used. It was, therefore, consoling to watch Robertson getting wet. He took every conceivable precaution but forgot to allow for the weight of ironmongery and the elasticity of nylon rope. By the time he had slid down to the halfway mark, his rucksack was submerged, and the water lapped about his ears. 'Pull, man!' he gasped. I responded energetically and landed him in a flurry of spray.

'The tide seems to be coming in much faster,' said Nunn doubtfully. 'I think we'll stick around here for a while. You can chuck us a traversing rope once you're up the first pitch.'

'What happens if we don't get up the first pitch?'

'That possibility has crossed my mind,' he admitted. At a guess, I suspect we'd have difficulty keeping afloat. But you can't grumble: few men are given such a powerful incentive to succeed.

This situation was tailor-made for Robertson's ideology. He recalled a passage from a favourite author – To go on was impossible, to retreat unthinkable – 'This is why we must go on!' he cried with Teutonic frenzy.

The first pitch resembled one of those gritstone problems where many fall and few prevail. The waves were breaking at Robertson's feet, and the seal began to bob up and down in excited anticipation of two new playmates. In these situations, I usually climb well. The landward face of the Old Man is plumb vertical for the first 100ft. I use this overworked term deliberately. Any sharp intake of breath would displace the climber from his holds, and there is no promise of a respite for a very long way.

I had only gained 12ft when I found myself edging leftwards by two horizontal cracks ideally spread for the hands and feet but separated by slightly bulging sandstone. A soft-shoe-shuffle took me to the corner of the face and out of sight of the audience; I stepped into warm sunlight and a hanging gallery of jug handles. Barely a minute later, I reappeared on the first ledge 60ft up. Robertson, with his usual bravura, tried to bludgeon a way straight up the wall, only succumbing after a struggle to the persistent lateral tension of the top rope. This enforced detour tore him from his principles, as he now reminded me: 'Let a seagull defaecate from the topmost pinnacle, and that is the route I must follow,' he solemnly announced, echoing the sentiments of his hero Comici – first Apostle of the Direttissima Doctrine.

Fortunes had been reversed, and now Nunn and Henderson were faced with the unpleasant prospect of ascending a 120ft Tyrolean rope to join the assault. Even with our height advantage, we found it difficult to throw the rope across to them so we tied on a hammer as a range finder. The first cast with the hammer shaved Nunn's head but he grabbed it on the recoil with a spectacular leap nearly ending in the fjord. The two experts, Robertson and Killer, then took charge of their respective ends in order to tension the Tyrolean rope by means of Jumars. This soon became a most complex operation, as each man was so committed to his own theories he had no time to check on the activities at the other end. But eventually, all was ready for the aerial crossing of the shore party. Robertson and I turned our attention to the next pitch. So compelling was the technical interest of the whole climb that the only implication I drew from the catastrophic possibility of the Tyrolean rope snapping was the personal inconvenience of being stranded indefinitely on the Old Man of Stoer. Two phantom lighthouse keepers waiting for the relief that never came...

This second pitch was undoubtedly the crux – perhaps Severe, although it looked much harder. Large safe handholds led up to the overhang, and Robertson could have been up it in a fraction of the time if he had not chosen to insert a security piton at the most awkward point. I found him sharing a cave with a young bird. 'A Fulmar Petrel,' I announced. 'Now, there's something peculiar about Fulmar Petrels; I remember reading about it in a bird book.'

'Anyway, this one seems to be choking,' remarked Robertson. The bird's neck had swollen hugely; its eyes glared. It looked as if it had something urgent to communicate. It had. Its beak shot forward, opened, and drenched us in foul-smelling slime. I recalled the fulmar's peculiarity. Robertson's cave was overshadowed by a considerable overhang which he had already inspected before deciding to belay. I lack conviction on considerable overhangs, so I decided to try my luck once more on the landward face, improbable as this had looked from the shore. The main bulge lay below me now, and the angle had reverted to the perpendicular. Vast jug-holds took me diagonally rightwards to a small eyrie. From this, a well-concealed chimney offered a one-way ticket to the summit. There was a good stance and plenty of cracks, but when I selected the appropriate blade, I realised I'd left my piton hammer down at the last belay.

Not caring to call attention to this rather elementary mistake, I pushed on, vaguely aware that I had accumulated a peel-potential of some 120ft. That figure did not worry me unduly, as I have a healthy disrespect for the rope, which I regard as a link with tradition and a reasonable assurance against the leader falling alone; the stronger the rope, the better his chances of company. The chimney was V-shaped, bottomless and harder than I had thought. It was also the communal cuspidor of a thriving community of fulmars on the ledge above. I went into it back-and-foot, hood pulled down, PAs skidding on avian regurgitations, breathless to the top. My exit was the signal for every fulmar in sight to release the final salvoes. Then up a short layback crack, and almost as the rope went taut I wriggled over the top onto the summit bed of sea pinks.

It took Robertson some time to realise he had reached the top. He reckoned he was entitled to the fourth pitch and was quite put out to find he'd run out of rock. For an hour we lay in the sun waiting for the others. We passed

Above: Tom Patey. Photo John Cleare. Below: The Original Route on the Old Man of Stoer. Photo Mike Hutton.

the time bombing the seals with divots. Seeing a small movement on the distant Tyrolean rope, I slung a divot at it and was surprised when it yelled back.

'What's happening?' we demanded.

Distant, unintelligible noises. Finally, Killer arrived at the end of the traverse rope and disentangled himself on the shore. His voice was suddenly audible.

'What the blazes held you up? We got fed up waiting and we're going back.'

'We've been shouting for you for the past hour.'

'Come off it! We heard you shouting at each other.'

'Nonsense,' I said. 'Come on up. It's a great climb. Three times bigger and better than Napes Needle.'

'Stuff Napes Needle!'

'How about a few pictures, then – for posterity ?'

'Sorry, mate-film's used up.'

'Englanders!' Brian muttered. 'You can't depend on them for anything.'

We baled out on a 150ft abseil which hung straight down to the first belay platform. Robertson came down last, and it was some consolation to the watchers on the shore when the rope jammed, and he had to climb back 150ft on Hiebeler clamps to free it. Before leaving he handed me a rope end.

'Tie this to a piton,' he said briefly.

'Which piton?'

'Any piton.'

I did as I was told, then set about clearing up the stance. About to extract the last piton, I noticed that the rope attached to it gave a twitch each time I tapped the peg.

'Don't you think you ought to leave that piton where it is?' Nunn shouted across.

'Why?'

'I think it would be a kind gesture, he said. 'It's holding one end of the double rope and your friend is climbing up the other.'

'Oh.' Reason dawned slowly.

'Don't worry about it,' he consoled. 'Everybody makes mistakes. Some make bigger mistakes than others.'

'Very true' I admitted.

And so the return was uneventful.

Scottish Stacks To Go At by Mick Fowler

My own interest in Scottish sea stacks has developed gradually, starting with ascents of the classic routes and progressing with increasing enthusiasm onto the less frequented challenges before moving on to new pastures. The Maiden was perhaps the most influential as it was the first Scottish stack that I approached by boat. As such, it represented a substantial advance in terms of approach ideas and opened the door to a whole new range of possibilities. Perhaps because of this new train of thought, my day on The Maiden remains particularly memorable. The idea materialised after close perusal of the SMC District Guide revealed the existence of not one, but two stacks off Whiten Head. It seemed that these were collectively known as 'The Maiden', and there looked to be a good chance that the westerly stack was unclimbed. The lure of a potentially untrodden summit (and another untouched for 18 years) proved too much, and it was clear that the trusty inflatable *Deflowerer I* (owned by Chris Newcombe and I, and purchased specifically for stack climbing) would soon have to make the first of many trips north.

The questionable sanity of living in London and driving to Scotland every winter weekend (to ice climb) and every summer weekend (to stack climb) was conveniently ignored. To be fair to ourselves, we'd been slightly cautious about our first foray into the Northern Seas and did make some efforts to contact local inhabitants and familiarise ourselves with the sea conditions. After phoning various Durness numbers at random, contact was finally made with an apparently knowledgeable individual who unequivocally advised us that a 30hp engine was the minimum recommended power. *Deflowerer I* had a 3.5hp engine which suddenly seemed a trifle inadequate. Lynn, Chris's wife, was particularly concerned by this, and the Newcombe household phone bill rose considerably as increasingly persistent efforts were made to hire a more suitable local craft – all to no avail. The choice was clearly limited to *Deflowerer I* or nothing... the final decision was never in doubt.

Having driven 720 miles and dispelled the myth that any Sutherland native will cheerfully lay on breakfast for a weary traveller, the team set about accidentally upsetting a Rispond fish farmer by mistakenly launching the boat in his prize breeding patch. Setbacks concerning copulating fish were thankfully short-lived, and in fine, calm weather, four empty stomachs were soon heading across the mouth of Loch Eriboll towards the distant and elusive Maiden(s).

Close perusal revealed that the SMC District guide had been entirely correct: two stacks – what more could one ask for? The eastern stack proved slightly higher, but as they are almost comparable in terms of impressiveness, it was immediately clear that a worthwhile unclimbed summit was up for grabs. We were blessed with a perfect day; the notorious swell was non-existent, and George the boatman could look forward to an easy day. Of the four climbers in the team, Chris Watts and Jon Lincoln started on the west maiden, whilst Chris Newcombe and I set about the east maiden. Photographic duties complete, the positions were then reversed. The end result was four new routes on perfect quartzite. We concluded that Scottish sea stacks certainly had a lot going for them.

This initial success induced increased enthusiasm and led to a more detailed assessment of the potential of the Far North. The next objective was chosen by reading between the lines of the SMC district guide. Clett Rock at Holborn Head (on the Caithness coast) sounded to be protected by deep water and serious currents. It also appeared to be unrepeated since the partially-aided ascent in 1970 and, therefore, awaited a free ascent.

To be honest, I'd never heard of it before coming across the name in the SMC guide, and we were all rather taken aback to find a stunning overhanging stack 200 yards long, 20 yards wide and 150ft high. Vertical or overhanging walls protected its grassy summit at every point, and active armies of squawking birds patrolled overhung ledge systems, apparently gleeful at the prospect of doing battle with a climbing team. At half height on the western end, we could just make out a rotting abseil sling left by the first ascensionists and trace the line of their ascent.

Above: A very young-looking Mick Fowler 'with my rather fetching recently-acquired NHS specs' at the stance after pitch one during the first ascent of Caveman at Berry Head. Photo Fowler collection.

After reading the passages earlier in this book chronicling the fowlers of the Scottish islands it might not be surprising that one of the most celebrated UK climbers, in fact 'the climber's climber', is a Fowler. Maybe his secret weapon is prehensile toes?

He should probably introduce himself: The name's Fowler, Mick Fowler, with a cheeky wink and a vodka martini, shaken not stirred, although I doubt cocktails are his cup of tea.

Facing page: The Old Man of Stoer. Photo Simon Richardson.

On the landward side, swirling current filled the 80ft-wide channel between Clett Rock and the dripping verticality of the Caithness coast. An impressive sight indeed, which was sufficient to persuade us to leave the excitement on offer until the nesting season had passed.

Returning in August, Jon Lincoln and I remained from the original crew, but Nicki Dugan and John Cuthbert replaced Chris Watts and Chris Newcombe to complete the team. Having landed with difficulty, but without mishap, the first pitch proved to be of a reasonable standard and led to a point beneath a band of overhangs at half height, which would clearly give the crux of the climb. Although the fulmars had now vacated their nests, the technical difficulty of the next section looked sufficient to cause plenty of interest without the added challenge of gushing bird vomit. Looks did not prove deceptive. Hard climbing with little protection led to a long... long reach for a prominent flat hold onto which a desperate mantelshelf was necessary to reach the end of the hardest section. More reasonable climbing then led to the top... untrodden for 18 years. The remnants of the first ascensionists' cairn were found, but in the absence of an in situ stake, our own abseil points were hauled to the top. These consisted of a metal fence post found on the drive up and half of one of the oars which had broken during some over-enthusiastic fending. Such implements would have proved perfect for the job had the summit soil been more substantial. Regrettably, though, this was not the case.

Although it was large in area, perfunctory prodding revealed the average depth to be 12" or so, hardly sufficient to hold a stake worthy of a 150ft free abseil. The swell below was now increasing considerably, as was the crowd of curious onlookers on the cliff top. Slate-grey clouds rolled in from the north, and a situation of considerable interest was obviously developing. At length, marginal anchors were placed, and, having endured a couple of hours of rising and falling swell, the Fowler body was feeling distinctly ill and not in a good condition for fielding the others. Apart from this, there were distinct practical problems. The end of the abseil was in a shallow zawn on the seaward side of the stack which resulted in the swell showing a marked tendency to break on the back wall and bounce back.

This state of affairs was particularly apparent as Nicki dangled on the end of the abseil rope attempting to release herself at the right moment as I struggled to manoeuvre the boat beneath her. At exactly the wrong moment, a rolling bank of swell exploded against the back wall, and simultaneously, the engine cut. The boat was awash, and Nicki was unsuccessfully struggling to escape the clutches of the sea by climbing hand-over-hand up the abseil rope. Substantial bailing and engine-drying operations resulted until further efforts had similarly frightening results before both Nicki, John and myself were in the boat. By the time we were ready to pick up Jon, the swell had increased further and, being

in need of manoeuvrability yet fearful of the engine being swamped, it seemed that the only option was to keep the engine running, bounce the nose of the boat off the rock and for Jon to jump at the instant the boat hit the rock.

Putting the plan into action was more problematical. The, by now, huge swell meant that a moment after impact, the prow of the boat scraped horrifically down the barnacle-covered rock and flopped neatly beneath a two-foot overhang before beginning to rise again. John and Nicki frantically pushed at the rock whilst I experienced the memorable sensation of being lifted out of the water, still grimly hanging on to the engine controls. Jon was forced to jump from about 15ft whilst Nicki and John pushed their hardest, trying to ignore the water now pouring over the sides. With a tremendous popping and rasping noise, the front was suddenly clear with such force that the roles within the boat were reversed. Those in the front were now dry whilst water poured into the rear, swamping the engine with distressing ease. Frantic use of the one remaining oar resulted in serious adrenaline flow but an effective escape to less violent waters where the engine could be restarted and a safe return made to Scrabster. Seldom do days on Stanage remain so deeply etched on the mind!

Elsewhere in the north, the coast bristles with potential for similar excitement. Be warned, though, the North Coast currents are strong, serious and unpredictable. Care is prudent, but the rewards are high. The exploratory phase is far from over. In few branches of British climbing can the future look so bright.

The Drongs by Hamish MacInnes

The Drongs have several heads like a Hydra, and we were faced with the perennial dilemma of all stack climbers: how to get to them.

They were beyond the ingenuity of Highland rope tricks and beyond the reach of all the ladders in Shetland.

The problem was solved by Don Whillans, who, as was his wont, gravitated to the local hostelry, where with singleness of purpose (even buying a drink), sussed out the only small boat owner in the bar, a local worthy who had won a leaky tub in a bet. Over 'reaming swats', he asked:

'Will you take us to the Drongs, Jimmy?'

The swaying potential Charon almost choked at the prospect, then started a drunken chant:

'You'll no get up the Drongs. You'll no get up the Drongs!'

Don and Joe Brown were probably the most powerful climbers of their generation, but until we went up to Shetland, they hadn't climbed together for 14 years. In such persuasive and illustrious company we did get up the Drongs.

Dry Bags and Fins by Simon Richardson

It was Guy Muhlemann, sometime in the late 90s, who introduced me to the idea of using dry bags for climbing sea stacks. Previously, one of us had to swim to the stack and then set up a Tyrolean to ferry the gear and the rest of the team across. This was a time-consuming process and limited us to objectives close to shore. But, incredibly, you could fill a kitbag-sized dry bag full to the brim with rope and hardware, and it floated. Coupled with wetsuits and fins, our horizons expanded considerably.

Our finest adventure was the first ascents of the Lorgasdal sea stacks on Skye one glorious September day in 1998. These 35m-high shapely slivers of basalt lie about 3km northwest of Idrigill Point in a bay where the Lorgasdal River plunges down vertical cliffs. The three-hour approach from Orbost past Idrigill Point and MacLeod's Maidens was followed by the challenge of how to reach the boulder beach 100m below. The only anchor was a large boulder perched directly over the waterfall, so we abseiled in our wetsuits before swimming 25m to the south stack. This was followed by a 100m swim to the north stack. The climbing was good, but the basalt was fragile, and the grades of HVS 4a and E1 4c tell a story in themselves. Unfortunately, Guy took a short fall on the second stack, and his ankle ballooned up. The swim back, jumar up the waterfall and long walk back with heavy sacks full of wet ropes and wetsuits are firmly imprinted in the memory.

Another success was Stac an Tuill on the wild and remote Minginish coastline on Skye. This spectacular 30m-high stack is characterised by a huge hole that cuts through its base and lies between Glen Brittle and Loch Eynort. It stands 20m offshore below 200m-high crumbling basalt cliffs. On our third visit, Mark Robson made an 80m abseil down the cliffs 800m to the north and swam under the towering cliffs to the stack. Wearing wetsuits, we climbed the west ridge via a series of cracks, which became steeper and increasingly rotten with height to the (more solid) summit block. Stac an Tuill is one of the most difficult-to-get-at sea stacks in Scotland, and we reflected afterwards that it would be rather unsporting to use a boat.

Mark became my regular stacking partner, and dozens of adventures followed using the new-found technique. We were particularly interested in unclimbed stacks, and the Scottish coastline offered countless opportunities. Some were all too obvious, such as Stoer South Stack, which lies 200m south of the classic Old Man. It is largely hidden from view in a steep geo, but even so, it was surprising it had not been climbed before as it was probably the highest unclimbed sea stack on the Scottish west coast. We set off from the same ledge used to approach the Old Man into a lively sea. Unfortunately, the water roughened up considerably during our ascent, and the return was challenging. I'm a strong swimmer but was very grateful when Mark pulled me out of the water by the scruff of the neck as I was about to be submerged by yet another large wave breaking onto our landing site.

Other objectives were more obscure. Fantasy Golf Stack is a rectangular grass-topped sandstone tower nestled beneath the west side of Sinclair Castle at Noss Head that was discovered after a lengthy search for a stack featured on a fantasy golf calendar, and Impossible Stack is the difficult-looking pinnacle 120m to the west. Fortunately, this turned out easier than it looked, but it was a typical battle with loose rock, mud and bird debris, which prompted us to use the ominous VSL grade (Very Severely Loose). Many stacks did not have anchors at the top, and over time, we perfected the dubious art of simultaneous abseiling.

A good find was the Stack of Old Wick, a superb 40m-high feature 100m southeast of the Castle of Old Wick. This was climbed one cold, dank day in early April, but at VS 4b, it was easier than it looked and had reasonable protection. On the summit, we found several old bike wheels tangled up in the long sea grass. This confused us at first – perhaps the stack had been climbed before after all – until we surmised that the local kids must have thrown them, like giant frisbees, onto the broad, flat summit. Eventually, sea stacking became as much about the adventure as the climbing. One October, Tom Prentice and I swam 200m to Wheat Stack on the Berwickshire Coast. This pyramid-shaped feature on the east side of Fast Castle Head resulted in a pleasant Moderate climb, but it was the return swim that made it memorable. Halfway back, we realised that two grey seals were escorting us back to shore. It was an unforgettable moment – the massive creatures were so close that we could almost touch them, and their power, bulk and efficiency through the water prompted wonder, fear and respect in equal measure.

The summer of 2020 was glorious, but Covid restrictions kept travel within five miles of Aberdeen City. This led to a focus on Tump bagging. In all, there are over 17,000 Tumps in the British Isles, and by definition, sea stacks that are 30m high are on the list. So, during the lockdown, increasingly longer cycle rides were made to ascend Aberdeenshire Tumps, and those further afield were set aside as projects for the future. Most intriguing of all was Maw Craigs, an unclimbed sea stack on the Moray Coast. I had seen it from a boat when looking for climbing opportunities many years ago and had dismissed it as a steep pile of choss, but now it was on the list of Tumps, it became an object of desire. Who could not resist an unclimbed summit on an official hill list? It took several attempts after the conclusion of lockdown for the stars to align, but in September 2021, the sea was calm, Mark was free, and the nesting birds were gone…

We're in position, about 2km west of Strahangles Point. The cliffs here are amongst the most impressive in Scotland and drop vertically over 120m to inaccessible pebble beaches. In between, sharp ridges shaped into fantastic pinnacles strike out to sea. The scenery is awe-inspiring, but, unfortunately, the rock is the softest of sandstone conglomerate, and the thought of doing any technical climbing sends shudders down the spine. A 100m swim leads to the stack. As we land, a loud roar from deep inside a sea cave startles us – a hidden sea monster seems appropriate for this primaeval place – but it turns out to be a large male seal, indignant at being disturbed.

Finally, we're in position below the south ridge, which appears to be the most likely line of ascent. Mark ties into the sharp end, takes the equipment – two ice hammers and three drive-ins – and immediately sets off up the initial steep wall leading to the ridge. I cower under a little roof whilst conglomerate crashes around me, and discarded feathers and bird poo dust fills the air. Our doubled 70m rope is not long enough, so heart in mouth, we move together for the final few metres. The pitch has no protection except for a single warthog that Mark has managed to place below a vertical step. It looks good, but I pull it out with my hand. Of more practical use is the foothold that Mark has cut into the rock using an ice hammer to assist the steep move up to easier ground.

The summit is painted white with centuries of guano. There is no belay, and a simultaneous abseil will follow, but we are relaxed and carefree. The climbing had been ungradable, poor even, but as an adventure, this is at the top of the scale. We are less than 60km from my home on one of Britain's unclimbed summits, looking across an azure blue sea on a beautifully clear autumn afternoon. Life rarely gets better than this.

Above: Mark Robson on the summit of Maw Craigs. Facing page: Simon Richardson on the first ascent of Lord Oliphant's Bicycle, Stack of Old Wick Below: Maw Craigs. The line of ascent was the ridge facing the camera. Photos Richardson collection.

Caithness Musings by Mark Robson

The east coast of Caithness, up to Duncansby Head by John o' Groats house, has continuous bands of sandstone cliff ranging from 100 to 200ft high, interrupted by shorter stretches of the Caithness plateau falling to sea level and sandy beaches as in Sinclair's Bay. Only at the top of this section has there been much climbing development.

Sea Cliff Climbing in Britain by John Cleare

Stacks: I'm not sure when my fascination began. Perhaps in the 1980s, during school years, high on Napes Needle, that most renowned of Lakeland climbs, still graded V Diff in those days. After all, isn't a Needle almost the same as a stack? The long, sweaty walk up from Wasdale requires far more effort than a dip in the ocean, coupled with a descent no less complex than retreat from a typical sea stack. Good schooling for the would-be sea stacker, for sure.

Or maybe it was closer to home in Derbyshire at the majestic Alport Stone? Towering a full 80m in height, this shapely free-standing pinnacle provided a perfect venue for teenage night-time adventure of all kinds. But perhaps Alport's finest is too much a man-made object, too solid in its composition to be considered a true stack.

Several years later, a spell spent living in the sunny South West offered enlightenment, and I discovered the world of stacks proper. No land-locked obelisks these. They were the real thing. Rising proud from the ocean, fashioned by wind, waves and by the fact that everything surrounding them had already fallen down. And, best of all, a rubber dinghy was obligatory.

Armed with the stacker's bible, Chris Mellor's *Stack Rock,* an altogether new side of climbing presented itself: the holiday atmosphere of Ladram Bay, where we departed a crowded beach in our small inflatable to enter a world of soft sandstone, guano and simultaneous abseils; The Parson with that terrifying railway tunnel approach. We'd been told by a previous suitor that, should a train pass, our best hope would be to lie flat on the tracks until the danger had passed. And that was even before the base of the stack itself had been attained. A grade of E3 4c, a pitch described as the 'Brown Spider' and protection in the form of tied-off six-inch nails and warthogs hinting at the excitement to come.

Willing partners were sometimes hard to come by in those years, but things changed after a move to Scotland when Guy Muhlemann, a good friend and veteran sea stacker, invited me to join him and Simon Richardson for one of their seaside adventures. Following success on Skye, where the three of us climbed the thrilling Vibrator, a small, slender stack in Loch Snizort that wobbled quite spectacularly when the loose summit blocks were hurled into the sea beneath, Simon and I became regular climbing partners. And so began a period from the early to mid-noughties when the far northeast coast of Scotland became our favourite hunting ground.

For sure, the biggest and the best of Scotland's sea stacks had been claimed long ago, but here, in the quiet backwaters extending south from the spectacular stacks of Duncansby Head to Berridale, marked by the striking Bodach an Uird, there lay a host of still unclimbed potential.

The Little Stack of Duncansby, or Tom Thumb Stack as Simon referred to it, was the first of the Caithness stacks to fall to us. I remember a bitterly cold day and a lonely beach despite the fact that it was the month of June and we were just a stone's throw from the tourist trap of John o' Groats. Simon only remembers me looking doubtful as we scrambled down the mainland cliffs hampered by rucksacks containing wetsuits, divers' fins, and an assortment of climbing paraphernalia before embarking on the first of many swims along this section of coastline.

As the partnership developed, so did confidence levels, and when we returned to that same beach a couple of years later to swim out to the Peedie Stack for the first ascent of Toil and Trouble, a 65m VS and probably the first time one of the big Duncansby stacks had been tackled without the use of a boat, we had things off to a tee. By then, we'd perfected the technique of approach using wetsuits, fins and drybags. Free from Tyrolean sieges or unsporting boat-based assaults, we were able to roam the coastline, often climbing several new stacks in a day. Not exactly 'light and fast', but certainly productive.

The next day, with a golfing theme clearly on our minds, we were successful in locating two fine stacks near Noss Head that Simon had spotted featured in a Fantasy Golf calendar. The first soon yielded to a 35m HVS which was named according to tradition The Original Route, before its smaller neighbour, The Tee, was quickly dispatched at a similar grade.

At this time, I was employed by the Ordnance Survey, and although we were sometimes lucky enough to come across a new objective purely by chance, most were discovered by painstakingly poring over aerial photography and the detailed topographic maps that I had access to through work. A little effort would usually be rewarded by something that warranted further investigation: Old Man of Wick, the Stack of Ulbster, and Stacks of Martin's Sclaite: all names on a map hinting that further exploration might prove fruitful.

Of course, sometimes we were less successful – an afternoon spent traipsing along the clifftop yielding nothing or the discovery of a tottering pile of choss that even we deemed unworthy of a swim. And sometimes something in between – our first ascent list containing several VSL or 'Very Severely Loose' ascents. Nice to look at, but not to touch.

Eventually, as is often the way, things moved on, and whilst enthusiasm remained, objectives became fewer and further between. Our hunting grounds shifted southwards to the Moray Coast, and trips to Caithness became less frequent.

But now, looking back through the notes of had-been and might-have-been objectives, I find that there are still stacks that we looked at and didn't climb. Things that we viewed from a clifftop and vowed to come back to on a day of calmer seas or after the seabird breeding season when storms had washed away the guano. I also have memories of one particular day, the sea sparkling and the sky blue, of a wall of solid sun-kissed rock and seabirds lazily wheeling around above. A wall that has still not yet seen the attention of others.

Maybe it's time to break out the fins again and head north.

Above: Mark Robson preparing for the swim to Stac an Tuill, Minginish, Skye. Facing page: Mark Robson setting off for the swim to Stac an Tuill, Minginish, Skye. Below: Mark Robson leading the first ascent of Stoer South Stack. Photos Richardson collection.

Mingulay

Writing in the late 1600s, Martin Martin, a Gaelic-speaking native of Skye, described fowlers climbing the 112m Liànamuil sea stack in his book *A Description of The Western Isles of Scotland*: 'The chief climber is commonly called Gingich, and this name imports a big man having strength and courage proportionable. When they approach the rock with the boat, Mr Gingich jumps out first upon a stone on the rock side, and then by the assistance of a horse-hair rope, he draws his fellows out of the boat and upon this high rock and draws the rest up after him with the rope until they arrive at the top.'

Kevin Howett was told by John Allen McNeil of Barra that the Mingulay islanders used to be dropped on the west face of this stack where they could climb into a huge sea cave and out the top, which is how they rigged a rope bridge to the main island to harvest the sea birds and graze sheep in the 1800s. An ascent by modern climbers in 2018 found the stack to go at a grade of Very Severe.

David Craig wrote about his trip to the Barra Isles in the *London Review of Books*[3]: 'Our climbing is a happy-go-lucky, latter-day rebirth of what the islanders, on Mingulay especially, did to earn a livelihood. Early in the 19th century, the Pabbay farmer-fishermen used to buy seabirds' feathers on Mingulay to sell on at Greenock and elsewhere. They still went fowling in the 1880s, when the craft had died out among the Mingulay folk, and roped down the first few hundred feet of those colossal bird-cliffs, the black groins of them streaked orange and green like tribal body make-up or Aboriginal cave paintings of dancing "demons", to catch fulmars with nooses on sticks. Feathers were a crucial commodity, sometimes accepted by landlords in lieu of rent. The best climber (i.e. fowler) on Mingulay was also the finest storyteller: Rory Rum the Story Man.'

Mingulay's most famous Gingich, Roderick MacNeil (1790–1875), was also known as Ruairidh an Rùma or Rory Rum the story man, so-called from a hogshead of rum he found on the shore and from the contents of which he nearly died. He was said to have a chest 'as round as a barrel, 48 inches in circumference' and was so powerful that 'whenever his fingers could get insertion in the crevices of the rock, he could move his body along the face of the precipice without any other support'. *Carmina Gadelica* author Alexander Carmichael visited Mingulay in 1871 in the company of his folklore mentor, John Francis Campbell, who commented: 'Rory Rum the story man, about 85, the best climber in Minglay till he got past work. He never wore shoes or stockings, never had a bonnet on his head till some years ago and now is crippled by the rheumatism and stoops over a long stick.'

The Story of Rory Rum by Kevin Howett

Come 1996, we found a local Barraigh man who could supply transport at a modest fee: John Allan McNeil, retired coxswain of the Barraigh Lifeboat, and whose family were amongst the last inhabitants of Miùghlaigh. A double island visit was planned. Pabaigh was blitzed with routes in good weather and calm seas. But it was not to last. The day before the transfer was due, the weather turned, and that night a force 10 storm blew in. The following morning it had not abated much, and the seas even in the lee of the island on the east were looking wild.

In the 90s, there was little chance of mobile reception, and John Allan had no portable ship-to-shore radios for us. Indeed, his little boat had little in the way of any health and safety essentials, so we doubted he would make the journey down to us in such terrible conditions. A number of us headed high up onto one of the northern hills of Pabaigh to try and get a phone signal, but to no avail. However, they appeared back at camp in a rush, saying they had spotted a tiny speck ploughing through the white horses heading our way. Was it John?

A very rapid de-camp was done, and we piled the kit on the rocks just as John arrived to make the transfer to Miùghlaigh. John's co-pilot Archie pulled the small boat into the bay almost in defiance of the elements. Kit and climbers were packed in like sardines, most of the kit in the hold and the climbers out in the open on the back decking, clinging to a pile of remaining kit. As Pabaigh disappeared into the mist, and we headed into the notoriously wild Sound of Mingulay, the swell began

Above: Kev Howett. Photo Howett collection. Facing page: Noel Craine abbing into Sròn an Dùin. Photo Grant Farquhar.

3   *London Review of Books* Vol 19, Number 21, 30/10/97.

Gordon Lennox and Kev Howett
on the first ascent of Hakkar,
E4, on Mingulay's Guarsay Mòr
Arena. Photo Howett collection.

to reach massive proportions, each accompanied by breaking white horses on their crest. Everyone clung to the gear in the centre of the boat, hoping for the best. John appeared with a life vest and passed it to Melanie, the lone female on the team, then resumed steering the vessel. There was, it seemed, only one life vest. John had to turn into and ride over each swell that was hammering through the sound from the west, then resume his course south. The relentless cold deluge of seawater that swamped the back of the boat each time we faced up to the swell was scary. I am sure I was not the only one doubting the wisdom of this journey, even if John was a lifeboatman in a past life.

He had no modern navigational instruments and seemed to navigate by the smell of grass and seagull shit, but I doubted this would work in these conditions. This was confirmed when the boat made two full circles, and John came out of the bridge and asked, 'Has anyone seen Miùghlaigh?' in his ever-calm and soothing Mingulayan tongue. There was a pause as we all pondered if he was joking or not before Graham took a compass bearing, and John instructed Archie to travel roughly in the right direction until Miùghlaigh appeared out of the clag. The only positive in this whole experience was that I was so terrified that I wasn't seasick.

It was quite a first introduction to the indomitable 'Mr JA' (as he became known to us) and was an indication that he could probably get us anywhere in any weather (and in the years to come, he didn't disappoint). The following day could not have been more different: a cloudless sky and a warm breeze. After lunch, I had my eye on the last remaining unclimbed area on Sròn an Dùin. So, with the sun turning onto the face, George and I headed south for the cliff, hyped up for some exposed and sea-lashed shenanigans. A fellow north-easterner relocated to Scotland, George had previously given up on trad after a ground fall at Back Bowden Doors some years before. He had transformed into one of Scotland's most prolific sport climbing developers, which he considered as safe highball bouldering, his first love. He was way stronger than me and could crank on holds I could only dream of. I was regularly put to shame on training sessions on his attic wall. However, he was not au fait with this sort of climbing any more.

The line I was after climbed through the shallow, roofed sea cave at the south end. It was immediately obvious that the swell was still enormous, and waves were crashing all along the base of the central area as well as under the intended line. But I thought we could come in from the right. We set up the abseil down one of my earlier routes, Dun Moaning, hoping to be able to start from its sea-level belay ledge. A slight miscalculation on the line of the abseil had me further left of that, and as I neared the base, it was confirmed that the area under the big roofs to my left was being battered. I had just descended below a small ledge and was hanging in space below a curving arched bulge. I had a choice: either try and pendulum some distance right to the belay of Dun Moaning or carry on down a little and find a belay. I bounced on down and found only a single nut on which to hang from. There was nothing here

resembling comfort, but at least the runner was bomber. More alarming was the fact that I was very close to the waves rushing past and shooting up the walls in the shallow cave only 10ft to my left, and the rock was dampened by sea spray. George came down looking pretty excited by the whole situation, so it was game-on. We hung from the runner and had a bit of a faff sorting the ropes, conscious that the sea could easily change its course slightly and we would be in trouble. At the time, I climbed to the mantra of 'total commitment' on all routes on the islands, and as soon as we were ready, I released the abseil rope which drifted several metres out to sea. Suitably psyched and lavishly committed, I set off to get through the bulge and out of wave range, optimistic that it would be easy enough as the rock is usually covered in holds.

I was wrong on that front. I tried to the left, but that was too wet. I tried directly through the largest part of the bulge and then traversing to Dun Moaning, all to no avail. None of the options sported any encouraging protection, so I was hesitant to go for it. We were kind of stuck as George reiterated his earlier comment on the error of letting the abseil rope go. But I had a backup plan: to fashion a bolas of interconnected slings weighted with a large cam with which to entangle the abseil rope tantalisingly out of reach and moving about in the swell. Luckily, one of the waves brought it within reach of our weapon, and we sighed with collective relief. Except we realised neither of us had brought anything to ascend the rope with. There was no alternative but to hand-over-hand up the rope. I reached the small quartz ledge buzzing and traversed to its left end to find an uncomfortable belay. At least now we could start over again.

The climbing was superb and led up into a small bower below the lower of twin upper roofs. I belayed and brought George up. He considered himself a VS climber (!) and was happy to relinquish the lead and settle into the comfy rock seat, basking in the late-afternoon heat. With the sun now full onto the crag and with no wind we started to cook, but we were alone on a stunning wall, going into the unknown, and we were having a ball. I headed for an enticing traverse left along a diminishing wall sandwiched between roofs as it had what I thought would be a jug-ridden horizontal slot, all the better for climbing in this heat. However, the holds changed from incut to slopey, and my now-redundant feet were swinging in space while I dangled on small sloping holds. By the end of the traverse, I was sweating profusely, utterly spent, and came off. Lowering to the stance to start again was not an option; I was too far left. In too much of a rush, I fell again, having not recovered. After a brief rest, I made it under the huge upper roof and took a belay.

George followed easily, clearly enjoying the wild position and the moves. We swapped belay, and I went through the body-length horizontal roof onto the easier upper wall. We sat on top in the most glorious 11pm sunshine having shared the best laugh ever on a climb. We felt that we had a suitable route that did justice to the legendary climbing exploits of Rory Rum the Story Man.

Oh No! Norman's in Reverse by Kevin Howett

Since starting to develop the sea cliffs on the Bishop's Isles in the Outer Hebrides in 1993, you would think there would have been many opportunities to get wet. But the remoteness of the islands tempered our enthusiasm for taking undue risks. Not surprisingly, the closest we got to having a face-to-face encounter with an angry sea was in the various exploits of the skippers of the boats who took us there. That's not to say we didn't occasionally get it wrong on the crag as well.

That first trip involved sailing from Barraigh into Miùghlaigh on Peter Daynes's yacht on a flat, calm sea. It came as a surprise, therefore, that whilst we and our kit were being transported onto the island, some of our team nearly ended up 'below deck'. The little dinghy rowed by Peter's son, Guy, was heavily laden with various kit and climbing luminaries (Graham Little, Mick Fowler and Chris Bonington) when a freak surge forced the dingy onto the rocky landing slab. I had visions of everything and everyone disappearing into the sea before we had even set foot on the island.

Our initial trip to Miùghlaigh gave Graham Little and I the first routes on Sròn an Dùin. These were climbed in stormy weather and were not without incident. We managed to avoid the huge seas that washed the base of the cliff but found ourselves utterly committed with no way out but upwards. Two years later (1995), we were back, this time to Pabaigh as well as Miùghlaigh. Chartering the converted lifeboat *Poplar Diver* crewed by Bob Theakston and deckhand Norman, we thought it might ensure a modicum of safety. But we encountered ferocious seas in The Minch, and we had misjudged the stability of a lifeboat; yes, it won't sink and will simply turn turtle and right itself, but it is so buoyant it rolls and bobs about like a lost cork. Everyone suffered various levels of sickness. I spent the whole trip above deck, cold and wet and fighting off that feeling.

This was followed by landing on Pabaigh, which was quite possibly the most frightening part of this trip. Norman was instructed to get the tender sorted. He seemed very nervous. We got the initial piles of kit on and several members of the team. Norman gunned the outboard and headed into a slim geo as instructed by Bob. The swell was significant, and waves were crashing into the little geo. Norman was clearly out of his comfort zone and having real problems controlling the tender, inadvertently switching into reverse at inopportune moments. Suddenly, there was a horrible crack, and the outboard died. A wave hit us hard and threw us towards the rocks at an acute angle, and but for the fast action of Graham Little leaping into the briny with the mooring line in hand, the climbing kit may have ended its useful life well below water. I dare say some of us might have survived, but it would have put paid to any climbing. Once disembarked, we got the outboard running again for Norman, but he had hit a rock and smashed a couple of the propellor blades off. We watched in amazement as he struggled back to *Poplar Diver*. Norman is now immortalised in various route names on Pabaigh.

It was the beginning of May 2000, and the Western Isles were experiencing a heat wave. Constant high pressure, blue skies and light winds had baked dry the usually sodden ground on Barra. We were back following up another hunch as to the whereabouts of an elusive huge geo we had spied through the mist and rain when escaping Miùghlaigh some years earlier. It just had to be Lingeigh, decided Graham, ever the enthusiastic optimist and the Ordnance Survey employee.
Lingeigh is a tiny, rounded lump of an island floating off Pabaigh that takes the full brunt of any Atlantic swell. John Allen (Mr JA) took four of us out (Graham, myself, Tony and Neil) in unnaturally calm seas to land in a tiny geo on the northern tip, where we struggled a mountain of kit up the carpet of sea pink. There was no fresh water and only one flat(ish) campsite. The sun warmed our backs as we devoured a meal, and then we set about exploring the gold-burnished evening. After only a 12 minute walk, we found ourselves, somewhat surprised, at the southern end of the island, having seen no crags at all. This looked like we would be doing nothing but sunbathe and drink whisky for three days. Returning dejected until within sight of the tents, we nearly fell into it: a deep slot, 55m high of immaculate gneiss, glittering like gold leaf.

Left: Helen Hall on the first ascent of Two Ewoks and a Wookie up a
Scout Walker's Trouser Leg, E3 6a, in Bigfoot Geo. Photo Tim Catterall.

Three days of frantic play followed. One climb after another, always immaculate, always an adventure, very often scary. With the island named Lingeigh and it being a slot, the search for route names ignited the once-lost teenage influences of toilet humour: four boys on an island re-enacting their childhood without the calming influences of parents or women. For this, we apologise, but perhaps this tells its own story.

On the day of departure, the sea was back to normal. Although the blue skies still hung in there, a huge swell had sprung up and was the portent of a big storm to come. Prospects of getting home were beginning to look remote as every part of the coastline of the island was now being strafed with massive waves. Mr JA and his co-pilot (another John) appeared out of the breaking swells, which were now almost devouring the little boat at every passing. Once they got 300m offshore, they quartered an area of sea looking for something. We realised they were searching for a safe anchor as they tried a couple of times before success.

After some time, the inflatable was launched and headed over to our original landing geo. One look at the waves crashing into the boulders at the back had them racing out. They searched back and forth for a better embarkation point. The winds were now getting stronger and the swell bigger, and it seemed like there was no other option but to try the geo. We started piling kit on the edge of it and watched in awe as John came into the tight confines surfing a wave; there was little room for mistakes on either side of the inflatable, and he had to rapidly set the outboard into reverse before being dashed on the rocks at the back. The other John flung us a few life jackets, which had never happened before, even in the most awful seas. I took it as a bad omen. On the next pass, he steadied himself to accept the kit which we flung at him whilst Mr JA kept the boat stable just long enough to accept a few rucksacks. As suddenly as they arrived, they departed on the back of the receding wave.

They were soon back again; this time, it was proving a little more difficult. As Mr JA manhandled the inflatable ready to insert it into the geo, we noticed that his boat had disappeared. It had vanished! We gesticulated to the two Johns, pointing to the desolate sea where it should have been. There was a moment of disbelief before they gunned off and disappeared around the east side of Lingeigh. We sat waiting for an age until they returned with the boat and found a better anchor. We could only guess at what they'd done to get the boat back. This was proving to be one of the more interesting sea-faring exploits so far.

Then Mr JA came bouncing towards us and into the geo, again surfing the wave but this time backwards, and shouting to throw stuff in and for one of us to jump on board. Getting the timing wrong would be a rather bad idea, and we had to make a group decision in seconds on who should jump in before Mr JA rode the wave back out again. One of us, I can't remember who, took the initiative and jumped in, unfortunately missing the sweet spot at the top of a wave and so fell through the air in

Above: Mr JA (standing) in his boat at Mingulay in 1996. Photo Kev Howett.

tandem with the inflatable for 3m to land in the base just in time to be whisked out the geo at full throttle.

There was an air of urgency and panic developing as this scenario was repeated until all four of us and our kit were off the island, each of us hoping the inflatable would be underneath us when we jumped. Only a man with 30 years of experience as a coxswain of capsizing lifeboats could have got us out that day. Back in the bar in Barraigh, the locals said that no one else would have even tried, and we would have been stranded for days.

On a trip in 2001, Mr JA again took us to the islands, this time dropping two brave souls off on Sanndraigh en route to Miùghlaigh. The weather was dry and sunny, but the lingering effects of one of the biggest storms of the year meant most of the west coast's big cliffs were being battered by high winds and large seas. Nearing the end of the stay, Hugh, Helen and I elected to check out a small and, as yet, unexplored zawn on the east. The north-facing wall of it gave up a reasonable E1 between the heavy showers, but the prize line for me was a unique pillar in the middle of the sea cave forming the zawn. A ledge below it was isolated by deep water on all sides, so the only access was by abseil (or swimming, which I didn't really fancy as the sea was too rough, even on this side, for that kind of approach). This ledge stuck out from the pillar like a foot, and so we named the zawn Bigfoot Geo.

The ledge was clearly tidal, but just how much needed to be worked out. So we watched for a while to judge how much time we would have for an attempt on the 'morrow, and it looked like we had a couple of hours to play with around low tide. Hugh and Neil fancied a go at the steeper back wall of the geo, so that left Helen to accompany me on what was her first visit to these islands and her first sea cliffs. There was a reasonable swell washing the big toes, but the ankle seemed free of water. I rigged the abseil and descended. It was quite a bit more overhanging than I expected, so I had to swing in and place a runner below the headwall clipped into the ab rope in order to guarantee landing on the foot rather than in deep water.

Before long, we were both secured to the ankle, and despite a big swell, most of it was funnelled into the caves on either side of us. The others removed the abseil rope to use to get into the north wall as the down-climbing option to gain access was out of the question because of the swell. The climbing was excellent at well-protected 6a, which was harder than Helen had climbed so far but was not impossible. My only worry was that if she got freaked out by the situation, it could change the outcome. I reached the top of the pillar when Helen shouted up that she was being splashed. I hadn't noticed the rise in the ferocity of the waves nor the time it had taken for us to get this far, and we, rather she, was rapidly running out of dry time. Rather than go to the top of the crag, I belayed, which meant I could see Helen and at least get her off the ledge.

By the time I set up a belay, most of the foot was being washed bar a small haven, unfortunately tight up against the right arête and not where the line started. Helen seemed unfazed by it all and set off up the left arête, now exposed to any crashing wave. She got wet (obviously) but did not complain and continued to the belay without further incident, climbing the now-wet lower section with ease. Perhaps she thought this was all normal? As I set off up the much steeper headwall (which succumbed easily), a glance down confirmed we had made the right choice. Two Ewoks and a Wookie up a Scout Walker's Trouser Leg (E3 6a) was therefore born whilst escaping Davy Jones's Locker.

We managed to get back to camp before Mr JA's boat arrived, and it was getting dark. He had recently upgraded the boat with the requisite lighting, ship-to-shore radio, full life vests, an inflatable life raft, and new navigation systems, and yet he seemed in a hurry to get back. As we passed Sanndraigh in the dark and could see the others' lights, we realised John was not stopping as he ploughed on in complete darkness. Suddenly, out of this blackness reared the Barraigh Lifeboat, all lights blazing. Only then did John put his port and starboard lights on. The island community were getting concerned about John's safety as he was getting noticeably older now, so we were escorted through the gap in the skerries off Bhatarsaigh to Barraigh. The incident was featured in the *Stornoway Press* later that week. The Sanndraigh team had to re-camp and be picked up the following day. Sadly, John had to retire as a boatman at the end of that year, leaving a void in the story-telling and wild sea exploits, which have made the exploration of these islands by climbers so enjoyable.

By 2004, many of the cliffs were looking climbed out regarding classic lines. However, this was not the case regarding some brilliant lines on Guarsay Mór which stood out for all to see. It was odd they had not been done. On a May trip, inclement weather initially meant catching rabbits and seal watching were favourite activities during the day, and long night jamming sessions took everyone's minds off gale force winds and disintegrating tents. Once it settled down, attention shifted to finding any overlooked lines in The Arena and The Grey Wall area of Guarsay Mór. Routes either

Above: Robert Mackenzie leading Too Young for a Gladiator, E1. Below: Kev Howett and Gordon Lennox on the first ascent of Hakkar, E4. Guarsay Mòr Arena. Photos Howett collection.

side of the Great Cave, first scaled by Mick Tighe and teams, were now regarded as amongst the best on the island.

The seas were still quite rough, and it was washing all over the slabby base of this 120m complex wall. As a result, it was difficult to judge where to position any abseils to hit the right place and whether there would be a ledge free of froth less than 15m above the sea from which to start. Gordon Lennox and I could see an unclimbed snaking corner system on the right-hand side of the cave which seemed to avoid being wave-washed and thought we would go for that. Robert Mackenzie and his dad Ruaraidh spied a line on the immaculate grey rock just right of us, but it didn't look hopeful for gaining a low belay.

Indeed, they had to belay on a tiny ledge 10m above the slabs, and even that was close enough to get a glubbing. Robert, aged 14, made an impressive lead of the second pitch, and the modern classic of Too Young for a Gladiator (E1) now gets lots of ascents, more usually from the bottom. Gordon and I descended towards the safe ledge. We followed the stepped corners to give Eye of the Storm (E1) without encountering anything wet. In doing this, we were able to get a good eye on the line we wanted to do all along – a line above the roof of the Great Cave itself.

We were back the following day. There were regular fronts moving in from the west, and we hung about the top of the wall in the rain, waiting for a dry spell. Seeing a band of blue sky approaching, we scrabbled to get down and back up before the next front rapidly approached. Rigging the abseil in the correct line was a bit touch and go, as we had to hit a small plinth at the base of a pillar that formed between the Great Cave and a smaller mini-cave to its right. By now, rather than diminishing, the sea was growing in ferocity as it had changed direction slightly to be head-on into the mouth of the cave. Huge surges of water powered into the back at amazing speed, and the backwash was accompanied by booming sounds of dread. The sea was a confused mess, boiling and spitting in every direction: a volcanic magma of white froth. This certainly upped the fear factor as I abseiled down. The wall overhung all the way, and I had to bounce to maintain contact. I stopped and placed a runner to keep me in enough to hit the plinth, but the outcome was still far from certain, and I had to bounce and swing around far too much before I just managed to grab the edge of the plinth and pull myself onto it.

This little ledge was actually missing the huge waves themselves, but their ferocity meant we were subject to masses of sea spray, as was the pillar above us. Despite this, I once again adhered to my personal mantra of 'total commitment' and let the abseil rope loose once Gordon was safely ensconced. It was his lead, thankfully, and he worked his way up the wet pillar until stopped by a particularly slimy section. The protection was poor and the chance of falling either onto the ledge or into the drink was high, so he backed off. Another massive wave thundered past us into the cave and spurted out

the right-hand corner adding another fine mist to the air. We swapped the lead, but I was not prepared to push it either. If we knew there was protection, then perhaps we would commit? This was a dilemma as there was no other way off the pillar. If only we could just jug up the abseil rope and inspect it? Luckily, I was prepared for such circumstances and re-enacted the bolas technique of interconnected slings weighted with a large cam with which to entangle the abseil rope perfected on the first ascent of Rory Rum the Story Man.

Gordon jugged up and found a runner placement. We briefly discussed ethics versus safety before a decision was made to leave it in place. This time, I kept hold of the abseil rope as he headed back up the pillar with renewed confidence and was able to make the moves to reach and clip the pre-placed runner. Watching Gordon climb smoothly upwards on very wet and uncertain ground reduced the sense of tension that had pervaded the proceedings so far, as it meant we had a chance of escape by climbing; the thought of jugging up 120m on the abseil rope with the risk of it being cut through was never an option I wished to have to take. I relaxed and started to appreciate and enjoy the maelstrom of spray, crashing waves and noise that filled the air around me. Within a few minutes, he was on the lip of the cave.

From my vantage point tethered to the base of the pillar, I could see more of the underside of Gordon's rock shoes than the rest of him. They would appear and disappear as he made his way along the very lip until he came fully into sight at the far end, arranging a belay. By the time I was following along the traverse, bigger waves were piling into the cave, forced up the back wall and spurting horizontally out from the roof centimetres below my feet. The creamy white water was released at high pressure and accompanied by a deep bass sound which shook my innards whilst the air all around was filled with an aerosol of salt water. It was awe inspiring, spectacular, and exciting. This was why I climbed out here on the Bishop's Isles.

The belay was a stunningly situated sentry box jutting out from the left edge of the sea cave and very comfortable. We swapped gear, and I headed upwards. Whilst the second pitch was easier, it was steep and sustained for 55m as it surmounted bulges, side-stepped roofs, and bridged up thin grooves before a final large corner led to a good ledge below very easy rock above.

The objective on our first visit to Mingulay in 1993 was The Builacraig. But this overhanging 210m goliath of a cliff thwarted our advances. Its name was reputedly the war cry of the undefeated Clan MacNeil from Barraigh. Our ascent of the route above the Guarsay Mór Great Cave had been accompanied by the war cry of the ocean and had filled me with a deeper appreciation of its power and beauty. It reminded me of the ceremonial dance of The Maori Haka which is a celebration of life. It seemed fitting to name the route as an amalgamation of these two cultures: the Barraigh McNeils and the Aotearoa Maori. We named it Hakka as both a celebration and a war cry.

Kincardine

The self-conferred nickname 'Silver City' was another over-reaching feat of turd-polishing euphemism. It was grey. Everything was grey. There was just no getting away from it. The buildings were all – all – made of granite and the sky was covered in a thick layer of permacloud. It. Was. Grey. If Aberdeen was silver, then shite wasn't brown.

Christopher Brookmyre. *A Big Boy Did It and Ran Away*

After his failed attempt to sail to Norway from Fraserburgh, Menlove Edwards spent a week staying at a house in Findon and climbing on the sea cliffs with his lifelong friend Dr Alec Keith, who was doing his house jobs at the Aberdeen Royal Infirmary. There is a photo in Jim Perrin's *Menlove* book of Alec Keith deep water soloing on Severe ground on the far east tip of Craig Stirling in the 1930s.

Rock Climbing Near Aberdeen by HG Drummond

Oh, the wild joy of living, The leaping from rock up to rock.

'David Singing Before Saul' by Robert Browning

The sport of rock climbing becomes more popular every year, and as the branch of mountaineering most readily obtainable in this country, one cannot wonder at it. Though probably the majority of the Cairngorm Club members look upon a good hill walk as the loadstone of their desires, there are yet a number to whom a scramble, necessitating the use of hands and feet, or a climb of not too great difficulty, gives an added zest to their day's outing. To the south of the Bay of Nigg stretches a long range of cliffs, as yet but little explored from a climbing point of view, whose value as a training ground cannot be overestimated. True, there are no peaks to be ascended if one excepts certain rocky pinnacles, but the essence of the sport lies not in bagging peaks but in struggling with and overcoming the difficulties and obstacles of nature. Continued practice gives an increase of skill and necessarily involves a corresponding increase in the difficulties that can successfully be grappled with and overcome. Among these cliffs, then, the happy climber may find corries, ridges, gullies, chimneys and cracks to his heart's content, all within easy reach.

Geologically, the region is somewhat of a puzzle and contains several different kinds of rock, some of which are good climbing material, others but indifferent. Generally speaking, the best climbing is obtained where a granite outcrop occurs, as there the rock is sound and, being much weathered, gives splendid hand and footholds. Cliffs composed of rotten rock, such as is found near Muchalls, or where the climbs end up in steep grass, earth and gravel, are better left alone, or at any rate, first done with a rope from above.

There are four sections of cliff where climbs are most easily obtained. The first stretches from the Coastguard station at Doonies Yawns to Long Slough; the second from Altens Havens to Souter Head; the third for about a mile and a half south of Cove quarries; and the fourth for some distance north of Newtonhill, but along the whole coast outlying climbs giving satisfactory sport can be found. The embryo mountaineer of Aberdeen who thinks of spending a climbing holiday in Skye, Glencoe or elsewhere has, therefore, only himself to blame if he quits his native heath without some practice and knowledge of perhaps the most important branch of his craft. Many climbs have been done on these cliffs, but multitudes are still waiting to be conquered, so that anyone having the 'wanderlust' may exult in the feeling of the explorer; that he has done some climb which has never before been attempted, even though it is not a Matterhorn.

It would be impossible here to give any list of those climbs that have been accomplished, but some be mentioned. Proceeding along past the farm of Doonies, a deep bay running right up to the railway line, called Long Slough, possesses a sharp knife-edged pinnacle detached from the southern wall of the bay. The ascent of this pinnacle is difficult, and as the rock near the top is rotten, not to be recommended,

Facing page: Climbers on Perdonlie Inlet on the Aberdeen Coast. Photo John Cleare.

although, with three on the rope, the leader can be safeguarded. Its ascent cost six separate expeditions and was made from the small col between it and the south cliff.

One remarkable through-chimney starts at the back of a cave on sea level about halfway along the ridge, giving pleasant back and knee work till the climber's head emerges, like a rabbit, through a hole on the summit ridge. To make a complete exit, however, some of his outer raiment may very possibly have to be removed in order to reduce his dimensions.

At Cove and Newtonhill, several fine climbs can be done. But enough has been said to indicate to the climber where he may seek pastures new. Let him look for himself. He that seeketh findeth, and not least among his pleasures will be the fine combination of rock and sea scenery opened out before him during his explorations.

The granite cliffs of Cornwall have long been exploited. Why not those of Kincardine?

Aitken's Tower by Edward W. Watt

This square pinnacle of rock at sea level is a well-defined landmark at Souter Head, and its ascent has been attempted on various occasions during the last 20 years but without success. Its height is about 35ft, and it is inaccessible by about 10ft at high tide. At low tide, however, one can reach its base on solid rock. The smooth face presented on the land side is obviously impossible, and the traverse on the left or north side will not go because there are not any hand or footholds. The right or south side is the only possible route, and this was the side ascended.

On Saturday, 23rd June 1934, JA Aitken, WN Aitken, Dr Martin, and J. McCoss were at Souter Head and the Tower was contemplated. After a very ingenious and difficult manipulation of the rope, WN Aitken, who was wearing rubbers, volunteered to have a go at the ascent. He was tied onto both ends of an 80ft rope. One end was laid over a crack, and the other end, which did service for half the climb, was then untied. One end was played out, and the other was taken in.

At the beginning of the traverse the hand holds, though good, are very small, and footholds do not exist; a scratch on the rock is all that is possible. There is an overhang at the first corner, and at this point, Aitken's fingers gave out. He was held in two directions, and nothing worse happened than wetting rather more than the soles of his shoes.

Round the corner, there is a magnificent slab handhold, and Aitken used it. After untying one end of the rope, he made short work of the ascent, being safeguarded all the time. A pole was sent up on the rope and fixed on the summit with a handkerchief tied to it. The descent was made down the smooth face on a doubled rope. The climb requires not less than three of a party, and it may be classed as exceptionally difficult and not for the novice.

The Jungle Book by Julian Lines

To each his own fear.

Rudyard Kipling. *The Jungle Book*

Only a few years ago, the notion of soloing above the cold North Sea and falling in was unthinkable and also a little crazy. But, attitudes have changed since the advent of deep water soloing. This isn't a sport; it is a passion. A passion fed by a drug that is only prescribed to those with adventurous blood and the will to be different. I became addicted, and when I returned to Scotland from my first trip to Dorset, I felt like Drake or Magellan, returning from the South China Sea with a galleon laden with precious jade, incense and spices.

The wall of golden granite of the Red Tower, 25 miles north of Aberdeen, is mystical in its seclusion. Its routes are perplexing to find and even more so to climb. Shere Khan and Bagheera aren't just characters from *The Jungle Book*. They are legendary climbing test-pieces, aspirations to the best, mere illusions to the rest. I arrived at the top of the wall and peered down the sheer 50ft cascade of rock that dipped into the blue... I felt at ease. The water was placid, and the rock was clean and honest. I love the granite and the way it instils confidence in its intricacy. Shere Khan steals the left side of the wall via a pair of perfect parallel cracks that appear to have been clawed at by a mythical tiger, producing a minimalist masterpiece. Six years before, whilst at university, I had failed to tame the tiger with a rope, but now the idea of trying without one seemed immeasurably more exciting.

A summer's afternoon was spent assessing the sea conditions – the swell, the tides and the currents – before I climbed the arête between Mowgli and Shere Khan. I named it Baloo – the bear surely deserved a seat on the wall amongst his friends. It was time to scale the tiger, but at sea level, he stood formidable. I pounced on, soon realising my mistake; his stripes were so deceiving. He vanished suddenly, leaving no trace behind but a searing pain rippling through my fingertips. I desperately tried to cling on, but like a disobedient cub, I was tossed away into the water – found wanting. Although the tiger had beaten me, the myth of the sea had been dispelled – a fall into the North Sea was survivable. I began to feel more at ease attempting to play with these fictional creatures.

My body was weary, and it refused to challenge Shere Khan for a second time, so instead, I turned my attention to Hole in the Wall: the original route, a classic, but out of touch with the other jungle beasts. Carefree, I passed by the eponymous hole, which didn't issue money but did issue brief finger respite. Up above, amidst the crux, I was glued flat to the wall, like a bank robber under arrest, restrained by the law of gravity. Nervous giggles welled from the depths of my stomach as I unwound and dropped into the cold embrace of the sea – a sanctuary where gravity has no jurisdiction.

Over a week passed before I could run wild and free again with the characters of the jungle. On my return to the vertical, granite jungle, I peered down the wall and gawped in surprise. Hole in the Wall was heavily coated in a trail of white chalk. The penny dropped, and I started to smile. Last week's conversation must have inspired my friend, Wilson Moir. He had obviously soloed Hole in the Wall whilst I had been away. I threw on my boots and followed the chalk. The crux glued me to the wall once again, in starfish mode, until I managed to summon up all the power I had to reach the elusive hold.

Three long years passed by before my next visit. The air was fresh, quintessentially Scottish, quite unlike the equatorial humidity and dust of the oriental shipyard that had been my home. The sea twinkled; Shere Khan and Bagheera were pleased to see me. It was likely that no one had visited them for years, apart from the odd passing fisherman who might have thrown them an occasional glance as he tended to his pots.

Facing page: Julian Lines on Shere Khan, F7c+ S1. Right: Julian Lines on Bagheera, F7b+ S2. Photos Dave Cuthbertson.

I sat down and gazed along the coastline, appreciating the warm granite, the bright yellow lichens, the sea pink and the background chatter of the razorbills, guillemots and kittiwakes. More so than ever before, this place had the scent of freedom and felt insulated from the real world's hustle and bustle. I threw a glance seaward – the sea was calm, like a mirror without reflection, only disturbed by the silky wake of a small, wooden fishing boat that had rounded the headland. The boat glided towards a red buoy and tickled alongside. The fisherman leant over and hauled in his creel before floating gently over to the next. An amiable feeling of peace filtered through the air; the day was tranquil and balmy.

Apathetically, I declined the ferocious challenge of Shere Khan, and instead, I turned to Bagheera – sleek and agile, a line of undiminished beauty. The base of the wave-splashed wall is smooth and black, just like the purring velvet cat. I climbed on stealthily, with an apposite mix of cheek and nerves, soon finding the rhythm. Without warning, the holds became as fictitious as the panther itself. My sinews stretched, outstretched paw, outstretched claw, tantalising. The curvature of rock taught me free flow, lithe muscles pirouetting me up and away. The hunt was nearly over, but the cat turned vicious as the water grew shallower and more distant. I reached the top, ecstatic, encased in the silence of symbiosis, visible only to the flashing, coal-specked eye of the resident kittiwake nesting in the Jungle Book corner.

The approaching summer solstice was surely a good omen, a pertinent time to revisit Shere Khan. I returned with Dave 'Cubby' Cuthbertson, who was keen to capture me at play with the big, graceful cats of Shere Khan and Bagheera. We found that the last wisps of haar had vanished, revealing a sun-drenched coastline above a perfect tide. The tension in the crisp, blue air was electric as I prepared – chalking holds and drinking Red Bull whilst Cubby checked his angles, film and lenses. Inches beneath me, the light danced like silver eels on the surface of the emerald-coloured sea. I cherished the colours, the life, the commitment and the perfect mixture of thoughts that gelled in my head. I launched on, pulling one way, then the next, spinning a helix in haste. I held my breath in case the cat awoke and shredded my suspension of disbelief. Time stood still until I found myself on top, simultaneously sad and delighted with my success, whilst Shere Khan lay beneath me, unperturbed, snoozing in the noonday sun, oblivious.

Left: Julian Lines on Bagheera, F7b+ S2.
Facing page: Jules on Super Cracks in Reality,
F7a+ SX. Photos Dave Cuthbertson.

Cracks in Reality by Julian Lines

A déjà vu is usually a glitch in the Matrix. It happens when they change something.

Trinity in *The Matrix* (film)

I was no longer content with normal deep water soloing; I wanted to find something close to my mental and physical climbing limit. I thought about an objective that would fulfil my requirements. Some crazy ideas formed until a solution resulted: Cracks in Reality in the Red Hole, three miles south of Aberdeen. On Christmas day, I wandered down to the Red Hole and sat on the cliff top, peering down into a narrow inlet that thrashed with angry white horses, their spray galloping halfway up the 120ft wall. The sky was without lustre; the rock was cold and damp, the light levels low and dim. The sun hadn't bothered to show face for hours, days even. The whole mood was dire, but I huddled, comfortable and pensive, contemplating the magnitude of this solo. After half an hour, I turned away uncertainly, with a desire to return at a later date, knowing that my fitness was the primary key to unlocking the insanity of my dream.

On my next visit, I crouched on my haunches at sea level on smooth wave-washed buttocks of rock, staring across the narrow inlet to the wall. I ascertained the depth of water, the line of fall, and my escape plan if I fell in; above all, I needed to persuade my subconscious mind that a solo was justifiable. Then I discovered a traverse over a small cave that might provide a direct entry into the route. I became excited at the idea – my commitment was sealed. I didn't feel ready in my heart of hearts to try Cracks in Reality, so I decided to solo the neighbouring Black Sleep first to test the water. Firstly, I made a rope ladder out of an old piece of rope by tying a number of alpine butterfly knots in it before securing it and throwing it into the sea. Without this ladder, a deep water solo would be foolhardy – the inlet was lined with velvet smooth rock, from which escape would be impossible.

I abseiled to the lip of the eight-foot roof that caps the wall; here, I had to kick out and swing in rhythm to my descent. Losing the pendulum effect would leave me dangling in space with no means of reaching the rock. I swung in and grabbed a hold on the belay 30ft above the water. Without considering the danger, I undressed from my harness, clipped it to the rope and let it go. The moment was electric as I watched my safety line spinning away into thin air, coming to rest a tantalising 10ft away from my grasp. I turned and faced the rock. Its texture was strange – smooth, red porphyry sprinkled with crystals of mica and rivets of quartz. I hoped that it was going to be fair. I started to climb a vague arête in a nervous state, fumbling over every hold in a pathetic manner. I needed a more positive hold to help my upward progress and boost my confidence, but none materialised. The arête vanished, and I climbed rightwards on further discreet finger holds, always trying to work out the sequence of moves before I had reached them to save strength.

I looked down at the sea to make sure it was there and could give me the confidence I needed to continue. I saw a positive foothold and climbed quickly onto it. At 60ft above the sea I could rest, surrounded by exits of varying difficulties, trying to absorb the experience, committed between the hot sun and the cold sea with a parched throat and dry lips. The remaining 20ft of wall teased me, daring me. I felt like a lowly hyena challenging a lion over a carcass as I set to the wall, but the lion grew nasty and sent me scuttling down to my foothold sanctuary. My next attempt had more purpose as I hunted out all the hidden finger holds that granted me safe passage into a welcoming groove at 80ft.

The psychological barrier of deep water soloing in the Red Hole had been broken, and now I was totally addicted to these solo adventures. Next was Cracks in Reality. I'd had aspirations to climb Cracks in Reality whilst at university, but they had not been realised. On one occasion, I went down to hold Matt 'The Cat' Ingham's ropes on the route. I hung on the belay, holding his 30ft fall in astonishment; we both jumared out.

Eight years later, I abseiled onto the belay just as a passing rainstorm unleashed havoc, but on this occasion, I had no jumars with which to escape. My friend Charlie

Ord, my climbing partner on that day, luckily hadn't abseiled, so he threw me a rope from the other side of the inlet. I swung around frantically, trying to grab the smooth, tilting slabs on the opposite side... what a circus!

Looking back, I realised that it was so fateful; it seemed to have been my destiny to solo the route, and success would be far more pleasurable than I could have anticipated. Furthermore, the direct entry would produce a better route: perhaps the finest on the whole Aberdeen coast – this alone was driving the cogs of my own obsession. Two days after soloing the Black Sleep, I returned with Cubby, who had driven from Glen Coe to capture the moment on camera. Apathy set in when we found the coast covered in cloud, but when the cloud soon evaporated, it left me with no further excuses.

The rock architecture at the start is weird and obtusely dimensional – a smooth hanging wall suspended above a water-worn slab angling into the sea; their relationship is awkward and not favourable to human ergonomics. I linked the two with a vulnerable and precarious split position; a slip here would be embarrassing. I pulled on and gulped at the sight of no water beneath me. My anxiety rode high, hands slipping on the ultra-smooth rock; the holds were blind and the path indeterminable, but this was the adventure I was here for. I reached a hanging rib of rock and looked around the corner... suddenly I was a fathom deep in cold water. I surfaced and trod water, squeaking: 'Can you throw the rope in?'

My heart was ripping at my chest like a caged rat whilst Cubby was fiddling with his camera. I was frustrated, and I just wanted to get out. Finally, the rope snaked into the water. I pulled up eagerly, the smooth rock making me slip and slide, my knees bashing against the slab, and my blood spilling down the rock and into the sea. I dried off, changed my chalk bag and returned. I passed my falling point, trying earnestly to piece a sequence together. I found one, but my hands were the wrong way round... splash! This time, my feet hit the slab of rock under the water. I pulled out, convulsing with the cold. I tried again for a third time, and this time I dropped down to the lip of the arch, jammed my body in and pulled over onto a smooth, black slab at the base of the orange wall. I just happened to glance up at the fearsome 120ft cliff above me, knowing that I was in no physical state to attempt it. I jumped in and retreated for the day.

The next morning was fine, and the sea was calm with a number of moon jellyfish now apparent. I tried not to focus on them. The sunshine lifted my morale as it revealed the wall in a morning glow set against the darker shadows of the overhangs that framed it. This time, the traverse passed fluently, and I continued upwards on chunky holds until they ran out. The rock became awkward, and the connection with the original line was still 20ft away. Looking down, I realised that I needed to jump back four feet to hit the water, and lactic acid began to rip through my forearms, turning them into a solid mass. I barely held on to reach the Procrastination belay ledge and respite. I was dwarfed

and held spellbound by the powers of the intense walls and overhanging corners that incarcerated me. I traced the 100ft, overhanging corner of Procrastination, curving insanely above my head. I could hear myself asking the question: 'Is there water below the corner?' No... don't even think about it!

'I guess not then.' The sun was intense, the rock was radiating heat furiously, and I was perspiring like a sumo in a sauna. In a dehydrated state, I climbed down to a small ledge just above sea level. Cubby's voice broke the silence: 'What are you thinking, Jules?'

'It is too hot; I'm going to wait for the sun to leave the face,' I replied. I had a deep suspicion that he thought I was going to give up and swim back across the inlet. The wall was getting the better of me mentally; my mortality was being bared, and the reality of it all was that I had to climb a 100ft vertical wall, capped by an overhang, with fingers and arms that hadn't enough strength left to peel an orange. My forearms were swollen, and I needed them to subside. I rested for as long as it took for the wall to go into the shade, and by then, my arms had depleted and my strength had returned. My tenacity surprised me as I climbed upwards to join the parent route. Then, at 60ft, I hit the psychological barrier – the point where deep water is no longer 'safe'. I was only halfway.

Above was the zone where danger increases and strength decreases exponentially – I was being cleaved in two by a double-edged sword. I started to grip harder as my safety net disappeared. My arms began to tire; my mind began to overload with risk assessments – subconscious calculations being performed at break-neck speed. I had to down climb into the safety zone or climb higher and risk all. Indecision made me look up instinctively, and I spotted a good hold way above. I dashed for it with committed eagerness and then up to a small foothold at 80ft, which was just enough to rest my arms and recollect. I looked down towards that lonely strip of surface tension, ready to snap me, before looking up at the roof, which had grown substantially. I felt very fragile and studious, trying to read the intricacies of the rock's literature, the hidden story leading me off the wall and onto a hanging slab under the mighty roof.

The rock under the roof was damp, and I had run out of chalk. My left foot was blistering in its boot, so I untied my lace and pulled my heel out, forgetting that I had no chance of taking both hands off to re-tie. I stretched into the jaw of the roof and swung out on hand jams and guano-coated holds 100ft above the water. In these circumstances, gravity feels as though it grows stronger, and I intuitively pulled harder, acutely aware of the penalty of letting go...

The neatly cut, final groove was easy. I took one last look down before pulling into the horizontal world. I stretched out and let the poisons of fear drain into the deep, deep, green grass.

Mercury Falling by Julian Lines

Deep water soloing in Scotland seems a ridiculous idea, but then again, so are other Scottish ideologies such as tossing logs and deep-fried Mars Bars. Temperature or the feeling of hot and cold is all in the mind; in the physical, it is just a column of mercury in a tube. In any case, cold water only makes you pull harder, hold on tighter, for longer, and makes you swim faster if you happen to fall in. When you get out (assuming that you ever do), you shiver enough to burn off any excess body fat. Then, on your next attempt, you'll be lighter and, therefore, more likely to levitate and succeed. But that's a dietary secret, and if it leaks out, every *Cosmopolitan* reader will be seeking out the shores of Scotland rather than jetting off to the Caribbean to achieve that slimline look. The Scottish coastline is vast, has immense variety in geology, and harbours some of the best DWS in the world, most of which have yet to be found. So here are a few to help you on your way.

Since the sun rises in the east, we will begin with Lean Meat (F6b+ S1) at Craig Stirling, a few miles south of Aberdeen. This cunning, overhanging flake crack bursts its way across a squiggly chunk of overhanging schist. Step off the psyche-up ledge and fidget for a while with the initial flake before moving up to grab the perfect (footless) handrail. Your heart will now be thumping hard; if you are made of lean meat, then this traverse will be boring and a foregone conclusion. However, if you are not so lean, then things become far more interesting because you will be pumped, and when you realise that you are not going to make the monster jug on the arête, you can smile, fall off and tuck yourself tightly into a bomb. Having done so, you will be safe in the knowledge that the headlines in the paper the following morning will read, 'Norway hit by a tsunami of unknown origin.'

Next up is Longhaven, so head into a tangle of jams of the Granite City and then exit towards a stark, treeless coastline 20 miles further north. The Red Tower has a collection of finger-frying paths of tenuous complexity. Hole in the Wall (F7a+ S1) is the supposed classic. It is one of the most sought-after E5 leads on the east coast. A roped ascent is like losing your virginity; on completion, you smile with relief, feeling elated but also wondering what all the fuss was about. Attempting it without a rope is a different encounter altogether; you feel a mixture of nerves and vulnerability, yet also excited and cheeky.

Sweeping around onto the Moray coast, you will reach the town of Cullen, home to both Cullen Skink, which could either be a lizard or a hallucinatory product if you are dyslexic, and Logie Head, which is a most friendly spine of sandstone that wades into the sea and finally truncates in the shape of Dark Star (F6a S0). If you like tomfoolery, then wait for a group of climbers to be climbing above the platform on the popular east-facing wall, then step off the end of the platform above a small, square-cut bay of floating kelp. You will now be burrowing your hands up some lovely overhanging cracks, and since you are out of sight, you can make all sorts of in extremis grunts. They will think you are a little unhinged and will be reaching for their phones to call the coastguard.

After sampling a bowl of warm haddock soup, continue along the Moray Coast – the land of the rain shadow and dust storms to reach the sunny village of Cummingston. Here, you will find some sandstone, sea stacks, caves and pebbly beaches. In amongst the caves is the shy, 25ft high hanging arête of Bunny Boiler (F7a+ S2/3), which is a little dubious because of the way the beach level changes beneath it. Rule of thumb: a sandy base and six feet of water at high tide is the best you are going to get. If you are up for it, first practice the art of the armchair landing. If you need a demonstration, ask Mike Robertson, but don't follow if he jumps from 60ft: five feet will suffice. Once you have mastered that, access the base of Sidle, drop down and let rip your best simian swings along a line of fragile-looking razor-edged pancakes and honeycombs, all very healthy, to reach the base of the arête. The holds are a little sandy, so stay cool. Dare to be different?

If you have survived thus far, then continue through the highland capital of Inverness. The Gulf Stream on the northwest coast might purvey slightly kinder temperatures. Driving down this part of the highlands, you will notice the gargantuan – black

Left: Julian Lines on Bagheera, F7b+ S2. Photo Dave Cuthbertson.

(sandstone) and white (quartzite) – primordial pieces standing defiant on their game-board of gneiss. By the time you are passing the pawn of Stac Pollaidh, you will need to be falling out of your thoughts and psyching up for your one-hour walk across boggy terrain to reach Baby Taipan Wall. The obvious, long, low-level break of the wall that is Land of Milk and Honey (F6b S0) isn't really at all like milk and honey; it's more like jams, slopes and rising heartbeats. So if there aren't any heart problems in your family, then you should give it a bash. It is close to the water, in fact too close, so you can do silly things like cut loose and break wind. From the twisting descent slab at the northern end of the wall, swing around the arête and check out the line; it's obvious – like a set of cat's eyes. Continue until you reach the pebbly beach - there is no easy way out, so it is best to reverse or wait for a St Bernard to arrive in a dinghy, by which time you may have died of pneumonia and become a fossil.

If the bog hasn't swallowed you, and the midges haven't driven you out of the country, then continue a little further to the Achiltibuie peninsula adjacent to the Summer Isles. If you have a large wad of notes in your pocket then take the tourist boat to the southern tip of Tanera Beag. If you haven't then you will have to muscle your way across in a kayak, or if you are completely insane then swim across with the Orcas; it's only two miles! The Cathedral Cave is the place to be if you like to spend most of your time upside down, in the water with the jellies, or both. The hanging left arête of Ambergris (F7a S2) is well-positioned and easy at the start. If you think you are good at wrestling hippos then roll up your sleeves and set to the hanging groove with muscles busting off the bone. If you survive, move up to a rail rest and continue. Don't look down – it's a long way, and you really don't want to punch a hole through the tourist boat, which has by now probably crept in unawares.

Whilst you are on the Summer Isles, it may be prudent to flag down the ferry to Stornoway as it passes. Head over to the west coast of Lewis and take the bridge over to the isle of Great Bernera. Snoop around the headland until you see something resembling the bow of the *Amoco Cadiz* punching out across the Atlantic. This is Mega Tsunami. At bang on 60ft high, it's a challenge for those who have no cells in the left side of their brain (or right side, for that matter). My advice to you is to wear foam-padded underwear, and don't solo alone; preferably don't solo it at all. Just look at it, beat your chest and imagine yourself up there being... err...brave. However, if you fancy it, make sure the sea isn't beating the gneiss to a pulp before gaining the perfect right-angled arête. At half height, make use of the 'solitary confinement' ledge, though – trust me – it doesn't do anything for the confidence. Climb upwards further into no man's land; at this point, if your mates are scaring you with pterodactyl noises, then gibber your way up an overhanging crack on the right side (F7a S2). Contrarily, if you want to scare your mates, then tackle the overhanging prow direct with sloth-like technique although hopefully immeasurably faster than said creature for the best deep water solo experience in the country (F7b S2).

Returning to the Isle of Lewis, head past Mangersta in search of a small tooth of rock protruding out of the sea. If there is a prevailing south-westerly (common), then you may never find this tooth as it will be under the waves. If calm, strip down, don dry bag and swim the narrow channel. The race is now on; squeeze into a Gollum-type niche, change, traverse hard left to the left arête of the wall: The Last Pirate in Persia (F5+ S0). Shake your way up this with cold, jump in and swim ashore. By now, you will be shivering enough to be of benefit to the National Grid.

Continuing with the island-hopping theme from Lewis to Harris and Harris to Skye, it's time to sniff out the chocolate-coloured dolerite of Neist. Luckily the rocks don't melt in the sun. Head down under the lighthouse into Foghorn Cove. Immiscible (F6c S0) is one of those routes that look lovely but turn out to be really frustrating. Saunter up the lower arête believing that it will be plain sailing all the way to the top. Wrong. The arête becomes more subtle at half height: you can see the holds, feel them (small), but can't commit to them, always thinking that a better hold surely must materialise. My advice is: stop pissing around because when you do decide to go, the tide will have gone out, which means the fall is twice as far and the water half as deep. Contort through the move and continue more easily, realising that climbing is better than chocolate sometimes.

Finally, the tour wouldn't be complete without going to Mull, and more specifically, Erraid, which leads us to Traigh Gheal, an old haunt of Robert Louis Stevenson. It's certainly a romantic place; however, the offwidth crack of Brine Shrine (6b S0) doesn't look so. Traverse down into a body bridge slot at the base – the brine shrine. Now, if you are a skeleton, then you can chimney up the inside with ease. If you are dumb then you can try all your Kung Fu offwidth skills and get spat out into the pure turquoise, but if you are of lateral thinking, you will find all the holds on the outside of the crack. Jumping from the top is obligatory. Then swim to the nearest harbouring yacht and use your best charm to cadge a lift to Eilean na Muc, a small island made of the most beautiful granite a little under a mile away. Here, on the southeast tip of the island, is a bijou hub for water rats. Teeter down a ramp and hop onto the overhanging wall; a topless flake seduces you onwards to a finger crack. If you are American, then jam this painful atrocity with much glee; other nationalities can evade the aforementioned pain by making cool shapes on the chickenheads out on the left wall, fashioning your own Origami (F7a+ S0).

When you have done all the above, or at least one or two of them, congratulations, you can now dispense with your water wings and award yourself a deep-fried Mars Bar.

The alternative look at the S grade system for deep water soloing: S0: Exciting if you can't swim. S1: Good for a go if you enjoy tombstoning. S2: For those who look good on camera whilst falling through the air. S3: Do you not have any friends?

Orkney's Truffles by Julian Lines

No edged weapon could harm him, and there was no strength that yielded not, and no thickness that became not thin before him.

The Orkneyinga Saga

Deep water solo venues are like truffles; they are craved, hard to find, and require specific conditions in which to come into existence; all great deep water solos are created within small windows of opportunity. Strangely enough, the area of Scotland that fits all the requirements is at Yesnaby on the west coast of Orkney. The rock is perfectly solid sandstone – a small oasis amongst Orkney's greater, disintegrating geological strata. It is kind on the fingertips, and there are long hours of daylight in the summer to utilise both morning and evening high tides. There are about 50 deep water solos, ranging from short and safe to cutting edge routes above marginal depths of water and thrown in for good measure, The Castle of Yesnaby sea stack. The main drawbacks lurking here are that it is exposed to westerly swells, and the water temperature will barely reach 12 degrees in the summer.

I had sniffed out the potential for deep water soloing when I had studied the topos in Gary Latter's guidebook, *Scottish Rock Volume 2*. I had a suspicion that Tim Rankin's route Dragonhead (E6 6b) could possibly undergo the deep water solo treatment. However, by the time I had a chance to check it out, that insatiable, travelling, deep water solo guru, Mike Robertson, had visited and soloed almost all the routes on Gardyloo Wall. The routes of Lum Crack (E2 5c S0), Thing of Dreams (E2 5b S1) and Gardyloo gold (E2 5c S2) work beautifully as deep water solos, being far less complicated to solo than to lead with ropes and with a rewarding experience to boot. After a few days of acclimatising, I engaged in the pure quality of moves on Dragonhead (F7a+ S0/1).

I then became fixated on the blank line to the right: Tim's 'piece de resistance' Yellow Hammer (E7 6c or F7c S1). In all-out mode, I lunged up the wall and miraculously struck and stuck the hidden crimps and had the finishing jugs in my hand; I was so stretched in extremis that I lost control of the use of my feet; my fingers began to uncurl like slow cogs. I laughed as there was absolutely nothing I could do apart from enjoy the 35ft ride into the water. I was smitten. I tried again on a number of occasions over the next couple of weeks; failure was down to not allowing myself to rest properly and putting my lower spine out of place whilst throwing a big heel-hook on a project. I seemed to be visiting the chiropractor in Kirkwall more often than the cliffs, and I was becoming increasingly more and more frustrated.

I had also become intrigued by a wide sea cave that faced the sea stack. It was split in two at the base by a central pillar in the shape of an overhanging arête. With my boots and chalk bag placed in a small dry bag, I waded/ swam across to the pillar and climbed up onto a small foot ledge in the arête, enough for me to perch on one foot at a time to don boots. The roof above me was massive, and the water behind me was deep. It was worth a try. A few moves up the arête led to an impasse and a seemingly blind dyno across the roof to a possible hold. I searched for other options, and with fortuitousness, I found a hold out left that led to an amazing handrail. I dangled and swung along this rail, which in turn dumped me on a triangular sit-down ledge below an overhung groove that stooped fearsomely. The sun shone in; I sat down, took off my boots and relaxed into the lotus position, cocooned in the cave's dimensions and invisible to the outside world. Being stuck in this niche of contentment all alone would frighten many, but I was relishing it; this spirit of adventure felt natural to me.

Above in the groove, initially baffled by the non-existence of holds and mean angles, I sweated, stretched and grappled with the weird contortions, rather more nervous of failing than falling. I finally extracted myself to complete one of the finest and most adventurous deep water solos in the country: Tantric Sex, F7a S2.

Left and facing Page: Jules on Ride the Houlihan, F6c S3. Photos Lines Collection: 'Todd Skinner did a route in Wyoming called Throw the Houlihan. Its something to do with cowboys lassoing. So the name Ride the Houlihan was a play on that as Scott was trying to lasso the stack and then riding the waves to get across.'

I sat on top and gazed over at the Castle, feeling that there was some unknown force driving me towards attempting the sheer, slightly overhung, unclimbed northwest face. It is probably a little less than 70ft high on a generous high tide and requires a short swim across a salty moat to access it. Joe Brown and party first conquered the Castle by its south face at E2 5b in 1967 before they headed to the nearby Isle of Hoy and put up two new climbs on the Old Man of Hoy, which was being televised live to the nation.

I turned up the following evening after 9pm when the tide was rising and the sun was setting. The slender stack cut a lonely shape, sticking up its giant, proverbial finger at the buoyant Atlantic. And because the high tide waits for no man, I quickly bundled my boots, chalk bag, clothes, towel and my down jacket into my dry bag, slung it on my back, abseiled down the mainland cliff and then slowly immersed myself into the cold water. Thankfully the swim wasn't too far for it to be able to suck all the heat out of me. I pulled out and traversed around to the platform on the south side. I ripped open my dry bag, towelled dry and put on my shorts and jacket to gain warmth.

The stack has a thin pillar at its southern end – a wee stone stiletto – forming the framework of an archaic stone window from which I stared out into the infinite room of sea and sky. After warming up for a while, the hour hand clicked inside my head. I threw off my jacket, doused my hands in chalk and pulled out onto the main body of the stack, lost in a sea of unknown sandstone verticality. At half height, storms of doubt loomed; there was a hard section ahead, and the rock above didn't look too trustworthy. I continued on nonetheless. In the next instant, the hold I was pulling on was still in my hand but no longer on the cliff, and I was torquing away... my cold and shivering sinews and tendons were stretched to their elastic limits as they tried desperately to keep me attached... I did, just, and then I tried a different avenue with further dubious holds, finally deciding to retreat before the Earth's rotation had been able to shut out all the daylight.

A few days later, I discussed with Scott Johnston, the local climber, kite surfer, ferry worker and milkman, a way to access the top of the stack. Scott came up with the idea of using his kite surfing line to lasso the top. I wasn't convinced. He gathered a handful of stones, tied one to the end of his line, took up a stance with a roll-up hanging out of his mouth and tried to slingshot the line over the stack. His eighth attempt worked. I swam over and pulled the line with two climbing ropes attached to its end. Scott swam over too, and we attached one set of rope ends to the base of the stack and jumared up the loose ends to commence cleaning. Shockingly, there were plenty of loose holds, and in retrospect, I was so glad not to have continued – it would have ended in disaster as one of the loose blocks that Scott prized off was the same size as me.

Climbing the stack had many logistical problems. I had to abseil, sort out the gear in my dry bag, strip off, swim to the stack, dry off and maintain warmth once there.

I also had to gain some composure and then start to climb. If and when I got to the top, I had to down climb or jump, gather belongings, swim back and jumar up to the cliff top.

On a further attempt, all was going well until I started to shiver uncontrollably two-thirds of the way up; the cold had finally breached my core. I clung to some small holds for ages, shaking, not wanting to commit, not wanting to jump and not wanting to retreat. All sorts of mind wars were going on until the ebbing tide made the decision for me. I reversed somewhat reluctantly.

I needed to change my strategy and warm up fully first. I found a good warm-up traverse wall on the mainland cliffs to the east of the stack. Having warmed up there, I picked up my dry bag and took the longer swim to the stack. I dried quickly and set to the wall. This time it all fell into place; I sat on the top, keeping a watchful eye on the resident fulmar. I was finally at peace with myself and the route: Ride the Houlihan, F6c S2/3 or E6 6a.

Daniel Laing and I added a fair number of new deep water solos in amongst the quirky sandstone caves and geos of Yesnaby. Whilst I was obsessed with climbing the stack, Yellow Hammer, and battling sciatica, Daniel was obsessed with a massive arête with a difficult entry above a comfy trench followed by the arête above a marginal amount of water that really required a spring high tide. After many falls and cold swims, he finally succeeded on Dharma Bums (F7a+ S3), a route that probably doesn't work at all for an ascent with ropes. After his ascent, I found him sitting on the top with that silent glow in his face that usually depicts the shedding of stress and the contentment of success. We'd had our fair share of truffles.

Facing page: Daniel Laing on Sea of Dreams, F6a+ S1. Above: Daniel Laing on Skullduggery, F6c S1. Orkney. Photos Julian Lines.

Fate decrees that the mountaineer should, sooner or later, fall a victim to the furor scribendi.

AF Mummery. *My Climbs in the Alps and Caucasus*, 1936

Sources

A Climber in the West Country by Edward C. Pyatt. David & Charles, 1968
Climbers' Club Journal and guidebooks
Deep Water by Mike Robertson. Rockfax, 2007. This is an outstanding and comprehensive DWS guidebook to the UK and other areas around the world
Fell & Rock Climbing Club Journal and guidebooks
Footlesscrow.blogspot.com
Scottish Mountaineering Club Journal and guidebooks
Sea Cliff Climbing in Britain by John Cleare and Robin Collomb. Constable, 1973
Stack Rock by Chris Mellor. PDF guide. First published 1994, fourth edition 2020
The White Cliff by Grant Farquhar. Atlantis Publishing, 2018
UKClimbing.com

The sources above have been used on multiple occasions in this book. The sources below are of previously published text; anything not listed is original writing for this book.

Front Matter

A Letter from Budapest by Damian Cook, Budapest, October 1995. Damian was living in Budapest at the time that Mike, Steve and Joff were writing the Dorset DWS guide, *Into the Blue*. They were keen to get as much of his input as they could, so amongst other sections and ideas, he contributed this.

Chalk

Scrambles Amongst the Alps by Edward Whymper. Murray, 1871
The Apprenticeship of a Mountaineer: Edward Whymper's London Diary, 1855–1859. Edited by Ian Smith, London Record Society, 2016
Climbing in the British Isles by WP Haskett Smith. Longmans, Green and Co, 1894
Rock Climbing in the English Lake District by Owen Glynne Jones. GP Abraham and Sons, 1911
This My Voyage by Tom Longstaff. John Murray, 1950
Aleister Crowley's 1894 application to join the SMC courtesy of Bob Aitken, Robin Campbell and the SMC
Chalk Climbing on Beachy Head by EA Crowley, *SMCJ* Vol 3, 1895
Chalk Climbing on Beachy Head by H. Somerset Bullock, *CCJ*, 1898
Mountain Craft by Geoffrey Winthrop Young. New York. Charles Scribner's Sons, 1920
Chalk – A Miniature Anthology by EC Pyatt, *CCJ*, 1952
Climbing and Walking in South-East England by Edward C. Pyatt. David & Charles, 1970
Beyond the brink: Beachy Head as a climbing landscape by Paul Gilchrist. *The International Journal of the History of Sport*, 2012
The Confessions of Aleister Crowley by Aleister Crowley. Jonathan Cape, 1929
Strange and Dangerous Dreams: The Fine Line Between Adventure and Madness by Geoff Powter. The Mountaineers Books, 2006
Aleister Crowley: The Biography by Tobias Churton. Watkins Publishing, 2011
The Brief Mountaineering Career of Aleister Crowley – The Great Beast 666 by Robin Campbell from *Mountain* magazine #11, 1970
Aleister Crowley parts 1&2 by Mike Cocker, *SMCJ*, 2020
The First Tigers by Alan Hankinson. JM Dent & Sons, 1972
Chalk and Cheese by Brian Wyvill, *Crags* #26, Aug/Sep 1980 and new text 2022
The Isle of Wight by John Cleare from *Sea Cliff Climbing in Britain*
Stacks To Go At by Mick Fowler from *Mountain* magazine, 1990s
Skeleton Ridge by Mick Fowler appeared in *Alpinist* magazine #25
A Battle at Hastings by Phil Thornhill from *On the Edge* magazine #14

Facing page: Ex-England wicketkeeper Bruce French soloing Riders on the Storm, HVS 5a, at Stennis Head, Pembroke. Photo David Simmonite.

Dorset

Into the Blue by Jonathan Cook. Climbers' Club, 1996
The Loss of the Halsewell by Rodney Legg in *Dorset Life*, Dec 2010
Remarkable Shipwrecks. Andrus and Star, 1813
An Introduction to Swanage by John Cleare from *Alpine Journal*, 1961
Göttertraversieren by Peter Gillman was originally published as The Traverse of the Gods in *CCJ,* 1967
A Trad and a Sports Climber and Where Worlds Meet by Damian Cook courtesy of Joff Cook
Freeborn Man by Toby Foord-Kelcey was originally published on his an englishman in squamish blog: englishmaninsquamish.blogspot.com in 2019
Above the Blue by Jerome Mowat was originally published on theprojectmagazine.com on 10/8/08
Head Above Water by Martin Crocker was originally published in *Climber* magazine, August 2001
Mark of the Beast by Pete Oxley edited from interview 24/10/22 and Wil Treasure Factor 2 podcast

South Devon

Coasteering by Edward Pyatt in *Alpine Journal,* 1978
Sea Traversing – The Truth by John Fowler from *Tor*, The Journal of the Exeter Climbing Club, Vol 2, 1970
Sea Mountaineering by Peter Biven from *Tor*, The Journal of the Exeter Climbing Club, 1973
Peter Biven (1935–1976) by Barrie Biven from an obituary in *CCJ*, 1976
The Great Picket Rock by John Cleare from *Alpine Journal*, 1974
Transcending by Harry Sales from *Alpine Journal*, 1974
The Wizard of Oz by Pete Saunders was first published as 'Britain's Finest Climb by Kafoozalem' on UKC, 2001
The Art of Climbing Without Climbing by Neil Gresham edited from original essay and dialogue from EpicTV Clips YouTube movie: 'Neil Gresham's Landmark Deep Water Solo', 2021
Little Big Man by Arnis Strapcans from 'Pat Littlejohn' in *Crags* magazine #9
Call to Arms and Journey Inwards by Dave Thomas edited from interview 17/3/22 and Wil Treasure Factor 2 podcast

Cornwall

The Road Between Tides, A Clockwork Orange and Bosigran by Des Hannigan from *The Almost Island* by Des Hannigan. Scryfa of Linkinhorne, 2017 with new text 2021
The Limpet Technique by Leslie Garnet Shadbolt from 'Sea Cliffs' in *CCJ*, 1912
The Diary of Virginia Woolf Vol 2 1920-1924 edited by Anne Olivier Bell. Harcourt Brace Jovanovich, 1978
To the Lighthouse by Virginia Woolf. Hogarth Press, 1927
The Cornish Cliffs by AW Andrews from *CCJs*, 1905 and 1937
Serpentine and Sunset on Zennor Hill by AW Andrews from *CCJ,* 1940
Tremedda Days: A View of Zennor 1900-1944 by Alison Griggs Symons. Tabb House, 1992
The so-called Tregerthen Horror by Antoni Diller on cantab.net, 2021
Aleister MacAlpine: Ataturk Crowley: Randall Gair: Count Charles Edward D'Arquires (1937–2002) by Des Hannigan was first published on artcornwall.org, 2015
West Penwith Coastal Traverse by AW Andrews edited from *CCJ*, 1938 and *Cornwall* by AW Andrews and EC Pyatt, Climbers' Club, 1950
Cornwall by AW Andrews and EC Pyatt. Climbers' Club, 1950
Samson. The Life and Writings of Menlove Edwards edited by Geoffrey Sutton and Wilfrid Noyce. Cloister Press, 1960
Menlove by Jim Perrin. Victor Gollancz Ltd, 1985
A Black Rainbow: The life and times of Menlove Edwards by Ken Smith, *Climber and Rambler*, September 1983
Mountain Warfare Training by Frank Smythe from *Alpine Journal* 273, 1946
The Art of Falling by AW Bridge from the *Mountaineering Journal*, June 1932
Commando Climbers edited from Commandos and Cornwall by Mike Banks from *CCJ*, 1956; Cornish Connections by Mike Banks from *High* magazine 153, 1995, and Cornish Perspectives by Mike Banks from *CCJ*, 1997. Excerpt from 'A Season' by AW Andrews
Commando Climber by Mike Banks. JM Dent & Sons, 1955
The Fighting Fourth by James Dunning. The History Press, 2013
Commandoveterans.org
Vertical Assault. The story of the Royal Marines Mountain Leaders' Branch by Major Mark Bentinck. Royal Marines Historical Society, 2008
My Shag Song by Nea Morin first published as 'Bosigran' in *Pinnacle Club Journal* #16, 1974–1976
Coasteering by Edward Pyatt edited from *A Climber in the West Country*
Project Penwith by Matt George from his Kernow Coasteering blog: kernow-coasteering.co.uk/coasteering-blog
From Sea to Shining Sea by Henry Barber edited from Desert Island Crags in *Crags* #25, June/July 1980
Déjà Vu from *Tears of the Dawn* by Julian Lines. Shelterstone, 2013, second edition Scottish Mountaineering Press, 2020

Carn Gowla by Des Hannigan from *The Long Deep* by Des Hannigan. Scryfa of Linkinhorne, 2019
A Family Holiday by Graham Hoey originally published as 'On Holiday with Howard' by Graham Hoey in *CCJ*, 2015–16
Dancing with Gravity by Stu Bradbury from the original article on UKClimbing.com 'The Gravity of Risk', 2022, and new text 2022
How cold is too cold? by Grant Farquhar was published in *Climber* magazine, 2023, and references How cold is too cold? Establishing the minimum water temperature limits for marathon swim racing. Jane Saycell, Mitch Lomax, Heather Massey, Mike Tipton. Department of Sport and Exercise Science, University of Portsmouth. *British Journal of Sports Medicine* 53(17), 2019

North Devon

This My Voyage by Tom Longstaff. John Murray, 1950
The Coast Scenery of North Devon by EA Newell Arber. JM Dent & Sons Ltd, 1911
Coastal Climbing in North Devon by CH Archer. Privately published in 1961 with supplements in 1963 and 1965
James Hannington Bishop and Martyr: the story of a noble life by Charles D. Michael. SW Partridge & Co, 1910
The Hidden Edge of Exmoor by David Kester Webb and Elizabeth Webb. Thematic Trails, 2011
Equipment for Amphibious Climbing is appendix number three from *Coastal Climbing in North Devon* by CH Archer
The Hidden Edge of Exmoor by Martin Crocker from *High* Magazine #210, May 2000
The 1978 Exmoor Coast Traverse by Terry Cheek from the ECT website: members.madasafish.com/~exmoorwalker/page100.html
The Hidden Edge – an attempt to complete Britain's longest climb in a single push by Dave Pickford was originally published in *Climb* magazine, 2013
The King of Exmoor by Martin Crocker was originally published as 'Just A Day in North Cleave Gut' on martincrockerclimbing.com

Lundy

Rock Climbing on Lundy edited from an article by EC Pyatt in *Lundy Field Society Annual Report* Vol 14 Part 12, and further articles by EC Pyatt and KM Lawder in *CCJs*,1962 and 1964
The Flying Dutchman by Neil Gresham. Preamble contains some information from UKClimbing.com and quotes an interview from Climbing.de

Wales

Extreme Rock & Ice by Garth Hattingh. New Holland, 2000, has a chapter about Total Eclipse of the Sun
Elegug Stacks by John Cleare from *Alpine Journal*, 1974
Pembroke Nirvana by Neil Gresham includes text previously published in *Rock & Ice* magazine #258, July 2019, an interview on Planet Mountain website 2012, and new text 2023
Little Big Crag by Martin Crocker was originally published on martincrockerclimbing.com
The Sea-level traverse of South Stack Island by Barbara James from *Pinnacle Club Journal* #16, 1974–1976
Things to do in Lleyn When You Are Daft by Martin Crook was first published on Footless Crow website, 2016: footlesscrow.blogspot.com/

Ireland

Fairy Legends and Traditions of the South of Ireland Part 2 by Thomas Crofton Croker, John Murray, 1828
The Duke's Head by Iain Miller first published in *Irish Mountain Log* #129, spring 2019
Ireland's Magic Isle by Alan Tees, 2023. An earlier version was published in the *Irish Mountain Log* #107, 2013
The Devil's Castle by Chris Harle was first published on outside.co.uk in May 2023

Scotland

Popular Tales of the West Highlands Vol 2 by John Francis Campbell. Alexander Gardner, 1890
Wonder Tales from Scottish Myth and Legend by Donald Alexander Mackenzie. Blackie & Sons Ltd, 1917
The History of St Kilda by Kenneth Macaulay. Becket and DeHondt, 1764
A Voyage Round the Coasts of Scotland and the Isles Vol 2 by James Wilson. Adam and Charles Black, 1842
A Description of the Western Isles of Scotland by Martin Martin. Andrew Bell, 1703
A Late Voyage to St Kilda by Martin Martin. D Brown and T Goodwin, 1698
Climbing in St Kilda by Norman Heathcote edited from *St Kilda* by Norman Heathcote and *SMCJ* Vol 6, 1901
St Kilda Past and Present by George Seton. William Blackwood and Sons, 1878
British Sea Birds by Charles Dixon. Bliss, Sands and Foster, 1896
The Ascent of Stack-Na-Biorragh by Richard M. Barrington from *Alpine Journal*, 1913

Evolution Without Natural Selection by Charles Dixon. Porter, 1885

St Kilda by Norman Heathcote. Longmans, Green & Co, 1900

Island on the Edge of the World: The Story of St Kilda by Charles Maclean. Canongate, 1972

Carmina Gadelica by Alexander Carmichael. Constable, 1900

First Crossing by WH Murray from *SMCJ* Vol 35, 1994

The Great Stack of Handa by Graeme Hunter was first published as 'The great Handa adventure' in *Scottish Mountaineer*, Autumn 2019

The First Great Climb by Dave MacLeod from his blog: davemacleod.com/blog, 2011

Sea Stacks by John Cleare edited from *Alpine Journal*, 1974 and *Sea Cliff Climbing in Britain*

The Old Man of Stoer by Tom Patey from *SMCJ* #158 Vol 28, 1967

Scottish Stacks To Go At from Stacks to go at Part 2 by Mick Fowler in *Mountain* magazine, 1990s

The Drongs by Hamish MacInnes from 'High Stacks' in *The Herald Scotland,* 1997

Rock Climbing near Aberdeen by HG Drummond from *Cairngorm Club Journal* #35, 1910

Aitken's Tower by Edward W. Watt from *Cairngorm Club Journal* #75, 1934

The Jungle Book and Cracks in Reality edited from *Tears of the Dawn* by Julian Lines. Shelterstone, 2013, second edition Scottish Mountaineering Press, 2020: scottishmountaineeringpress.com

Mercury Falling by Julian Lines from *SMCJ*, 2015

Orkney's Truffles by Julian Lines was written as promotion for *Tears of the Dawn* and first published in *Scottish Mountaineer,* May 2013

Special thanks to the contributors:

John Alcock, Henry Barber, Lee Bartrop, Steve 'Crusher' Bartlett, Ricky Bell, Richard Bingham, Nick Biven, Stu Bradbury, Nick Buckley, Marc Calhoun, Frank Cannings, Simon Cardy, Jess Carr, John Cleare, Damian Cook, Jacob Cook, Jon 'Joff' Cook, Liam Cook, Martin Crocker, Martin Crook, Dave 'Cubby' Cuthbertson, Andy Donson, Max Dutson, Tim Emmett, Toby Foord-Kelcey, Mick Fowler, Matt George, Peter Gillman, Mark Glaister, Pete Greening, Neil Gresham, Nick Hancock, Des Hannigan I've stood on Cape Cornwall in the sun's evening glow. Down Chywoone Hill at Newlyn to join the fishing fleets go. Watched the sheave wheels at Geevor as they spun around, And heard the men singing as they go underground. Many new routes I have been completing. And, some people were definitely cheating. For this is my Eden, and I'm not alone. For this is my Cornwall and this is my home. We will not have bolts placed or holds chipped. Just because they were too gripped. I've stood on the cliff top in a westerly blow. And heard the wave thunder on the rocks far below. First thing in the morning, on Chapel Carn Brea, To gaze at the Scillies in the blue far away. For this is my Cornwall, and I'll tell you why: I love it here, and here I shall die. Brian Hannon, Chris Harle, Sue Hazel, Craig Hiller, Graham Hoey, Graeme Hunter, Mike Hutton, Gordon Jenkin, Jethro Kiernan, Rob Lamey, Tom Last, Julian Lines, Pat Littlejohn, Dave MacLeod, James Mann, John McCune, James 'Caff' McHaffie, Alan 'Richard' McHardy, Iain Miller, Jerome Mowat, Pete Oxley, Ken Palmer, Iain Peters, Robbie Phillips, Dave Pickford, Greg Pittam, Richard Pollard, David Price, Simon Rawlinson, Stephen Reid, Glenn Robbins, Mike Robertson, Mark Robson, Carl Ryan – Black Planet Photography, Pete Saunders, Colm Shannon, Ben Silvestre, David Simmonite, Andrew Steinberg, Tony Stone, Steve Taylor, Alan Tees, Dave Thomas, Phil Thornhill, Scott Titt, Crispin Waddy, Andrew Walker, Lukasz Warzecha, Frankie Woods, and Brian Wyvill.

Thanks also to:

The Alpine Club, The Cairngorm Club, The Climbers' Club, The Scottish Mountaineering Club, Bob Aitken, John Appleby, Mike Adams, Andrew Burr, Mike Cocker, Duncan Critchley, Jim Crowley, Graham Desroy, John Dunne, Andreas Hechenberger, Dave Henderson, Dr David Hillebrandt, Mark Kemball, Rob Lovell and the Scottish Mountaineering Press, David Medcalf, Colonel Ian Moore, Calum Muskett, Jim Perrin, Eloïse Pitts Crick, Guy Robertson, Tony Scott, Gav Symonds, Wil Treasure, Mike 'Twid' Turner, Mike Weeks, Charlie Woodburn, and anyone I have inadvertently omitted.

Grant Farquhar, Bermuda, 2023: grant@atlantis.bm.

What if it tempt you toward the flood, my lord,
Or to the dreadful summit of the cliff
That beetles o'er his base into the sea,
And there assume some other horrible form
Which might deprive your sovereignty of reason
And draw you into madness? Think of it.

Hamlet by William Shakespeare